T0299748

Routledge Revivals

Active Lavas

Originally published in 1993, *Active Lavas* looks at the practical aspects of monitoring uncontrolled streams of molten rock and how field data can be applied for theoretical modelling and forecasting the growth of lava flows. It describes the basic features of common subaerial lava flows and domes – both on Earth and on other bodies in the Solar System – before discussing the logistics of measuring lava properties during eruption and how these measurements are used to develop simple theoretical models for forecasting flow behaviour.

Active Lavas

Monitoring and Modelling

Edited by Christopher R.J. Kilburn and
Giuseppe Luongo

Routledge
Taylor & Francis Group

First published in 1993 by UCL Press

This edition first published in 2022 by Routledge
4 Park Square, Milton Park, Abingdon, Oxon, OX14 4RN
and by Routledge
605 Third Avenue, New York, NY 10158

Routledge is an imprint of the Taylor & Francis Group, an informa business

Publisher's Note
The publisher has gone to great lengths to ensure the quality of this reprint but points out that some imperfections in the original copies may be apparent.

ISBN 13: 978-1-032-35070-7 (hbk)
ISBN 13: 978-1-003-32517-8 (ebk)
ISBN 13: 978-1-032-35075-2 (pbk)
Book DOI 10.4324/9781032350707

Active lavas:
monitoring and modelling

Edited by

Christopher R. J. Kilburn & Giuseppe Luongo

First published in 1993 by UCL Press

UCL Press Limited
University College London
Gower Street
London WC1E 6BT

The name of University College London (UCL) is a registered
trade mark used by UCL Press with the consent of the owner.

ISBN: 1-85728-007-5 HB

British Library Cataloguing-in-Publication Data
A CIP catalogue record for this book is available from the British Library.

Typeset by Westfield Typesetting, Norfolk.
Printed and bound by Biddles Ltd, King's Lynn and Guildford, England.

CONTENTS

Contents

Contents

Contents

Preface

This book hopes to raise more questions than it answers. Active lavas are exciting but exasperating phenomena. They fascinate with their natural beauty but, despite the wealth of observational data gathered over the last decades, they remain poorly understood in anything but general terms: it is easier to forecast the weather than the growth of a flow.

The chapters collected in this volume cover most aspects of lava flow studies, from basic field descriptions to future monitoring techniques. Eighteen authors have contributed and, while a general consensus exists, diverging opinions also occur (e.g. on modelling lava resistance). This is a healthy sign and hopefully signals stimulating debates in the future.

The book falls naturally into four parts. Part I describes flow structures and their development, Part II examines monitoring techniques, Part III looks at flow models, their applications and limitations, and Part IV discusses the responsability of the volcanologist during volcanic emergencies.

Many people other than the contributors and editors have helped produce this book. Several colleagues have agreed to review chapters and thanks are due to Steve Baloga, Norman Banks, Steve Blake, David Chester, Joy Crisp, Ron Greeley, Pete Lipman, Hank Moore, Peter Mouginis-Mark, Tina Neal, Bill Rose, Herb Shaw, Don Swanson, Jean-Claude Tanguy and Geoff Wadge. Thanks, too, to Benedetto De Vivo, Grazia Giberti, Giuseppe Rolandi, Antonio Pozzuoli, Roberto Scandone and Agostino Zuppetta (in Naples) and to Pete Burrows, Raphael Carbonell and Denis Timm (in London) for their logistical support at critical moments in preparing the text. Last, but not least, nothing at all would have been possible without the patience of the publishers.

To Claudia Rametta

Contributors' addresses

Barca, D., Dipartimento di Scienze della Terra, Università della Calabria, 87036 Arcavacata di Rende (CS), Italy.

Borgia, A., Istituto Nazionale di Geofisica, Via di Vigna Murata 605, 00143 Roma, Italy.

Crisci, G. M., Dipartimento di Scienze della Terra, Università della Calabria, 87036 Arcavacata di Rende (CS), Italy.

Di Gregorio, S., Dipartimento di Matematica, Università della Calabria, 87036 Arcavacata di Rende (CS), Italy.

Dragoni, M., Dipartimento di Fisica, Università di Bologna, Viale Berti, Pichat 8, 40127 Bologna, Italy.

Fink, J. H., Department of Geology, Arizona State University, Tempe, AZ 85287, U.S.A.

Guest, J. E., University of London Observatory, Mill Hill Park, London NW7 2QS, U.K.

Hardee, H. C., Department of Mechanical Engineering, Box 30001, Dept 3450, New Mexico State University Las Cruces, Las Cruces, NM 88003, U.S.A.

Kilburn, C. R. J., Environmental Science Division, IEBS, Lancaster University, Lancaster LA1 4YQ, U.K.

Linneman, S. R., P.O. Box 440, Winchester, ID 83555, U.S.A.

Lopes-Gautier, R. M. C., Jet Propulsion Laboratory, California Institute of Technology, MS 183-601, 4800 Oak Grove Drive, Pasadena, CA 91109, U.S.A.

Luongo, G., Dipartimento di Geofisica e Vulcanologia, Università di Napoli, Largo San Marcellino 10, 80138 Napoli, Italy.

Nicoletta, F., Dipartimento di Fisica, Università della Calabria, 87036 Arcavacata di Rende (CS), Italy.

Peterson, D.W., U.S. Geological Survey, Branch of Volcanic and Geothermal Processes, 345 Middlefield Road, MS-910, Menlo Park, CA 94025, U.S.A.

Pieri, D. C., Jet Propulsion Laboratory, California Institute of Technology, MS 183-601, 4800 Oak Grove Drive, Pasadena, CA 91109, U.S.A.

Pinkerton, H., Environmental Science Division, IEBS, Lancaster University, Lancaster LA1 4YQ, U.K.

Rothery, D.A., Department of Earth Sciences, The Open University, Milton Keynes, MK7 6AA, U.K.

Sørensen, S.-A., Department of Computer Science, University College London, Gower Street, London WC1E 6BT, U.K.

Tilling, R. I., U.S. Geological Survey, Branch of Volcanic and Geothermal Processes, 345 Middlefield Road, MS-910, Menlo Park, CA 94025, U.S.A.

Part 1

MORPHOLOGY

Preface

The most abundant information on lava flows concerns their morphology. Indeed, in cases such as planetary studies, this may be the *only* information available. Part I therefore examines the structural evolution of flows and the associated dynamics of emplacement. Chapters 1–3 concentrate respectively on silicic, blocky and aa lavas, while Chapter 4 sets terrestrial studies in a planetary framework.

Notable omissions are descriptions devoted to subaqueous and pahoehoe lavas (although the latter feature in sections throughout the book, particularly Chapter 5). The first reflects the difficulty of monitoring underwater lavas (extraterrestrial studies are simpler in comparison) and the second that most observed flows have been aa or blocky types. Owing to the current activity in Hawaii, however, interest in pahoehoe emplacement is increasing again and new insights can be expected in the near future (e.g. Hon et al. 1993). Similarly, undersea projects using remote-sensing submersibles (e.g. the GLORIA and TOBI projects) promise substantial advances in our knowledge of submarine volcanism. As a compromise in the meantime, two short bibliographies follow for readers seeking background material on underwater and pahoehoe lava flows.

Underwater lavas

Applegate, B. & R. W. Embley 1992. Submarine tumuli and inflated tube-fed lava flows on Axial volcano, Juan de Fuca ridge. *Bulletin of Volcanology* **54**, 447–58.

Ballard, R. D. & J. G. Moore 1977. *Photographic atlas of the Mid-Atlantic Ridge rift valleys.* New York: Springer.

Ballard, R. D., R. T. Holcomb, T. H. van Andel 1979. The Galapagos Rift at 86°W: 3. Sheet flows, collapse pits, and lava lakes of the rift valley. *Journal of Geophysical Research* **84**, 5407–22.

Holcomb, R. T., J. G. Moore, P. W. Lipman, R. H. Belderson 1988. Voluminous submarine lava flows from Hawaiian volcanoes. *Geology* **16**, 400–4.

Jones, J. G. 1968. Pillow lava and pahoehoe. *Journal of Geology* **76**, 485–8.

Jones, J. G. & P. H. H. Nelson 1970. The flow of basalt lava from air into water - its structural expression and stratigraphic significance. *Geological Magazine* **107**, 13–9.

Lonsdale, P. & R. Batiza 1980. Hyaloclastite and lava flows on young seamounts examined with a submersible. *Geological Society of America, Bulletin* **91**, 545–54.

McBirney, A. R. 1963. Factors governing nature of submarine volcanism. *Bulletin Volcanologique* **26**, 455–69.

Mills, A. A. 1984. Pillow lavas and the Leidenfrost effect. *Journal of the Geological Society of London* **141**, 183–6.

Moore, J. G. 1975. Mechanism of formation of pillow lava. *American Scientist* **63**, 269–77.

Moore, J. G. & J. P. Lockwood 1978. Spreading cracks on pillow lava. *Journal of Geology* **86**, 661–71.

Moore, J. G., R. L. Phillips, R. W. Grigg, D. W. Peterson, D. A. Swanson 1973. Flow of lava into the sea, 1969–1971, Kilauea volcano, Hawaii. *Geological Society of America, Bulletin* **84**, 537–46.

Walker, G. P. L. 1992. Morphometric study of pillow-size spectrum among pillow lavas. *Bulletin of Volcanology* **54**, 459–74.

Pahoehoe lavas

Greeley, R. 1987. *The role of lava tubes in Hawaiian volcanoes*. US Geological Survey Professional Paper 1350, 1589–602.

Hon, K., J. Kauahikaua, K. McKay 1993. Emplacement and inflation of pahoehoe sheet flows - observations and measurements of active Hawaiian lava flows. *Bulletin of Volcanology* (in press).

Macdonald, G.A. 1953. Pahoehoe, aa and block lava. *American Journal of Science* **251**, 169–91.

Peterson, D. W. & D. A. Swanson 1974. Observed formation of lava tubes during 1970–1971 at Kilauea volcano, Hawaii. *Studies in Speleology* **2**, 209–22.

Swanson, D.A. 1973. Pahoehoe flows from the 1969–1971 Mauna Ulu eruption, Kilauea volcano, Hawaii. *Geological Society of America, Bulletin* **84**, 615–26.

Walker, G. P. L. 1989. Spongy pahoehoe in Hawaii: a study of vesicle-distribution patterns in basalt and their significance. *Bulletin of Volcanology* **51**, 199–209.

Walker, G.P.L. 1991. Structure and origin by injection under surface crust, of tumuli, "lava rises", "lava-rise pits", and "lava-inflation clefts" in Hawaii. *Bulletin of Volcanology* **53**, 546–58.

Wentworth, C.K. & G.A. Macdonald 1953. Structures and forms of basaltic rocks in Hawaii. *Bulletin of the US Geological Survey* **994**, 1–90.

CHAPTER ONE

The emplacement of silicic lava flows and associated hazards

Jonathan H. Fink

Abstract

The major risks from silicic lavas derive from self-generated pyroclastic flows and surges and associated debris avalanches, block and ash flows, and lahars. The degree of risk reflects the style of lava emplacement and useful distinctions can be made between lava domes and flows, composite and simple extrusions, endogenous and exogenous growth, and elongate and circular feeding systems. Laboratory and theoretical studies provide a basis for forecasting rates of flow advance, while field data identify surface morphologies which may act as precursors of explosive conditions, particularly endogenous pyroclastic phenomena.

Introduction

The relatively high viscosities of silicic magmas give them a greater tendency to erupt explosively than their mafic counterparts. Hence, assessments of the dangers posed by volcanoes that produce rhyolite and dacite tend to focus on explosive rather than effusive processes. Because of their generally low velocities and eruption frequencies, silicic lavas rarely threaten lives or property by engulfment, as commonly happens with more mafic flows. Nonetheless, explosions, pyroclastic flows and surges generated from within advancing silicic lava flows and domes, along with debris avalanches, block and ash flows, and lahars, can all result in significant hazards for local populations, as well as for scientists studying such eruptions.

In order to place the hazards associated with effusive silicic volcanism in perspective, this chapter first outlines some of the morphological and structural characteristics of silicic lava flows and domes. It then reviews some laboratory and theoretical studies which allow evaluation of the factors that control the rate of advance of viscous lavas. Finally it considers some of the processes responsible for endogenous explosive activity. It does not directly explore larger scale explosive processes and their associated hazards, nor the related problem of lahar generation caused by the mixing of silicic lavas with surface water or snow.

Styles of silicic lava emplacement

Silicic lavas commonly make up a relatively small fraction of the volume of any volcano. The exceptions are large (up to 50 km^3) sheet-like rhyolite deposits that have many of the characteristics of welded tuffs, but texturally and structurally resemble lava flows (Bonnichsen 1982, Twist 1985, Bonnichsen & Kauffman 1987, Hausback 1987, de Silva et al. 1988, Duffield 1989, Henry & Wolff 1992). Such deposits probably reflect a continuum of large scale eruptive styles for silicic magmas ranging from effusive to explosive. Extremely large silicic lava flows were probably fluidized or super-heated in some manner to allow them to attain their great areal coverage before freezing. The flow of such lavas would constitute very significant but infrequent hazards.

At the other end of the volumetric spectrum, some of the smallest silicic extrusions are found along chains of domes fed by dikes. These blocky piles may be as small as 10 m^3. Others may completely fail to emerge, solidifying before they break the surface. These "crypto-domes" may only be indicated by raised-up mounds along dike alignments marked by normal faults, ground cracks, and the positions of other domes and flows. Small domes are also found on composite volcanoes where dike emplacement is not apparent (e.g. 1739 Pietre Cotte obsidian flow on the island of Vulcano (Keller 1980)).

Flows versus domes

"Typical" silicic extrusions have volumes of about 0.1–1.0 km^3, and consist of a central high-standing portion (the "dome") surrounded by a distal region with lower relief (the "flow" or "coulée") (Fig. 1.1). The length-to-width ratios of such extrusions are controlled largely by underlying topography, with steeper slopes leading to more elongate flows and flat topography producing more circular flows. Blake (1990) has shown that, for extrusion of a lava with Bingham rheology on a slope, the transition from equant dome to elongate flow (or "coulée") will occur when the radius R satisfies the inequality:

$$R > \tau_0/(\rho g \sin^2\theta),$$

where τ_0 is the lava yield strength, ρ is the lava density, g is gravitational acceleration and θ is the underlying slope. The shape of the central dome may also be influenced by a balance between the volume flow rate and the lava viscosity. High flow rates and low viscosities tend to promote lower central constructs. Finally, the temperature-dependent yield strength prevents flows from thinning beyond a critical depth, and thus also

6

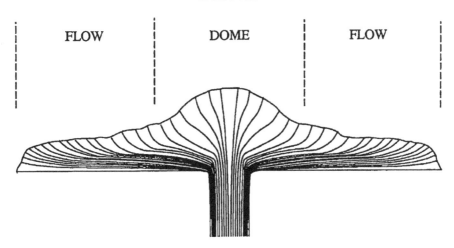

Figure 1.1 Schematic diagram of common form of silicic extrusion showing central dome, outlying flow, and flow banding pattern.

influences the overall morphology.

In relatively large silicic extrusions the raised central portion forms primarily by the upward movement of lava emerging from the vent, whereas the outlying part of the flow is emplaced largely by lateral spreading. In some smaller domes like Mount St Helens that lack contiguous lava flows movement vectors may be more complex. Evidence of these trajectories may be seen in the orientation of flow layers, which tend to be vertical near the vent and horizontal in the distal areas (e.g. Bryan 1966, Macdonald 1972, Fink 1983) (Fig. 1.1). Flow layers, which range from less than a millimeter to a centimeter or more in width, are common in silicic lavas and are made of varying concentrations of microlites or vesicles. Identification of the central dome helps define the position and possible elongation of the vent, which in turn can assist in evaluating conduit and vent geometry and the state of stress at the time of eruption.

Composite versus simple flows

A second classification of silicic extrusions contrasts composite and simple or monogenetic extrusions. This distinction refers to whether a given flow is emplaced by a single continuous outpouring, or if it starts and stops intermittently. This aspect of emplacement can provide insights into the size of a magma reservoir, the long-term eruption rate, triggering mechanisms, and the volume of future eruptions. For example, a 1 km^3 composite dome made up of 20 lobes of equal volume implies a near-surface reservoir whose "eruptible" capacity is approximately 0.05 km^3. If the same dome formed during a single event, a much larger reservoir would be required. For domes that erupt at regular intervals in response to a continuous process such as crystallization-induced volatile release (e.g. Cashman 1988) or tectonic extension, distinguishing between composite and simple eruption styles can help place quantitative constraints on the triggering mechanism.

Most of the silicic extrusions which have been witnessed while active (e.g. Mount St Helens, Santiaguito, Mont Pelée, Bezymianny, Augustine) are composite domes. Without observations during growth, it is difficult to determine whether a given extrusion is composite or simple. A monogenetic flow may produce more than one lobe if confining marginal levees rupture, or if resistance to forward advance causes upstream diversions. The highly complex outline of the Glass Mountain rhyolite flow on Medicine Lake Highland volcano in northern California is thought to have formed in this manner. Similarly, the dome that grew continuously at Mount Unzen, Japan in 1991 had a complex shape with several lobes (Nakada & Fujii 1992). On the other hand, a composite dome may have a molten core that persists during repose periods, causing the entire edifice to deform as a single mass, even though magma only enters and leaves the surface of the dome at irregular intervals (e.g. Swanson et al. 1987).

The episodic emplacement of composite extrusions is an example of the cyclic nature of silicic eruptions. Longer term cycles, which commonly include alternating explosive and effusive phases, may range from months (e.g. Swanson & Holcomb 1990) or decades (e.g. Rose 1987), to tens of thousands of years (Sheridan et al. 1987). Identification of such eruption patterns is a major goal of volcanic hazards assessment.

Endogenous versus exogenous growth
Another aspect of dome growth which has implications for eruption style is whether new magma proceeds directly from the conduit to the surface, or whether it resides in the dome interior for an extended period. When all new lava appears on the surface of a dome, the style of growth is referred to as exogenous, whereas if the crust fails to fracture and the dome grows by internal inflation, then the growth is endogenous (Fig. 1.2). Active domes may show a combination of the two styles. For the Mount St Helens dome, early growth was mainly exogenous while later growth had a large endogenous component (Swanson et al. 1987, Fink et al. 1990). As with composite and simple flows, exogenous and endogenous growth cannot generally be distinguished in prehistoric lavas. The distinction is most commonly made for domes, but portions of flows which spread laterally may also exhibit endogenous growth if their outer crust remains largely intact during emplacement.

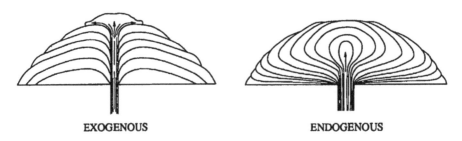

EXOGENOUS   ENDOGENOUS

Figure 1.2 Schematic diagram showing endogenous and exogenous addition of lava to a growing dome.

σ_1

σ_1

Figure 1.3 Typical pattern of dikes over a buried silicic magma body. The regional stress pattern dominates at some distance away from the body, whereas a radial pattern appears closer to the influence of the magma body. (After Bacon et al 1980.)

In order for a transition from endogenous to exogenous growth to occur, the tensile stress at the flow surface must exceed the tensile strength of the lava crust. Factors which promote cooling of the crust, such as precipitation and long repose periods between extrusions may favor endogenous growth. Factors which act to stretch the flow surface, such as high eruption rates and flow down steep slopes, promote exogenous growth.

These two styles of growth also lead to different modes of gas loss from the lava. During exogenous growth, much of the new magma is exposed to the atmosphere, allowing it to release volatiles by vesiculation. In contrast, during endogenous growth new magma remains under pressure within the dome interior, causing more gas to stay in solution. Rapid fracturing of the carapace of the dome may cause explosive decompression of the interior and subsequent pyroclastic activity. Such explosivity, discussed in more detail below, is thus favored in domes that undergo endogenous growth.

Central pipe versus dike conduits
The distribution pattern of silicic vents can provide important insights into the factors controlling eruption style, as well as the types of hazards that might be expected. Vent locations generally reflect the geometry of conduits which connect the magma source with the surface. In basaltic eruptions, spatter ramparts and other low-relief vent constructs are commonly destroyed or obscured when subsequent lava flows bury, erode or carry them away. In contrast, most silicic lava flows are thicker and less susceptible

to later modification. Thus, evidence of conduit geometry can be inferred more easily for silicic lavas.

The geometry of intrusions reflects the state of stress in and around a magma body. In general, dikes will align themselves parallel to the direction of maximum tectonic compression. However, the influence of regional stresses may locally be overwhelmed by pressure in the magma chamber, leading to a more radial pattern (Bacon et al. 1980, Bacon 1985) (Fig. 1.3). In three dimensions, this radial geometry may result in arcuate cone sheets (Anderson 1938). Passage through a rotating stress field may cause a dike to divide into collinear segments which subsequently reorient themselves, resulting in the commonly observed *en échelon* geometry (Delaney & Pollard 1981) (Fig. 1.4). Segmentation and rotation may also occur when a rising dike passes from a relatively hot region where deformation is dominantly ductile to a cooler zone characterized by brittle deformation (Reches & Fink 1988) (Fig. 1.5). Lateral flow of dikes across

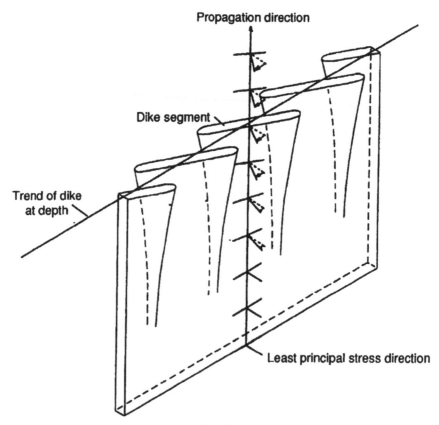

Figure 1.4 Diagram showing the response of a dike to a stress field whose principal axes rotate with depth about a vertical axis. As the dike rises, it adjusts to the changing stress orientation by breaking into segments which rotate in order to remain normal to the direction of maximum extension. (After Delaney & Pollard 1981.)

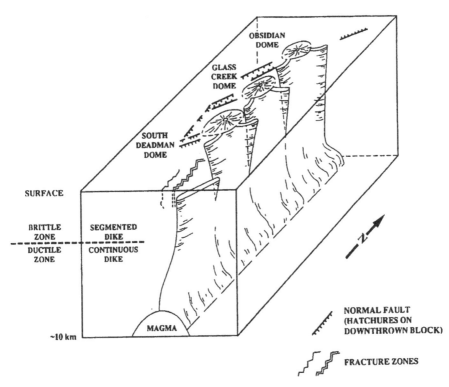

Figure 1.5 Diagram of the proposed dike beneath the Inyo domes, showing the development of rotated segments in the rising dike as it passes from a ductile to a brittle region of the crust. (After Reches & Fink 1988.)

lithological boundaries may also lead to segmentation and changes in orientation (Baer & Reches 1987).

The geometry of buried silicic dikes may be inferred from a variety of structures preserved adjacent to and on the surfaces of silicic extrusions (Fink & Pollard 1983, Fink 1985). The most obvious indication that domes were fed by dikes is an overall alignment. Sets of up to two-dozen silicic domes commonly form linear or *en échelon* trends that extend for up to 20 km. Individual domes along these trends may also be elongate parallel to the dike direction.

As a dike approaches the Earth's surface, it may produce a series of paired sets of fractures whose trends parallel that of the dike and whose separations are proportional to the depth to the top of the dike (Pollard et al. 1983) (Fig. 1.6). Mapped variations in the closest spacings of these ground cracks can thus be used to infer the locations and orientations of buried dikes. These cracks may merge to form elongate graben immediately prior to the eruption of magma. A rising dike may also heat groundwater, resulting in an aligned set of phreatic craters, as is seen in the three craters at the south end of the Inyo dome chain (Fig. 1.7).

On the surface of a dome or flow, the orientations of flow foliations may reflect the

11

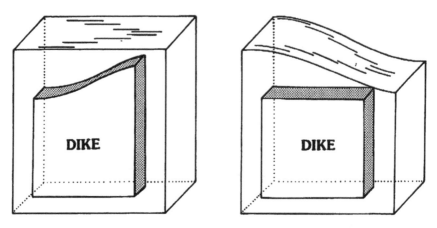

Figure 1.6   Schematic diagram showing the geometry of fractures which form over a rising dike as it approaches the surface. The spacing of the paired cracks is proportional to the distance from the top of the dike to the ground. (After Fink & Pollard 1983.)

alignment of the feeding conduit. Elongate depressions and ridges preserved on the flow surface may also parallel the dike trend. In some cases, however, the vent may widen explosively into a funnel-shaped depression (Eichelberger et al. 1986, Sampson 1987), so that smaller-scale structures near the center of the flow may no longer reflect the dike orientation.

Observations of basaltic eruptions, mapping of exhumed basaltic intrusions, and theoretical calculations demonstrate that smooth planar fractures provide the most efficient means for magma to force its way through the crust to the surface. However, once a pathway is established, subsequent flow of magma will favor a pipe-like conduit with circular cross section (Delaney & Pollard 1981). This process is illustrated by the transition, commonly observed in Hawaii, from a laterally extensive, dike-fed fissure eruption to more restricted, pipe-fed fountains. Comparable transitions have not been observed for silicic eruptions, but can be inferred from the mapped distribution of explosive and effusive products from recent activity at such locations as the Inyo–Mono volcanic chain in eastern California (e.g. Miller 1985, Sieh & Bursik 1986) and South Sister volcano in Oregon (Scott 1987). At the Inyo site, initial activity 550 years ago produced tephra and pyroclastic flows from vents distributed along an 11-km long trend, soon followed by the eruption of lava flows from three widely separated locations (Fig. 1.7).

From a hazards standpoint, the delineation or prediction of dike geometry is particularly important for impending silicic eruptions in populated areas. In places such as the Inyo domes, where seismic and hydrothermal activity in the past decade have suggested the possibility of renewed volcanism, the probability of a dike-fed eruption is especially alarming, since it offers the prospect of near-simultaneous explosive eruptions along a 10 km trend. Some indication of where such a dike (or dikes) might breach the surface could be provided by observing patterns of newly formed ground cracks, or measuring

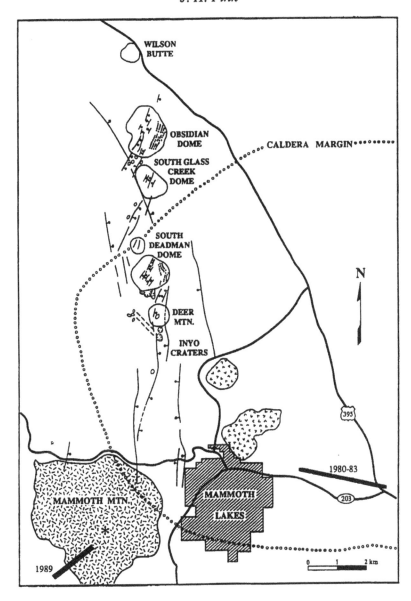

Figure 1.7 Map of the Mammoth Lakes region in eastern California, showing the location of the town, the three large 550-year-old Inyo domes, the phreatic Inyo craters, major faults parallel to the dome trend and bounding Long Valley caldera, Mammoth Mountain, and the epicenters of major earthquakes from 1983 and 1989. Note that a line connecting the 1989 earthquake epicenters with the summit of Mammoth Mountain intersects the Inyo dike trend (Fink 1985) and the trend of the 1983 south moat earthquakes in the same location, just north-east of Mammoth Mountain. Also shown are State Highway 203 and the "scenic loop" road added as an escape route in 1984 to the north of Mammoth Lakes.

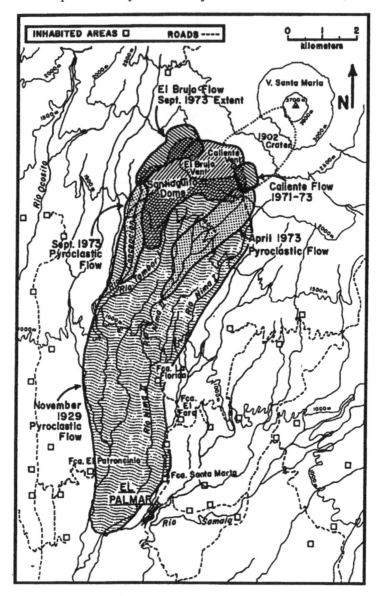

Figure 1.8 Map of the Santiaguito dome complex, showing the two most recently active vents (El Brujo and Caliente), lava flows they had produced as of 1973, as well as the areas covered by pyroclastic flows caused by dome collapse in April and September 1973, and in 1929. (From Rose 1987.)

heat flow with infrared sensors. Studies of the 1980 Long Valley earthquake swarm suggested the possible intrusion of an east–west-oriented dike at the south end of the caldera, just east of Mammoth Mountain (Savage & Cockerham 1982). Later earthquakes in June 1989 were located immediately south of Mammoth Mountain, aligned with the vents and associated structures of the Inyo domes. Concerns about the possibility of laterally extensive eruptions were partly responsible for construction of a new escape road for the town of Mammoth Lakes (Fig. 1.7).

The hazards associated with collapse of silicic domes, whether dike fed or not, may be enhanced by topography. For example, activity at the Santiaguito complex in Guatemala has persisted since 1922 along a 5 km long chain of domes perched high on the south-west flank of Santa Maria volcano (Rose 1973, 1987). Because of the great relief, periodic collapse of the domes and shedding of debris produces significant hazards to towns many kilometers downstream (Rose et al. 1977) (Fig. 1.8). Predicting which towns and facilities are most threatened requires knowledge of which vent is most likely to be active. Delineation of the factors controlling which sites along a silicic dike are most likely to have persistent activity would help in the assessment of hazards at sites like Santiaguito.

Dynamics of silicic flow advance

Although silicic lava flows have rarely threatened towns or property by direct engulfment, increasing population density (e.g. Vulcano, Italy) and geothermal development (e.g. Coso volcanic field, California) in various volcanic regions make such hazards more likely in the future. In addition, domes which form within a summit crater such as at Mount St Helens provide much more of a hazard from pyroclastic flows and gravitationally driven mass movements after the lava reaches the crater lip. Finally, the distance travelled by mass movements generated when lava flow fronts collapse will depend on the heights of those fronts. In order to predict where and when a particular silicic flow or dome is liable to either encroach upon a downstream site or emerge from a crater, it is necessary to model how such flows advance. In this section we briefly review some recent studies of the dynamics of very viscous flows.

Factors influencing the advance of lavas

The three most important factors controlling the area likely to be covered by a given flow are underlying topography, lava rheology and total erupted volume. Estimating how quickly this area will be inundated requires further information about eruption rate. Digital topographic maps can be used in conjunction with simple models of flow rheology to predict the paths of lava flows (e.g. Ishihara et al. 1990). In such studies, rheological assumptions must first be made which describe the way that lava deforms in response to applied stresses.

Laboratory measurements indicate that the rheology of silicic magmas is both temperature and strain rate dependent. Shaw (1963) and others have shown that the viscosity (η) of silicic magma is an exponential function of the inverse of temperature (T):

$$\eta = \eta_0 \exp\left(E/RT\right),$$

where η_0 is a constant, E is the activation energy, R is the gas constant and the strain rate is assumed to be fixed. Spera et al. (1982, 1988) have demonstrated that rhyolite magma also has pseudoplastic behavior:

$$\sigma = A \mid \varepsilon \mid^{1/n},$$

where σ is the applied stress, ε is the strain rate, A and n are constants, and n has an experimentally determined value of about 1.25. For the purpose of modelling, these behaviors have been approximated variously as isothermal Newtonian (Huppert 1982, Huppert et al. 1982), isothermal Bingham (Blake 1990), temperature-dependent Newtonian (Fink 1980), temperature-dependent Bingham (Dragoni et al. 1986, Fink & Griffiths 1990, Ishihara et al. 1990) or a combination consisting of a ductile interior with a brittle solid crust (Denlinger 1990, Iverson 1990). The Newtonian models are most useful when velocity data are available, whereas the Bingham models are more appropriate for interpretations of the morphology of remotely observed flows and others whose emplacement was not directly witnessed. The pseudoplastic and temperature-dependent models require more accurate strain rate and thermal data than are commonly available in field studies of lavas.

The effects of lava flows on man-made structures in their path depend on their rheological properties. Advancing Bingham materials will spread laterally and thin until their marginal thickness (h_p) equals a minimum value permitted by the yield strength (Hulme 1974):

$$h_p = \tau_0 / (\rho g \sin \theta).$$

Once this condition is attained, the flow will follow the maximum gradient down-slope, leaving marginal levées of thickness h_p which help define the depth of the active channel. Thus the higher the yield strength of a flow, the higher the topographic barrier it will be able to overtop as it moves downhill. Furthermore, because the thickness (h_p) of the non-deforming plug on the upper surface of a flow is also proportional to the yield strength, flows with greater strengths will carry more massive surface layers capable of knocking down obstacles rather than overtopping them.

Predicting the advance rate of future flows requires extrapolation from careful measurements of effusion rates. Observations of active dacite domes at Santiaguito (Rose 1973), Bezymianny (Bogoyavlenskaya & Kirsanov 1981) and Unzen (Nakada & Fujii 1992) have shown that all exhibited relatively steady average flow rates. Swanson & Holcomb (1990) used digital topographic data and calculated that although the overall volume flow rate for the Mount St Helens dome decreased from 1980 to 1986, within this time it remained steady for periods of up to 3 years. For flows not observed while active, only crude estimates and predictions of past and future flow rates are possible (e.g. Sheridan et al. 1987).

Models for the spreading of lava flows

Several recent studies have attempted to relate the radial spreading of viscous lavas to their rheological properties and eruption rates. Huppert et al. (1982) developed a model based upon laboratory experiments with Newtonian oils and field measurements made on an active andesite dome at Soufriere volcano. They found that, for an assumed isothermal Newtonian rheology and constant flow rate (Q), the dome radius (r) could be expressed by

$$r = 0.715(g'Q^3/3v)^{0.125} t^{0.5},$$

where $g' = g(\rho-\rho_a)/\rho$ is the reduced gravity, ρ_a is the density of the ambient fluid (air for subaerial lava flows), $v = \eta/\rho$ is the kinematic viscosity of the lava and t is time since extrusion began. This model predicts that in the absence of a topographic influence, viscous lavas should spread at a rate proportional to the square root of time. In order to explain observed advance rates, Huppert et al. concluded that the lavas had viscosities several orders higher than would be expected from laboratory measurements. This discrepancy was attributed to the resisting influences of a cooled surface crust and talus blocks in the flow front.

Blake (1990) developed a model for spreading of isothermal Bingham domes which he applied to the Soufriere data of Huppert et al. (1982) as well as to morphological measurements from dozens of other silicic domes and to a set of laboratory experiments using kaolin slurries. Blake found that these data sets could be best explained by the relationship:

$$r = (Q^2 \rho g'/\tau_0)^{0.2} t^{0.4}.$$

When applied to the Soufriere dome in conjunction with observed transition times (see next paragraph) this model predicted viscosities more consistent with laboratory values than did the Newtonian model. For radial spreading data, these two models differ primarily in the exponent of their time dependence.

Blake also found that during the emplacement of viscoplastic domes at constant flow rate, the radial spreading rate underwent a transition from Newtonian to Bingham time dependence after a time T, given by:

$$T = (g'^{0.75} \eta_b^{1.25} \rho^{0.75} Q^{0.25})/\tau_0^2,$$

where η_b is the plastic viscosity. At early times, lava would tend to pile up over the vent to a depth well in excess of h_c and the spreading dome would remain Newtonian in character, with the radius increasing as the square root of time. For $t > T$, the effect of yield strength on the spreading rate became progressively more important, with the radius eventually increasing as $t^{0.4}$.

In order to evaluate the role of cooling on flow advance, Fink & Griffiths (1990) modelled lavas as temperature-dependent Bingham materials, and simulated them by extruding liquid polyethylene glycol (carbowax) into a tank of cold sucrose solution. When the wax came in contact with the cold liquid in the tank, a crust formed which served to retard the advance rate of the flow. Fink & Griffiths were able to replicate many of the processes observed in natural lava flows and domes, including surface folding, fracturing, endogenous growth, and pillow formation. Morphological changes from one surface structure to another could be related to the rate of formation of surface crust, and these relations were successfully scaled from laboratory to field situations. Furthermore they found that the spreading rates of their simulated domes exhibited a similar transition from Newtonian to Bingham behavior as discovered by Blake (1990). In this case, however, it was the growth of a cooled crust which caused an effective Bingham rheology to develop.

In practice, these three models all show that in the absence of significant subflow topography (a likely condition within a summit crater but not in areas of dike emplacement), silicic lavas extruded at a constant rate will spread with a velocity roughly proportional to the square root of time. Thus, even though the actual rheological behavior of an active silicic lava may not be precisely known, the rate at which it will spread across a flat surface can be approximated if the density, volume flow rate, and viscosity or yield strength can be estimated.

Once lava starts moving downslope, the form of its velocity profile will depend on whether it is advancing across a planar surface or within a channel. For flow of a Bingham material of thickness H moving across a planar surface, the maximum (surface) velocity is given by (Dragoni et al. 1986)

$$v = [(\rho g H^2 \sin \theta)/2\eta_b][1-\tau_0/(\rho g H \sin \theta)]^2.$$

The Newtonian case can be obtained by setting $\tau_0 = 0$:

$$v = (\rho g H^2 \sin \theta)/2\eta_b.$$

From these two equations it is clear that the presence of a yield strength reduces the velocity, and if the flow depth is less than the critical value h_c the flow will stop advancing. For flow in a semicircular channel of radius R, the maximum velocity may similarly be expressed as (Dragoni et al. 1986)

$$v = [(\rho g R^2 \sin \theta)/4\eta_b][1-2\tau_0/(\rho g R \sin \theta)]^2.$$

The above three equations do not take into account the effects of cooling, which causes a rapid rise in both viscosity and yield strength (e.g. McBirney & Murase 1984). Dragoni et al. (1986) present a series of graphs showing the dependence of flow depth and velocity on effusion rate and underlying slope for the case of strengths and viscosities which increase exponentially with cooling. In order to calculate the time necessary for an advancing flow to travel a specified distance, temperature-dependent relationships like those presented by Dragoni et al. (1986) must be integrated to take into account downstream variations in slope and channel width. Several examples of this method applied to mafic flows may be found in Ishihara et al. (1990).

In order to determine which of the various rheological models are most appropriate for an active lava flow, detailed measurements of flow front advance rates, cross-channel velocity profiles, penetrometer data (e.g. Pinkerton & Sparks 1978) or longitudinal variations in flow margin thickness and levée width are needed.

Endogenous explosive activity associated with silicic lava flows

The principal hazards associated with advancing silicic lavas are caused by pyroclastic flows and explosions generated from their interiors. Depending upon the local topography, such flows can travel long distances and, if they mix with water, may generate lahars capable of travelling even further. These pyroclastic flows are particularly dangerous because they can occur several weeks or months after a lava flow starts advancing, and may be derived from the front, several kilometers from the vent. Since many eruptive cycles from silicic magma bodies can be related to stratification and a downward decrease of volatiles within those bodies (e.g. Eichelberger & Westrich 1981, Blake 1984, Fink et al. 1992), it is common for explosive activity from a vent to decrease sharply once effusion of lava flows begins. In contrast, endogenous explosions can occur with little warning at times when they are least expected.

Although this type of hazard was identified at Mount Merapi in Indonesia more than 50 years ago (Neumann van Padang 1933), recognition of its significance came primarily from a study by Rose et al. (1977) of eruptions at the Santiaguito dome complex in the early 1970s. Collapse of the front of a dacite flow more than 1 km from its vent led to formation of a pyroclastic flow that swept 5 km down the valley of the Rio Concepcion, killing at least one person. More recent analysis of the products from a much larger eruption of Santiaguito in 1929 which killed over 5,000 people indicate that it too probably originated from within an advancing lava flow (Mercado et al. 1988). The pyroclastic flows produced during this eruption were preceded by a pyroclastic surge, implying a much greater amount of energy release.

Two models have been proposed to account for the generation of this type of explosion (Fig. 1.9). According to the first (e.g. Mellors et al. 1988), collapse of a flow

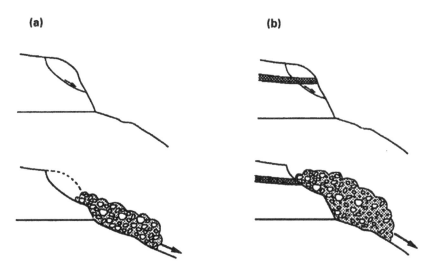

Figure 1.9 Diagram showing two models proposed for generating pyroclastic flows by dome collapse: (a) expansion of air mixed with hot blocks leads to fluidization of the fragmental debris; (b) a volatile-enriched zone within the flow (shown by cross-hatching) is rapidly exposed by collapse of the flow front, causing explosive decompression.

front causes rapid mixing of fine-grained hot rock with entrained air, and this mixture converts potential energy to kinetic energy as it advances downslope as a block and ash flow (McTaggart 1960). This model is consistent with observations of small pyroclastic flows generated from collapse of the front of the Mount St Helens dome, and implies that cooling during flow advance should minimize endogenous activity, while over-steepening of the flow front should accentuate it.

A second model (Fink & Manley 1987, 1989) states that volatiles may become concentrated within an advancing flow, and exposure of the resulting water-rich zones by collapse of a flow front can cause explosive decompression. Furthermore, if the volatile content becomes high enough, explosions may be generated without flow front collapse, resulting in formation of explosion craters which are commonly observed on the distal surfaces of young silicic flows. This model requires a mechanism for volatiles to become concentrated within dry lavas that characteristically have less than 0.2 wt% water.

Evidence of volatile concentration in silicic flows is provided by drill cores taken from rhyolite flows in the Long Valley and Valles calderas. Samples from these cores had water contents that were nearly all around 0.1 wt%, except for anomalous zones of coarsely vesicular pumice (CVP) found approximately one third of the way below the surface which had water contents of up to 0.5 wt%. Mapping of the surfaces of many rhyolite flows (e.g. Fink 1983, Fink & Manley 1987) reveals that these CVP zones were capable of rising buoyantly to the flow surface, and that they were probably associated with explosion craters (Fink & Manley 1989).

Fink & Manley (1987) proposed that volatiles could become concentrated by migration along microcracks as a flow advanced. Shear stresses associated with move-ment could promote such fractures in the lower portions of the flow. Upward migration of gases would be arrested when they reached the base of the non-deforming surface crust of the flow. This mechanism could lead to increasing concentrations of volatiles, and consequently increasing endogenous explosive hazards, as a flow advanced.

In some cases, volatiles may increase during extrusion without being concentrated by surface flow. Anderson & Fink (1990) analyzed water contents of samples from various lobes of the Mount St Helens dacite dome. They found that magmatic water contents increased during each extrusive episode. Volatile contents typically increased from values of 0.1 wt% near the flow front to 0.4 wt% near the vent. Such increases would cause the likelihood of endogenous explosive activity to increase during the course of an eruption. At the Mount St Helens dome, the potential danger of this volatile increase has been mitigated by the relatively flat topography of the near-vent region and the relatively low height to diameter ratio which makes rapid slope failure and slumping less likely.

Clearly the principal hazard associated with silicic extrusions comes from en-dogenous pyroclastic phenomena. Precursors of such explosive behavior may be provided by changes in the surface texture of a flow. In glassy rhyolite flows such as those of the Mono–Inyo chain, the appearance of very dark scoriaceous outcrops within the otherwise light-colored pumiceous carapace may indicate that volatiles have become concentrated enough for a gravity instability to develop (Baum et al. 1989). In dacite domes like those of Mount St Helens, changes from smooth to scoriaceous surfaces generally reflect an increase in overall volatile content (and therefore potential explosivity) of the extruding lava. In either case, the formation of an oversteepened active front creates the most dangerous situation, particularly if the extrusion is located high up on a volcano. Unfortunately, better assessment of this type of hazard will probably require more observations of the destructive potential of actual flows.

Summary

In this chapter various aspects of the emplacement of silicic lavas have been considered. Distinctions between domes and flows, composite versus simple extrusions, elongate versus circular conduits, and endogenous versus exogenous styles of growth need to be made and understood before hazards can be successfully assessed. Eruptions from silicic dikes which range up to 10 km or more in length can expose a large area to pyroclastic and phreatic hazards, and collapse of domes in summit regions of steep volcanoes can threaten a 360° sector. Knowledge of the location and geometry of dikes can come from mapping of structures on and around recently emplaced silicic domes in conjunction with various geophysical techniques. Rates of advance of silicic lavas can be predicted using simplifying assumptions about flow rheology, along with topographic data and theoretical models based in part on laboratory simulations. The principal dangers from small to moderate silicic extrusions occur when interior volatile-enriched zones are exposed during collapse of flow fronts. The resulting "Merapi-type" block and ash flows may travel rapidly down valleys, causing destruction many kilometers from the vent. Advance warning that this type of volatile concentration is taking place may be provided by observed increases in the scoriaceous character of the flow surface.

Acknowledgements

Thanks are due to Don Swanson and Bill Rose for helpful reviews. The research was supported by National Science Foundation grants EAR 86-18365, EAR 88-17458, and EAR 90-18216.

References

Anderson, E. M. 1938. The dynamics of sheet intrusion. *Proceedings of the Royal Society of Edinburgh* **58**, 242–51.

Anderson, S. W. & J. H. Fink 1990. The development and distribution of lava surface textures at the Mount St. Helens dacite dome. In *IAVCEI Proceedings in Volcanology*. Vol. 2, *Lava flows and domes: emplacement mechanisms and hazard implications*, J. H. Fink (ed.), 25–46. Berlin: Springer.

Bacon, C. R. 1985. Implications of silicic vent patterns for the presence of large crustal magma chambers. *Journal of Geophysical Research* **90**, 11,243–52.

Bacon, C. R., W. A. Duffield, K. Nakamura 1980. Distribution of rhyolite domes of the Coso Range, California: implications for the extent of the geothermal anomaly. *Journal of Geophysical Research* **81**, 2425–33.

Baer, G. & Z. Reches 1987. Flow patterns of magma in dikes, Makhtesh Ramon, Israel. *Geology* **15**, 569–72.

Baum, B. A., W. B. Krantz, J. H. Fink, R. E. Dickinson 1989. Taylor instability in rhyolite lava flows. *Journal of Geophysical Research* **94**, 5815–28.

Blake, S. 1984, Volatile oversaturation during the evolution of silicic magma chambers as an eruption trigger. *Journal of Geophysical Research* **89**, 8237–44.

Blake, S. 1990. Viscoplastic models of lava domes. In *IAVCEI Proceedings in Volcanology*. Vol. 2, *Lava flows and domes: emplacement mechanisms and hazard implications*, J. H. Fink (ed.), 88–126. Berlin: Springer.

Bogoyavlenskaya, G. E. & I. T. Kirsanov 1981. Twenty-five years of Bezymianny volcanic activity. *Vulkanologiya i Seismologiya 1981–2*, 3–13 (translated by D. B. Vitaliano 1981).

Bonnichsen, B. 1982. Rhyolite lava flows in the Bruneau-Jarbidge eruptive center, southwestern Idaho. In *Cenozoic geology of Idaho*, B. Bonnichsen & R. M. Breckenridge (eds). *Idaho Bureau of Mines and Geology Bulletin* **26**, 283–320.

Bonnichsen, B. & D. F. Kauffman 1987. Physical features of rhyolite lava flows in the Snake River Plain volcanic province, Southwestern Idaho. In *The emplacement of silicic domes and lava flows*, J. H. Fink (ed.) Geological Society of America Special Paper 212, 119–45.

Bryan, W. B. 1966. History and mechanism of eruption of soda-rhyolite and alkali basalt, Socorro Island, Mexico. *Bulletin Volcanologique* **29**, 453–79.

Cashman K. V. 1988. Crystallization of Mount St Helens 1980-1986 dacite: a quantitative textural approach. *Bulletin of Volcanology* **50**, 194–209.

Delaney, P. T. & D. D. Pollard 1981. Deformation of host rocks and flow of magma during growth of minette dikes and breccia-bearing intrusions near Ship Rock, N.M. US Geological Survey Professional Paper **1202**.

Denlinger, R. P. 1990. A model for dome eruptions at Mount St Helens, Washington based on sub-critical crack growth. In *IAVCEI Proceedings in Volcanology*. Vol. 2, *Lava flows and domes: emplacement mechanisms and hazard implications*, J. H. Fink (ed.), 70–87. Berlin: Springer.

de Silva, S., S. Self, P. Francis 1988. The Chao dacite revisited. *Eos, Transactions of the American Geophysical Union* **69**, 1487 (abstract).

Dragoni, M., M. Bonafede, E. Boschi 1986. Downslope flow models of a Bingham liquid: implications for lava flows. *Journal of Volcanology and Geothermal Research* **30**, 305–25.

Duffield, W. A. 1989. Fountain-fed silicic lava flows. *New Mexico Bureau of Mines and Mining Research Bulletin* **131**, 76 (abstract).

Eichelberger, J. C. & H. R. Westrich 1981. Magmatic volatiles in explosive rhyolitic eruptions. *Geophysical Research Letters* **8**: 757–760.

Eichelberger, J. C., C. R. Carrigan, H. R. Westrich, R. H. Price 1986. Non-explosive silicic volcanism. *Nature* **323**, 598–602.

Fink, J. H. 1980. Surface folding and viscosity of rhyolite flows. *Geology* **8**, 250–54.

J. H. Fink

Fink, J. H. 1983. Structure and emplacement of a rhyolitic obsidian flow, Little Glass Mountain, Medicine Lake Highland, northern California. *Geological Society of America, Bulletin* **94**, 362–80.

Fink, J. H. 1985. The geometry of silicic dikes beneath the Inyo domes, California. *Journal of Geophysical Research* **90**, 11,127–33.

Fink, J. H. & R. W. Griffiths 1990. Radial spreading of viscous-gravity currents with solidifying crust. *Journal of Fluid Mechanics* **221**, 485–509.

Fink, J. H. & C. R. Manley 1987. Origin of pumiceous and glassy textures in rhyolite flows and domes. In *The emplacement of silicic domes and lava flows*, J. H. Fink (ed.). Geological Society of America Special Paper 212, 77–88.

Fink, J. H. & C. R. Manley 1989. Explosive volcanic activity generated within advancing silicic lava flows. In *IAVCEI Proceedings in Volcanology*. Vol. 1, *Volcanic hazards: assessment and monitoring*. J. Latter (ed.), 169–179. Berlin: Springer.

Fink, J. H. & D. D. Pollard 1983. Structural evidence for dikes beneath silicic domes, Medicine Lake Highland Volcano, California. *Geology* **11**, 458–61.

Fink, J. H., S. W. Anderson, C. R. Manley 1992. Textural constraints on effusive silicic volcanism: Beyond the permeable foam model. *Journal of Geophysical Research* **97**, 9073–83.

Fink, J. H., M. C. Malin, S. W. Anderson 1990. Intrusive and extrusive growth of the Mount St. Helens lava dome. *Nature* **348**, 435–7.

Hausback, B. P. 1987. An extensive, hot, vapor-charged rhyodacite flow, Baja California, Mexico. In *The emplacement of silicic domes and lava flows*, J. H. Fink (ed.). Geological Society of America Special Paper 212, 111–8.

Henry, C. D. & J. A. Wolff 1992. Distinguishing strongly rheomorphic tuffs from extensive silicic lavas. *Bulletin of Volcanology* **54**, 171–186.

Hulme, G. 1974. The interpretation of lava flow morphology. *Geophysical Journal of the Royal Astronomical Society* **39**, 361–383.

Huppert, H. 1982. The propagation of two-dimensional and axisymmetric viscous gravity currents over a rigid horizontal surface. *Journal of Fluid Mechanics* **121**, 43–58.

Huppert, H., J. B. Shepherd, H. Sigurdsson, R. S. J. Sparks 1982. On lava dome growth with application to the 1979 lava extrusion of the Soufriere of St. Vincent. *Journal of Volcanology and Geothermal Research* **14**, 199–222.

Ishihara, K., M. Iguchi, K. Kamo 1990. Numerical simulation of lava flows on some volcanoes in Japan. In *IAVCEI Proceedings in Volcanology*. Vol. 2, *Lava flows and domes: emplacement mechanisms and hazard implications*, J. H. Fink (ed.), 174–207. Berlin: Springer.

Iverson, R. M. 1990. Lava domes modeled as brittle shells that enclose pressurized magma, with application to Mount St. Helens. In *IAVCEI Proceedings in Volcanology*. Vol. 2, *Lava flows and domes: emplacement mechanisms and hazard implications*, J. H. Fink (ed.), 47–69. Berlin: Springer.

Keller, J. 1980. Pietre Cotte flow, Fossa di Vulcano. In *Rediconti della Società Italiana di Mineralogia e Petrologia* **36**, L. Villari (ed.), 369–414.

Macdonald, G. A. 1972. *Volcanoes*. Englewood Cliffs, New Jersey: Prentice-Hall.

McBirney, A. R. & T. Murase 1984. Rheological properties of magmas. *Annual Review of Earth and Planetary Sciences* **12**, 337–57.

McTaggart, K. C. 1960. The mobility of nuées ardentes. *American Journal of Science* **258**, 369–82.

Mellors, R. A., R. B. Waitt, D. A. Swanson 1988. Generation of pyroclastic flows and surges by hot-rock avalanches from the dome of Mount St. Helens volcano, USA. *Bulletin of Volcanology* **50**, 14–25.

Mercado, R., W. I. Rose Jr, O. Matias, J. Giron 1988. November 1929 dome collapse and pyroclastic flow at Santiaguito Dome, Guatemala. *Eos, Transactions of the American Geophysical Union* **69**, 1487, (abstract).

Miller, C. D. 1985. Holocene eruptions at the Inyo volcanic chain, California – implications for possible eruptions in the Long Valley caldera. *Geology* **13**, 14–7.

Nakada, S. & T. Fujii 1993. Preliminary report on the activity at Unzen volcano (Japan), November 1990–November 1991: Dacite lava domes and pyroclastic flows. *Journal of Volcanology and*

Geothermal Research **54**, 319–33.

Neumann van Padang, M. 1933. De uitbarsting van den Merapi (Midden Java) in de jaren 1930-1931. *Ned. Indies, Dienst. Mijnbouwk. Vulkan. Seism. Mededel.* **12**, 1–135.

Pinkerton, H. & R. S. J. Sparks 1978. Field measurements of the rheology of lava. *Nature* **276**, 383–6.

Pollard, D. D., P. T. Delaney, W. A. Duffield, E. T. Endo, A. T. Okamura 1983. Surface deformation in volcanic rift zones. *Tectonophysics* **94**, 541–84.

Reches, Z. & J. H. Fink 1988. Mechanism of intrusion of the Inyo dike, Long Valley caldera, California. *Journal of Geophysical Research* **93**, 4321–34.

Rose, W. I. Jr, T. Pearson, S. Bonis 1977. Nuees ardentes eruption from the foot of a dacite lava flow, Santiaguito volcano, Guatemala. *Bulletin Volcanologique* **40**, 53–70.

Rose, W. I. Jr 1973. Nuee ardente from Santiaguito volcano, April 1973. *Bulletin Volcanologique* **38**, 365–71.

Rose, W. I. 1987. Volcanic activity at Santiaguito volcano, Guatemala, 1976–1984. In *The emplacement of silicic domes and lava flows*, J. H. Fink (ed.). Geological Society of America Special Paper 212, 17–28.

Sampson, D. E. 1987. Textural heterogeneities and vent area structures in the 600-year-old lavas of the Inyo volcanic chain, eastern California. In *The emplacement of silicic domes and lava flows*, J. H. Fink (ed.). Geological Society of America Special Paper 212, 89–102.

Savage, J. C. & R. S. Cockerham 1982. Earthquake swarm in Long Valley caldera, California, January 1983: evidence for dike inflation. *Journal of Geophysical Research* **89**, 8315–24.

Scott, W. E. 1987. Holocene rhyodacite eruptions on the flanks of South Sister volcano, Oregon. In *The emplacement of silicic domes and lava flows*, J. H. Fink (ed.). Geological Society of America Special Paper 212, 35–55.

Shaw, H. R. 1963. Obsidian-H_2O viscosities at 1000 and 2000 bars in the temperature range 700 to 900°C. *Journal of Geophysical Research* **68**, 6337–43.

Sheridan, M. F., G. Frazzetta, L. La Volpe 1987. Eruptive histories of Lipari and Vulcano, Italy, during the past 22,000 years. In *The emplacement of silicic domes and lava flows*. J. H. Fink (ed.), Geological Society of America Special Paper 212, 29–34.

Sieh, K. & M. Bursik 1986. Most recent eruption of the Mono craters, eastern central California. *Journal of Geophysical Research* **91**, 12539–71.

Spera, F. J., A. Borgia, J. Strimple, M. Feigenson 1988. Rheology of melts and magmatic suspensions, 1. Design and calibration of concentric cylinder viscometer with application to rhyolite magma. *Journal of Geophysical Research* **93**, 10,273–94.

Spera, F. J., D. A. Yuen, S. J. Kirschvink 1982. Thermal boundary layer convection in silicic magma chambers: effects of temperature dependent rheology and implications for thermogravitational chemical stratification. *Journal of Geophysical Research* **87**, 8755–67.

Swanson, D. A. & R. T. Holcomb 1990. Regularities in growth of the Mount St. Helens dacite dome, 1980-1985. In *IAVCEI Proceedings in Volcanology*. Vol. 2, *Lava flows and domes: emplacement mechanisms and hazard implications*, J. H. Fink (ed.), 1–24. Berlin: Springer.

Swanson, D. A., D. Dzurisin, R. T. Holcomb, E. Y. Iwatsubo, W. W. Chadwick Jr, T. J. Casadevall, J. W. Ewert, C. C. Heliker 1987. Growth of the lava dome at Mount St. Helens, Washington. In *The emplacement of silicic domes and lava flows*, J. H. Fink (ed.). Geological Society of America Special Paper 212, 1–16.

Twist, D. 1985. Geochemical evolution of the Rooiberg silicic lavas in the Loskop Dam area, southeastern Bushveld. *Economic Geology* **80**, 1153–65.

CHAPTER TWO

The blocky andesitic lava flows of Arenal volcano, Costa Rica

Scott R. Linneman & Andrea Borgia

Abstract

Since 1968, Arenal has produced more than 500 blocky basaltic andesite lava flows. Study of these and earlier lavas indicates a heirarchy of geological units which structurally link the simplest flow to the volcano as a whole: unit flows, composite flows, lava fields and lava armor. This hierarchy yields insights into the growth of Arenal and identifies key field observations for better understanding blocky flow dynamics and associated hazards.

Introduction

In this chapter a variety of field observations made on active and inactive blocky lava flows at Arenal volcano, Costa Rica, are presented. These descriptions are intended to help in the understanding of lava flow dynamics and monitoring, essential to volcanic hazard assessment. This chapter may be read in one of two ways. First, the reader may be drawn to the figures and their extensive captions which supply an efficient characterization of the morphology and dynamics of the lava flows at Arenal. Alternatively, a more complete understanding of the lava emplacement process may be attained through the comprehensive text descriptions and discussions.

Arenal volcano is a composite cone situated in northern Costa Rica between the Central American volcanic arc segments of the Cordillera de Guanacaste and the Cordillera Central (Fig. 2.1). Arenal is small; its nearly symmetrical cone rises only about 1,100 m above the surrounding topography (Fig. 2.2) with a total volume of about $1.5 \times 10^9 \, m^3$. Arenal is young; the oldest associated eruptive unit is dated at less than 3,000 years BP. The growth of the cone has been distinctly episodic. Each eruptive

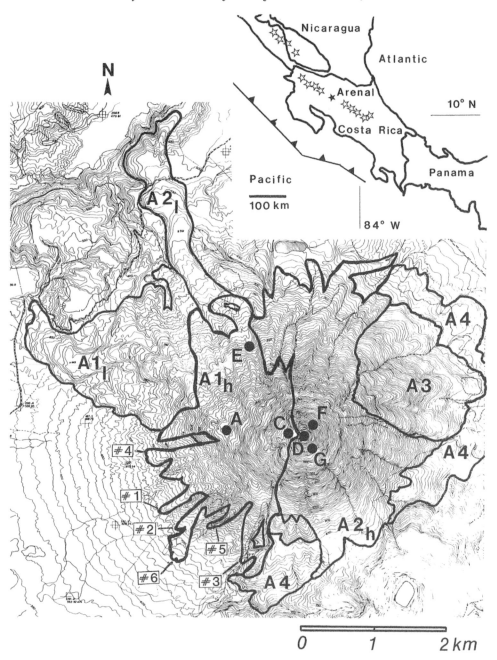

Figure 2.1 Topographic and geological map of Arenal volcano with the areal distribution of the various lava fields. Note the breaks in slope and parabolic intersections at lava field contacts. Key to lava fields: A1$_l$ (erupted from crater A, beginning in 1968); A1$_h$ (crater C, beginning 1973); A2$_l$ (crater E); A2$_h$ (crater D); A3 (crater F); and A4 (crater G). Positions of measured composite flows (#n) are also shown. Chato volcano is at the lower right. Note the angular contact between the Arenal A4 lava field and Chato volcano that could be wrongly interpreted as formed by erosion along conjugate faults. (After Borgia & Linneman 1990.)

Figure 2.2a Arenal volcano May, 1982: looking east, several features of Figure 2.1 can be seen. From the top down: the old summit (crater D), the active lava cone (crater C), lava field $A1_h$ (dark grey), lava field $A1_l$ (lighter grey) and the devastation zone from the 1968 blast (lightest grey, DZ). Chato volcano is the truncated cone at the right.

Figure 2.2b July 1984: looking south, the young upper lava field ($A1_h$) forms the upper right horizon and further to the right the shallower horizon is the lower lava field ($A1_l$). Note the abrupt break in slope at the contact between the two lava fields. The fully vegetated lava field to the left is A3. The deep incisions in the centre (ET) were formed by erosion of abundant prehistoric tephra and breccia, not by faulting of the edifice. Prehistoric lava flows ($A2_h$) with sparse vegetation are visible just to the lower right of the erosional incisions.

episode produced a large volume flow field, each flow field being approximately 15% of the present volume of the cone (Borgia et al. 1988). The compositional variation of the lavas which make up these flow fields is remarkably small. Although more-evolved tephra units have been found, virtually all of the lavas of Arenal are high-alumina basaltic andesite (SiO_2 54–57 wt%) (Malavasi 1979, Reagan et al. 1987, Borgia et al. 1988).

The nearly 20 years of continuous effusion of lava flows at Arenal volcano have offered a unique opportunity to study in great detail the development of the two lava flow fields erupted from this small basaltic andesite lava cone. This study has high-lighted features that are distinctive of Arenal (and perhaps of other small andesitic cones) and different from other volcanoes where lavas are more fluid or more viscous. By both monitoring active lava flows and studying prehistoric flows at various erosional levels, a complete model of the mechanics of lava flow emplacement has been assem-bled that extends from the most simple unit flow to the construction of the volcano. The Arenal example illustrates that only through the integration of all available information on active and ancient flows may one achieve a valid lava flow model. Such a model may, in turn, form the basis for reliable volcanic hazard determination and consequent mitigation.

In this chapter the detailed descriptions of Arenal flows are divided into three parts. In the first part, descriptions of the active flows pertain specifically to the period of continuous effusion of lava flows between January 1980 and June 1983. The geometric and kinematic characteristics of Arenal flows reflect a remarkably simple lava distribu-tion system. The simple structure of the lava flows is derived from the limited range of three variables: low effusion rate, high viscosity and nearly constant density, due to sparse vesiculation. In the second part, observations on prehistoric flows that bear on the morphological, stratigraphic and structural interpretation of the volcano are presented. Finally, in the third part, some directions for monitoring future andesitic eruptions and for assessing volcanic hazards are provided.

Some of the observations of the Arenal flows reported here form the basis of a mathematical model for volume-limited unit flows (Borgia et al. 1983) and an important extension of that model to volcanic form (Borgia & Linneman 1990). Because this chapter emphasizes the authors' original field observations of Arenal flows, readers interested in the modelling approach are directed to those papers.

Field observations of active arenal lava flows

The current episode of effusion at Arenal has produced more than 500 blocky basaltic andesite lava flows almost continuously since its explosive initial eruption in July 1968 (Melson & Saenz 1968, 1973, Bennett & Raccichini 1977). High effusion rates (3 m^3 s^{-1}) characterized the period 1968–74 when the flows issued from a crater low on the western flank. Since effusion shifted to the present near-summit crater in 1975, the effusion rate has remained nearly constant at 0.3 m^3 s^{-1} (Wadge 1983). Based on these rates of effusion, the current episode has erupted a total volume of almost 0.4 km^3.

The observations of constructive volcanic features at Arenal are organized according to the same hierarchy of geological units which makes up the volcano: the *unit flow* is the most fundamental component; each *composite flow* includes a number of unit flows; a *lava field* is built of several composite flows; the *lava armor* (i.e. the volcanic edifice) consists of superposed lava fields. Both the unit flows and composite flows are elongate features generally possessing mirror symmetry and consisting of distinct channel zones and frontal zones, each with unique form (geometry) and function (dynamics). The lava field consists of a conoid formed by the composite flows erupted from a single crater; therefore different parts of the lava field lack fluid dynamic continuity. The lava armor is characterized by the superposition of the lava field conoids and by their parabolic intersections.

The birthplace of a lava flow: description of the active crater

Since the beginning of effusive activity at the near summit crater in 1974, the lava flows at Arenal have issued from a continuously growing lava cone located in the horseshoe-shaped crater opened in 1968 (crater C). This cone is nearly symmetrical with steep sides (up to 40°) (Fig. 2.3a) and, during the period 1982–90, its elevation increased by approximately 10 m per year. By 1989 the cone had grown to about the height of the old (pre-1968) summit (crater D) (Fig. 2.3b). However, during the period 1980–83, the active crater was bounded on the east by the arcuate scarp produced during the 1968 explosive activity. This crater geometry limited the possible directions that a new lava could advance to the west flank of Arenal.

During normal effusive activity, lava rose from the conduit into the vent and flowed through one side of the lava cone into the active channel (Fig. 2.3c & d). This crater flow pattern would continue until disrupted. As the height of the lava cone increased, its steep flanks periodically would become unstable and collapse outward. Such a collapse would allow lava to drain into the new breach, changing the direction that the lava exits the crater. The instability which created the breach was caused by either continuous constructive factors (lava effusion rate or cone growth rate) or episodic destructive factors (explosions or earthquakes) leading to landsliding of the cone. During the period 1980–83, the authors observed these changes in crater outflow direction to occur every 6–8 months. The importance of such shifts, as discussed below, arises from the observation that they mark the end of lava supply to one long-lived composite flow and the beginning of a new one.

The elementary lava flow: descriptions of the unit flow

The *unit flow* at Arenal is a finite quantity of lava produced during continuous effusion (Borgia & Linneman 1990) (Fig. 2.4). It is composed of a continuous fluid body of lava

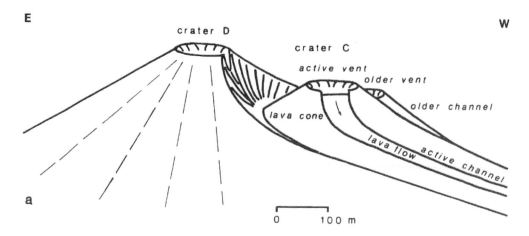

Figure 2.3a The active crater of Arenal. This sequence (a-d) illustrates a change in flow direction which marks the end of lava supply to one composite flow and the birth of another. Sketch of the lava cone looking south as it appeared in February 1983. The cone depicted here eventually collapsed to the west forming a new composite flow. (From Borgia & Linneman 1990.)

Figure 2.3b Photograph of the summit of Arenal during the explosive phase in May 1988 looking south-east. The old summit (crater D) is visible to the upper left. The young lava cone was almost at the same elevation as the old summit. Large amounts of lava breccia surrounded the cone, probably deposits of recent hot avalanches. A small flow is flowing out to the north-west (towards the observer). (Photograph by Guillermo Alvarado.)

c

Figure 2.3c Photograph of the lava cone in February 1983 looking south (as in part a). Lava flowed towards the observer with the line of fumaroles marking the right margin of the flow channel. During this interval, lava filled the vent completely (to the top of the lava cone). Such a condition led to the instability which caused the collapse of the flank of the cone to the west and formation of a new opening. In this case, the collapse was triggered by a large regional earthquake.

d

Figure 2.3d Photograph looking south of the summit of the lava cone after the change in the direction of outflow in late April 1983. The lava flowed to the right and the level was then about 2 m *below* the crater rim. Notice that the top surface of the lava is scoriaceous aa.

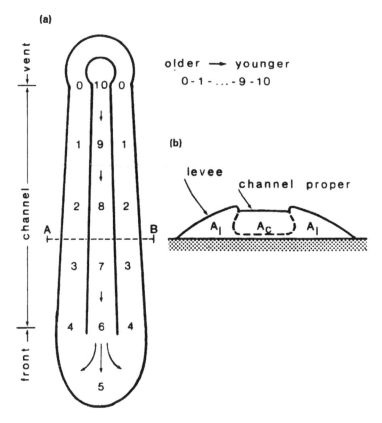

Figure 2.4 (a) Map view and (b) cross section of unit flow. Numbers decrease with increasing age of the lava. A_l is the cross-sectional area of the levées; A_c is the cross-sectional area of the channel proper. (From Cigolini & Borgia 1980.)

enclosed by debris. An idealized vertical section consists of top debris, vesiculated crust, unvesiculated core and bottom debris (Fig. 2.5) (Cigolini & Borgia 1980). A unit flow is divisible into a channel zone and a frontal zone, each of which is described in detail below. Illustrations of the different parts of the unit flow are presented in Figures 2.6–8.

The channel zone

The channel zone reaches from the vent down to the frontal zone (Fig. 2.4). Obviously, the length of the channel zone of an active unit flow increases with time. We further divide the channel zone into the channel proper, in which the lava actually flows downhill with parallel streamlines, and the levées, which are the stationary lateral boundaries of the flow. The levées are covered by debris which creates uniform slopes of approximately 42°. The channel proper has a slightly convex upward shape and is always covered with debris. The nature of the debris layer changes significantly from the vent to the front as a unit flow lengthens. Near the vent, the top debris layer can be as thin as 0.5 m and consists predominantly of aa lava with slabs of lava (Cigolini et al.

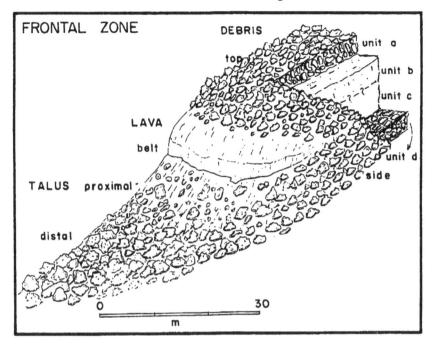

Figure 2.5 Sketch of a typical frontal zone. In cross section, four different units may be distinguished: unit a is formed by the top debris; unit b is the vesiculated crust of the continuous lava; unit c is the dense nucleus of the continuous lava; and unit d is the basal debris formed at the front over which the lava flows. (From Borgia et al. 1983.)

1984) (Fig. 2.9a). By 1 km downstream from the vent, the top debris layer can be several metres thick, consisting principally of blocky lava (Fig. 2.9b & c).

At Arenal, the levées of a unit flow are established in the frontal zone. In general, then, the path of the lava forming a unit flow is from the vent, through the channel proper to the front and from the front to the levées. Because the levées are a substantial fraction of the volume of the channel zone, the consequences of this path for the relative age of lava in a unit flow are significant (e.g. for petrological studies): the lava increases in age from *the vent to the front* in the channel proper, but from *the front to the vent* in the levées (Fig. 2.4). In addition, the general flow path also requires that the age of the lava in the channel proper increases from the vent to the front in the upper part of the flow and from the front to the vent in the bottom debris. This situation is most easily envisioned by analogy to a conveyor belt or bulldozer tread.

During its lifetime, every unit flow experiences a developing phase and a collapsing phase (Cigolini et al. 1984). Because the front advances throughout both phases, the effect of collapse is most noticeable in the channel zone. Lava from the vent continuously supplies the front during the developing phase. When the continuity with the vent is lost (i.e. effusion is cut), the unit flow begins its collapsing phase. Surface velocity profiles show clearly the difference between the developing and collapsing phases as well as the development of a thermal structure within the unit flows (Fig. 2.10). The

a

Figure 2.6a Sequence of photos (a-e) of the active flow in early March 1982. The full *channel zone* of the active flow is highlighted by the white dust formed by shearing along the longitudinal margins of the flow.

b

Figure 2.6b The complex *rear frontal zone* of this small composite flow was formed by a new flow front merging with the preceeding ones and generating a wave-like pattern. Some RFZ waveforms result from the addition of new flow fronts, while others are created by compressional deformation of the crust due to the arrival of the new front or because the front encountered a topographic obstacle.

S. R. Linneman & A. Borgia

c

Figure 2.6c The *flow front* of the same flow as it encountered a topographic obstacle (an older flow) and was changing direction (turning to the observer's left) to follow a deep gully. The core of the flow was exposed as the light grey section about one-third from the top above the long talus slope of debris.

d

Figure 2.6d The *nucleus* of the flow was exposed at the front. Notice 1–2 m of debris was on top of the core (the large block near the top is 1.5 m in diameter). The temperature in the crack was 960°C as measured by a Cr–Al thermocouple.

e

Figure 2.6e   The *talus slope* of the front was produced by blocks from the core and top debris cascading downslope in front of the flow and stopping at an angle of repose of about 40°. The cloud of dust was produced by grinding between the glassy blocks.

profiles measured near the vent during the developing phase show no evidence of the "plug flow" which characterizes profiles away from the crater. The authors interpret this difference to the growth of a cooler, high-viscosity crust away from the crater. The velocity profiles of collapsing flows show the same uniform shape of "plug flow", but at a significantly slower velocity.

The collapsing process can be compared to squeezing a tube of toothpaste although, instead of being pushed, the fluid lava is drained downward by gravity (Fig. 2.7d & e). As the fluid lava in the channel continues to drain into the front, it leaves behind a collapsed and extended crust and top debris which fills a significant part of the channel proper (Fig. 2.11a & b). The extending crust occasionally breaks into elongated sections 2–3 m thick, exposing the hot interior, which vesiculates forming glassy aa lava (Fig. 2.12a & b).

The inner surfaces of the levées are often visible in the collapsed channel zone. These surfaces are typically oxidized to a reddish color and often show flow lineations in the downstream direction (Fig. 2.11c–e). The efficiency of the drainage process, measured by the depth of collapsed channels, depends upon the steepness of the slope over which the flow is moving. Evidence for this relationship is presented below. The geometry of the collapsed channel zone of a unit flow cannot be determined uniquely because every channel at Arenal examined in detail was apparently the host to more than one unit flow (i.e. composite channels). Therefore, the shape of collapsed channels will be discussed below in the section on composite flows.

The frontal zone
The part of the unit flow farthest from the vent is the frontal zone. It is where the levées are constructed: the lava flows from the channel proper through the frontal zone and into the levées. When no more lava can drain into the frontal zone, the front stops, despite the fact that it maintains its volume and is still fluid. The frontal zone is, in many cases, the only uncollapsed section of unit flows which has stopped. The frontal zone (Fig. 2.6 & 7) consists of: (a) a distal accumulation of debris (blocks) with a slope of 38–42° and an areal extent directly proportional to the ground slope; and (b) a parabola-shaped protrusion of lava that emerges from the debris, usually 1–1.5 m in height with prominent vertical cracks. This exposed lava is surrounded on the top and sides by debris; typically the blocks are 50–80 cm across.

The multiple lava flow: descriptions of the composite flow

When several unit flows flow down the same channel system during a relatively short interval of time (days to weeks) they form a *composite flow*. Although the unit flow is a fundamental element of lava emplacement at Arenal, such simple flows are rarely

a

Figure 2.7a   Sequence of photographs (a-e) of the active flow in late February 1983. A new front (light grey) is shown flowing down the collapsed channel of a composite flow. The front of this totally confined flow eventually merged into the rear frontal zone of the composite flow about 200 m downstream (also see Fig. 2.15b). On the right is the composite levée with an accreted layer from the flow which preceeded this one by about 6 days. The new flow front was relatively small, not filling the collapsed channel zone. The largest block near the top of the new front is 2 m in diameter.

Figure 2.7b   Close-up of the same new flow front shows that very little of the continuous nucleus cropped out. In this confined flow, the nucleus was enveloped by debris, much of it sand sized, from the top and the sides. The two large blocks (about 2 m in diameter) near the top were recently detached from the core and still slightly incandescent.

Figure 2.7c   Side view of the same flow front, taken 20 minutes after (b), shows that one of the large blocks has fallen. The dark wall on the right is the interior of the levée of the collapsed composite channel which the active flow was following.

S. R. Linneman & A. Borgia

Figure 2.7d The *collapsing zone* of the same unit flow is shown progressing from left to right. The linear white areas delineate the zones of high shear on the flow margins. The channel was nearly full at the right edge of the photograph and nearly collapsed at the left edge. A central cleft of fractured crust (dark grey) marks the middle of the collapsed flow.

Figure 2.7e The nearly collapsed channel just upstream from the scene in (d) shows the same white shear zones, the central cleft of broken crust, the two convex-upward sections on the side of the cleft, and the channel walls. Another new front passed down this same channel within a few days of this photograph.

Figure 2.8 Photograph of exposed cross section of a prehistoric unit flow from lava field A2$_h$ near crater D. All of the enveloping debris has been removed by erosion, revealing the unsheared massive nucleus above the sheared (foliated) lower part of the flow. In general, the unsheared zone tends to be more vesicular than the sheared zone.

preserved. Indeed, no indisputable collapsed unit channels have been observed. That is, all observed channels apparently accommodated more than one unit flow. The importance of composite flow emplacement for the growth of lava fields at Arenal cannot be overemphasized. During February to May 1982, 13 unit flows were observed to flow down the same channel at time intervals of 5–9 days. No interruption of effusion was observed during the formation of the 13 flows, nor during the 4 year (1980–83) period of the author's fieldwork at Arenal. This constant production of new flows coupled with their low velocity meant that several unit flows were often active at any one time (Fig. 2.13a & b). The morphology of pre-1980 composite flows, including prehistoric lava flows, shows that they, too, formed by repeated emplacement of unit flows (Fig. 2.13c). An estimate for the current eruptive phase is 40–60 composite flows formed by 500–1000 unit flows. As with the unit flow, in composite flows a composite channel zone and a composite frontal zone are designated (Fig. 2.14).

The composite channel zone
When subsequent unit flows flow down the same channel they each affect the geometry (and structure) of that channel. Some general features of composite flow channels are illustrated in Figure 2.15. Two observations indicate that the most important alteration is the addition of a new layer of lava to the levée of the composite channel zone (Fig. 2.14). In several collapsed composite channels partially disintegrated levées in which

parts of earlier levée walls are exposed were observed (Figs. 2.7a & 16); the layers tend to be very thin at the crater (Fig. 2.16a) becoming as thick as 1 m further downstream (Fig. 2.16c) and closely mimic the convex (towards the channel) shape of the pre-existing levée wall (Fig. 2.7a). In between the layers of lava a layer of debris which varied in clast size from 20 cm to sand was observed (Fig. 2.16b). Also observed was the overflow of levées, in which a 0.5 m thick layer of lava was deposited on the outer debris slope of the levées. The age of lava within these composite levées decreases outwards from the core of the levée (Fig. 2.14)

The geometry of composite channel zones also suggests that layer addition plays an important role in determining their final morphology. During the period February to May, 1982, new flows exited the vent area southward by way of one of three primary composite channels. The southerly of these three was collapsed and inactive and thus its geometry measurable. The near-vent composite channel is notably deeper relative to composite channels farther down the volcano. Apparently, lava addition to the top and channel-ward portion of the levées, coupled with more efficient draining of the nucleus during the collapsing stage of each unit flow, makes the composite channel deeper. The levées are, in fact, relatively immobile to flows of even greater volume of lava than those which initially established them, allowing subsequent flows to build the levées higher.

a

Figure 2.9a Photograph of contorted slabs of "lava aplastada" (Cigolini et al. 1984) surrounded by aa (scoriaceous) lava near the crater. The vent is obscured by gasses to the left and the downstream flow direction is to the right. This photograph was taken in February, 1982, a period of relatively high effusion rate. Compare with Figure 2.3d (April 1983), where a lower effusion rate produced only aa lava.

Figure 2.9b Photograph of blocky lava typical of most flows near the base of the volcano (>1 km downstream from the crater). The blocks shown here are from a large composite flow (labelled #6 in Fig. 2.1) of the upper lava field (A1$_h$) and have diameters of 1–2 m. Unit flows on the mid-flanks have blocks 30–70 cm in diameter (compare with Fig. 2.6).

Figure 2.9c Photograph of some very large (4 m) blocks on the edge of large composite flows of the lower lava field (A1$_l$). These form pyramid-like structures protruding from surrounding blocky lava.

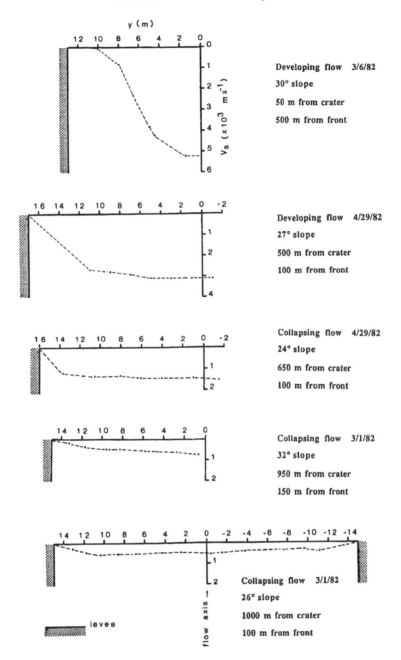

Figure 2.10 Surface velocity profiles measured on different unit flows. Date of measurement, flow phase (developing or collapsing), slope, and estimated distances from the vent (upstream) and the flow front (downstream) are presented to the right of each profile. Velocities are measured by timing markers on the surface of the flow over a known distance (10–20 m). The error in the value of the velocity is estimated at 10%.

Measurements of the dimensions of a variety of composite channels reveal that the channels vary in width (at the top of the levées) from 20 to 60 m. Wider channels were observed but not measured. The measured collapsed channels vary in depth (from a line even with the top of the levées) from 3 to 11 m. There is apparently no correlation between channel width and slope of the flow for the data on all flows (Fig. 2.17a). This means that other factors besides the hydrostatic pressure of the magma, such as topography and cooling, control the width of the flow.

In fact, when pairs of measurements on the same flow are connected in a graph a consistent slight positive correlation is noticed. That is, for individual flows, a decrease in slope results in a decreased channel width. This observation, which is the opposite to what one may expect, is a consequence of the geology of the volcano. The lava field A2$_h$ that was emplaced during the last prehistoric eruption (in Fig. 2.1) blanketed the western flank of the volcano half way down the slope. Below this only a thick sequence of unconsolidated tephra was present. Surface runoff had not eroded the lava armor, but deep stream cuts were present in the tephra (Fig. 2.13a). The limiting line of greatest channel widths in Figure 2.17a suggests that the inverse width–slope relationship holds for unconfined flows. Thus, while the flows tended to be unconfined close to the summit, they become strongly controlled by topography further down the slope. The expected

Figure 2.11a This symmetrical collapsed channel is 10 m wide at the base and is about 500 m from the crater. Note the longitudinal cleft at the centre of the collapsed flow and the convex upward shape of the broken crust on either side of the cleft. The cleft forms as the lava is drained out of the nucleus of the flow; collapse progresses from the centre outwards towards the sides.

Figure 2.11b This asymmetrical collapsed channel is on the south-west side of Arenal about 300 m from the crater. Such asymmetry in levée size and shape may indicate uneven structural support from topography such as older flows or may simply be due to post-collapse slumping.

Figure 2.11c Collapsed channel about 60 m south of the crater shows the characteristic shape of the levée walls: convex towards the centre of the channel and becoming subvertical near the base.

The blocky andesitic lava flows of Arenal volcano, Costa Rica

Figure 2.11d The verticality of the same levée wall (in (c)) illustrates how it might be misinterpreted as a fault surface in prehistoric flows. Notice, however, the irregular, flow-generated bulges.

Figure 2.11e A close-up of the surface of the same levée wall (in(d)) shows the flow striations which may also be confused for fault slickensides. The arrow indicates the flow direction.

Figure 2.12a This extended crust was produced by the collapse of a flow on a steep (38°) slope approximately 1.5 km south of the crater. As the crust broke into elongated sections, the newly exposed core vesiculated. No levées were observed in this case, but similar features were found in collapsed channels with well-defined levées. Such extensional collapse is rare at Arenal.

Figure 2.12b A close-up of the scoriaceous surface produced during extensional collapse. The length of the clinometer is 12 cm and the flow direction is down.

relation of wider flows on shallower slopes is overruled by the topographic constraints. Therefore, extreme care must be taken when interpreting geometric measurements of lava flows for which the underlying topography is unknown.

On the other hand, if measurements of collapsed channel depth versus slope from the same flow are plotted we see clearly the expected positive correlation (Fig. 2.17b). Such a correlation probably reflects the increased efficiency of the drainage process during

Figure 2.13a Aerial photographs (a-c) of Arenal showing distributary composite flows. (Photographs from the Instituto Geographico National.) 4 September 1971: lava was effusing from crater A (compare with Fig. 2.1). Notice how the various pulses of lava formed the composite flows. The arcuate ridges and prominent levées can be identified on the prehistoric flows (on the left side of the photograph) as well as on the new flows. To the right of crater A is a linear feature formed by preferential erosion of the tephra units below the lava armor. Such a lineament could easily be mistaken for a fault. The site of crater C is marked by fumaroles near the summit of the volcano.

Figure 2.13b   5 March 1980: a complex composite flow was forming on the south flank. Notice in this distributary system that levée overflows produce new channel branches followed by collapsing of the old channel downstream.

Figure 2.13c   5 March 1980: composite flows on the north-west flank. Compare the similar forms of the recent (lava field A1$_h$) and prehistoric (lava field A2$_h$) flows including composite channels with accretionary levées and multiple unit flow fronts.

49

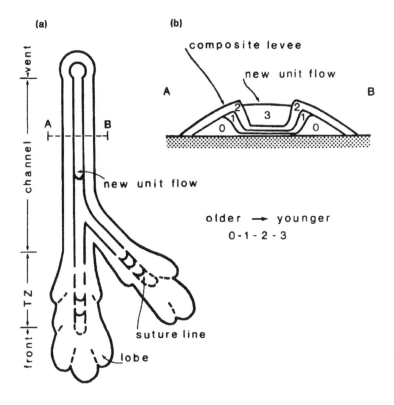

Figure 2.14 (a) Map view and (b) cross-sectional sketches of a composite flow. Numbers decrease with increasing age of the lava. Notice the composite nature of the levée, the transition zone (TZ) , and the composite front with lobes and suture lines. Compare with aerial photographs of Figure 2.13. (From Borgia & Linneman 1990.)

collapse on steeper slopes. Also noticed was that the continuous accumulation of unit flows to form a composite flow makes the top surface of the composite channel zone approach with time a constant slope.

The composite frontal zone
The unit flows which descend the composite channels feed into the composite frontal zone (Fig. 2.14). The composite frontal zone is an accumulation of unit flow fronts; most of which can remain fluid enough to deform upon the arrival of a new front from behind. The composite frontal zone are generally much larger than their unit flow equivalents, often having various lobes which may move discontinuously and independently of one another. The unit flow fronts arriving at the rear of the composite frontal zone appear to "sink" into its back (Fig. 2.18). This produces parabolic contacts on the surface of the composite flow. Once this process has been observed directly, such features become obvious on aerial photographs (Fig. 2.13b). Because of the morphological complexity of the composite frontal zone, it is often difficult to distinguish a

boundary between frontal and channel zones. The terminology of Kilburn & Lopes (1991) is adopted by dividing the composite frontal zone into an upstream portion (rear frontal zone, RFZ) which acts as a transition zone between the channel and front, and the snout which is characterized by a readily observable three-dimensional flow pattern.

The topographic profiles of several snouts on relatively shallow slopes (6–9°) show a uniform shape despite a factor of 2 difference in size (compare Figs. 2.7a, 18a & 19a). These snouts are 20–40 m in length from the start of continuous debris to a slope change which approximates the ground slope. The point marking this change of slope for these measured flows can be 10–20 m above the ground surface (Fig. 2.19a, b, d, & f).

The shape of the RFZs of composite flows reflects the complexity of the composite frontal zone, the underlying topography and the transition from front to levéed channel. Topographic profiles across the RFZ (Fig. 2.19) reveal the composite levées as longitudinal ridges flanking the flow above the characteristic 40–42° slope of the side debris. The RFZ is convex upward between the levées. The symmetry of the profile is dependent upon the underlying topography as shown by the profiles of frontal zones banked against a hillside (Fig. 2.19a & c). The profiles also demonstrate the complexity of composite frontal zones. The contacts between unit flows form *suture lines* on the upper surface of composite flows. In the centre of the flow, the age decreases upwards towards the crater (Fig. 2.14).

a

Figure 2.15a Composite flows. 13 December 1983: this active composite flow had a composite channel defined by the parallel white lines. The fumaroles near the bottom of the flow indicate where unit flow fronts have merged. The nearly unconfined composite flow front is growing in height and width by the addition of unit flow fronts.

b

Figure 2.15b   26 February 1983: the same composite flow front 2 weeks later, confined by a valley formed from prehistoric and recent flows, was almost dead. The surface of the front was nearly flat and it had advanced only 300 m in the 2 weeks preceeding this photograph. Notice the dark "belt" of exposed nucleus lava with talus beneath it.

c

Figure 2.15c   30 March 1983: the same composite channel is shown after the flow out of the crater was diverted to the west (left), an event described in Figures 2.3c & d. In this photograph the channel is collapsed and empty. The large blocks of lava in the foreground are at the edge of the composite flows of the lower lava field (Fig. 2.9c).

The RFZ often preserves waveform surface features. These topographic waveforms are typically oriented with their crests and troughs perpendicular to the levées (and the direction of flow); however, exceptions to this rule have been observed. The waveforms vary greatly in wavelength (5–40 m) and amplitude (0.3–2.0 m). However, only a weak correlation exists between these two dimensions (Fig. 2.20a). Similarly, little or no correlation is observed between wavelength and slope (measured from crest to crest) or amplitude and slope. The waveforms are basically symmetric as demonstrated by the one-to-one correlation of trough-to-crest measurements with half-wavelengths (Fig. 2.20b). The asymmetry of the waveforms which fall above the error envelope may identify these as composite flow fronts and not as folding due to compression and shortening of the crust. The symmetric waveforms result from the deformation of the crust in the RFZ in response to the compression produced by normal stresses transmitted from the channel to the front or by obstacles, such as a hillside or reduction in the ground slope, which cause the frontal zone to turn or slow and inflate, building compressive longitudinal stresses in the rigid crust of the RFZ.

The product of an eruptive episode: description of the lava field

The result of a single eruptive episode at Arenal is a number of composite flows erupted from the same crater. Collectively, these flows form a *lava field*. In contrast to the unit and composite flows, lava fields do not have channel or frontal zones. Two distinctive features of the Arenal lava fields are noteworthy: their shape and the general pattern of age relationships within a field.

A well developed lava field tends to have radial symmetry and a conoidal shape with the axis at the crater common to all of the composite flows. In profile, the lava field will have a constant slope near the crater, but will be concave upward in the distal portion. The uniform slope proximal to the crater results from the emplacement of the "last" layer of composite flows. The concave part of the profile is produced by the second noteworthy feature, a general reduction of flow length with time. The distal edges of a lava field tend to be formed by the earliest emplaced composite flows. Prominent collapsed composite channels form radial topographic barriers for subsequent flows. The isochrons within a single lava field will tend be older at shallower lower elevations to younger on the steeper higher elevations.

An analysis of the six lava fields of Arenal and two from the adjacent Chato volcano, each 10–15% of the present volume of the volcano, reveals an inverse correlation between the horizontal length and the slope of the lava fields (Fig. 2.21a; Borgia & Linneman 1990). Simply stated, lava fields are shorter on steeper slopes. Though this concept is counterintuitive, it has been verified by observation of the two modern lava fields (cf. lava fields Al_l and Al_h in Fig. 2.1).

a

Figure 2.16a   Accretionary levées. These thin, vesiculated and discontinuous accretionary levées are at the vent. Such features could be misinterpreted as dikes which intruded the shear zone between the levées and the crust of the flow.

b

Figure 2.16b   Further downstream, the accretionary levées become thicker and have an unvesiculated core.

c

Figure 2.16c Multiple accretionary levées are exposed by erosion near the active vent. The dense cores of these levées are 60–80 cm thick. A partially collapsed accretionary levée in a more distal part of a composite channel is shown in Figure 2.7a.

Figure 2.16d Prehistoric accretionary levées crop out in a stream bed near the summit crater. The stream has eroded through the lava breccia and exposed three distinct layers of lava separated by breccia.

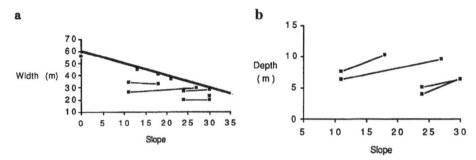

Figure 2.17 Plots of collapsed composite channel geometry. (**a**) Effect of slope on channel width. The plot reveals that the channels vary in width (at the top of the levées) from 20 to 60 m. Wider channels were observed but not measured. There is apparently no correlation between channel width and the mean slope of the flow at which that width was measured for the data on all flows. When pairs of measurements on the same flow are tied together, we notice a consistent slight positive correlation. That is, for individual flows, a decrease in slope results in a decreased channel width. This observation, which is the opposite of what one may expect, is a consequence of the geology of the volcano which allows flows to be less topographically constrained on the upper flanks of the volcano relative to the deeply dissected lower flanks (see Fig. 2.13a). The expected relation of wider flows on shallower slopes is overruled by the topographic constraints. The limiting line of greatest channel widths suggests that the inverse width–slope relationship still holds for unconfined flows. (**b**) Effect of slope on collapsed channel depth. The measured collapsed channels vary in depth (from a line even with the top of the levées) from 3 to 11 m. When measurements of *channel depth* versus *slope* from the same flow are plotted, the expected positive correlation can be seen clearly. Such a correlation probably reflects the increased efficiency of the drainage process during collapse on steeper slopes.

The structure of the volcano: description of the lava armor

The sum of the lava fields constitutes the *lava armor*. At Arenal, the intersections (contacts) between lava fields are parabolic (Fig. 2.1). This shape is the result of the superposition of lava field conoids which are not coaxial (Fig. 2.21b). That is, the summit craters which form the individual lava fields tend to be displaced from one another. The authors suggest that this occurs because the solidified conduits are more difficult to re-open than to break through in another place.

Interpretation of geological features

The observation and interpretation of geological features of non-active (dead) flows is essential in understanding lava flow emplacement dynamics, which in turn is fundamental to hazard determination. From another point of view, some of the aforementioned observations about the morphology and emplacement mechanisms of blocky lava flows at Arenal may be beneficial for the field- and photogeologist trying to recognize lava formations in volcanic terrains and the relations between them. In most volcanic terrains, *outcrops* represent a limited, covered and/or eroded view of the original lava flow. Such outcrops could be interpreted incorrectly if one does not consider the original mechan-

(a)

Figure 2.18a Composite flow fronts. The irregular upper surface of this profile of an unconfined composite front illustrates the superposition of unit flows. The basal slope is 10°.

(b)

Figure 2.18b Quarrying exposed this cross section of a unit flow nucleus within a composite flow front from the lower lava field A1₁.

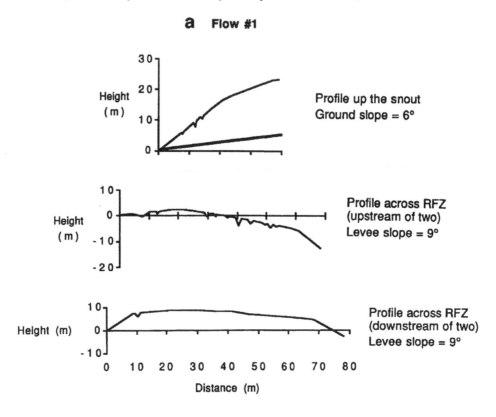

Figure 2.19a Topographic profiles of frontal zones (a-f). The locations of these six flows are shown in Figure 2.1. The profiles "up the snout" were measured parallel to the apparent flow direction. The ground surface is shown schematically as a line with a slope projected from downstream of the front. The profiles across the RFZs were measured perpendicular to the apparent flow direction. The "levée slope" expresses the angle of slope along the top of the levées parallel to the flow. Flow #1 was designated #8 by the Instituto Costarricense de Electricidad (ICE #8). The asymmetry of the upstream profile demonstrates the effect of the underlying topography, in this case a ridge on the left side.

ism of their emplacement. In the following sections a number of geological situations are presented, at both outcrop and aerial photograph scale, which may require specific knowledge of the mechanisms of lava flow emplacement for correct geological interpretation.

This discussion of field and photographic interpretation of volcanic features is by no means complete. It is the result of the authors' field experience and applies to Arenal. It should not be extrapolated to other volcanoes without considering the appropriate differences in lava chemistry, mechanism of emplacement and climate of the study area. For general references on volcanic facies the appropriate sections in Cas & Wright (1987) are recommended. The "rule of thumb" for an active volcano like Arenal is that constructional processes dominate over erosional processes. For instance, streams tend to be confined by lava flow emplacement rather than by erosional entrenchment.

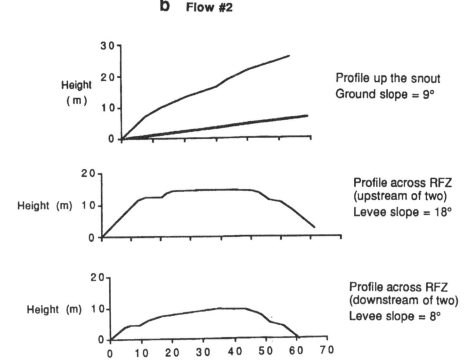

b　**Flow #2**

Profile up the snout
Ground slope = 9°

Profile across RFZ
(upstream of two)
Levee slope = 18°

Profile across RFZ
(downstream of two)
Levee slope = 8°

Distance (m)

Figure 2.19b　Flow #2 was designated ICE #9. The change in slope in the snout profile may reflect the emplacement of another unit flow into the composite frontal zone.

Distribution of unit flows, composite flows and lava fields

Contacts
Unit flows form composite flows. The contacts between unit flows form *suture lines* on the upper surface of composite flows. In the centre of the flow the age decreases upward towards the crater. Unit flows also produce the accretionary levées of composite flows with concentric structure. The age of lava within these levées decreases outwards from the core of the levée (Fig. 2.14).

Composite flows form a lava field. The contacts between composite flows tend to be *elongate* from the crater to the lava field margin. But composite flows tend to become shorter with time; older composite flows are found at lower elevations and on shallower slopes than younger flows which are emplaced at higher elevations and on steeper slopes (Fig. 2.21a).

A lava field is emplaced as a series of composite flows during eruption from a single vent. The shape of a lava field is conoidal: an upper conical part and a lower part with a concave upward logarithmic profile. Subsequent eruptions from different vents which emplace different lava fields have different axes of symmetry. Thus, the intersections between the various lava field conoids define *parabolic contacts* (Figs. 2.1 & 21b).

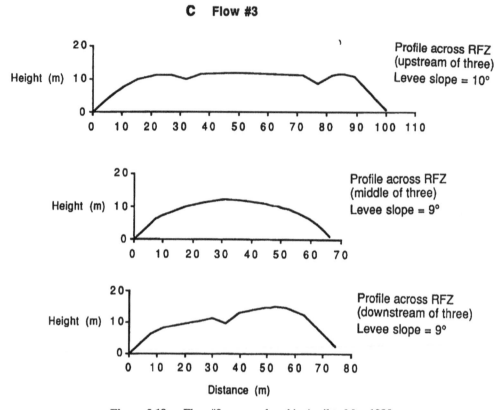

Figure 2.19c Flow #3 was emplaced in April to May 1980.

Flow direction indicators
The principal flow direction of a lava flow is not always the local maximum slope as seen in outcrop. Very commonly overspills form extensive sheet flows over the blocky talus of the levées. The flow direction of these sheet flows (which should be considered as only secondary flows) might be very different than the principal direction of flow (up to 90°). The sheet flows will generally have uniformly steep slopes of about 40° (Fig. 2.16c & d). The best indicators of flow directions are: (a) the slope of the levée ridge; (b) the suture lines of composite flows (convex downward); and (c) the orientation of elongated vesicles in the central part of the flow.

Levées versus flow fronts
The channel is drained during the collapsing stage on slopes above 20°. Thus, above 20°, the majority of massive lava outcrop consists of levées. The two levées have only blocks and rubble between them. Although it seems obvious that two levées form a flow, outcrops of single levées can easily be mistaken for a whole flow. In some cases, the accretionary levées of composite channels form structures that could later be mistaken for dikes (Fig. 2.22). Flow fronts are, on the other hand, more common on slopes less than 10°. They typically consist of a solid core covered by lava debris (Fig. 2.18b).

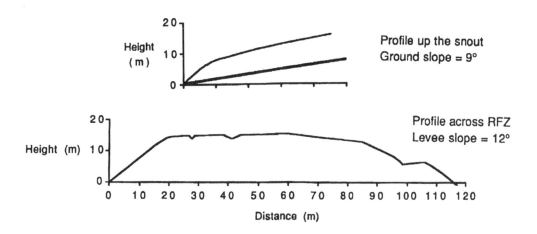

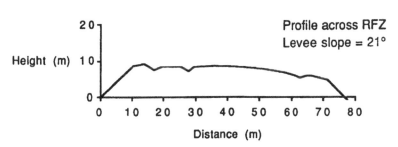

Figure 2.19d & e Flow #4 was designated ICE #19. The asymmetry of the downstream profile was produced by a topographic obstruction on the right side of the flow. Flow #5 was designated ICE #34.

Constructional–erosional features
Lack of recognition of constructional–erosional features may lead to incorrect interpretation of certain geological situations because the application of some fundamental principles is not trivial. For instance, the principle of superposition in which topographically lower units are older is not always easy to apply in the volcanic setting. This is particularly evident in the cases where erosion is responsible for removing part of the older units (Fig. 2.23 & 24).

Splitting of valleys
The emplacement of a lava flow in a valley has some consequences for later geomorphologic interpretations. First, what was originally one valley becomes two or three

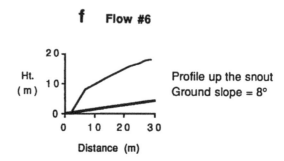

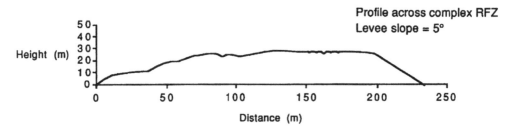

Figure 2.19f Flow #6 was designated ICE #10. Note the change of scale on the profile across this large, complex frontal zone.

after the emplacement of a flow. The drainage will begin again on the sides of the lava flow and occasionally in the centre (if the channel is collapsed) (Fig. 2.24a). In addition, the bottom and top of the lava flow are excellent aquifers with high permeability, enhancing erosion in these units. Secondly, consider the special case of a structurally controlled valley. If the valley was formed by erosion along a fault axis, emplacement of a lava flow in the valley would create the illusion that the fault is to one of the sides of the flow instead of beneath, or that there are two parallel faults (Fig. 2.25).

Across-stream erosion
Across-stream erosion produces age relations and outcrop contacts which are not always intuitive. In streams one may find one, two or three lava types, depending upon the level of erosion (Fig. 2.24b). Non-uniform erosion rates occur in different directions because the flow breccia at the bottom of the flow is eroded most easily, leading to the collapse of the flow nucleus. Erosion is slower from the top to the bottom of the flow because the nucleus is hard to erode. Such non-isotropic erosion leads to a rapid downcutting from the lateral margin to the centre of the flow perpendicular to the flow direction (Fig. 2.16d).

Along-stream erosion
Stream erosion down the flanks of a volcano may produce two patterns which are morphologically similar but which expose different stratigraphic levels. Thus, resulting outcrops have different age relationships. In the first pattern, streams cut into the same

a

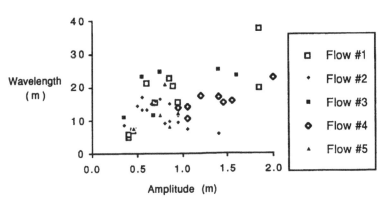

b

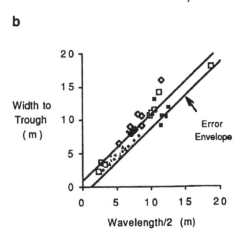

Figure 2.20 The geometry of the waveforms on the surface of rear frontal zones. These are typically oriented with their crests and troughs perpendicular to the levées. (a) This plot of wavelength versus amplitude reveals the waveforms vary greatly in wavelength (5–40 m) and amplitude (0.3–2.0 m). However, only a weak positive correlation exists between these two dimensions. (b) A plot of trough-to-crest measurements versus half-wavelengths demonstrates the symmetry of the waveforms. The asymmetric waveforms which fall above the error envelope may be composite flow fronts and not simply folds due to compression and shortening of the crust.

lava field, producing a stream profile that is a top-of-flow waterfall (Fig. 2.26). The same general morphology is produced when the downcutting stream crosses different lava fields. The major difference is the possible presence of tephra between the lava field in this second case. Both of these forms are generated because erosion is negligible at the lava surface, but it is large at the base because of the unconsolidated permeable basal rubble.

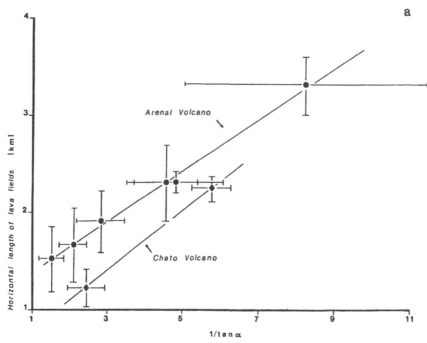

Figure 2.21a Plot of the horizontal length of lava fields at Arenal and Chato volcanoes versus the inverse of the slope. Although the intercepts for the two volcanoes are different, the slopes are equal within error. Error bars are one standard deviation.

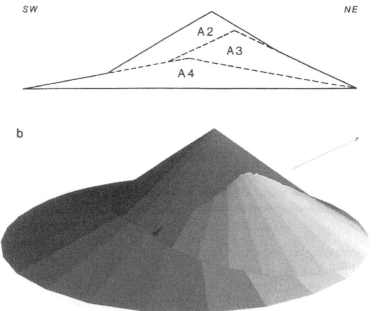

Figure 2.21b Schematic cross section and oblique perspective of Arenal depicting the superposition of non-coaxial lava flow fields. Note that the shorter flows occur on steeper slopes and that older flows crop out at the distal base of the cone. Notice also how the parabolic contacts between lava fields in the sketch correspond to the parabolic geological contacts of Figure 2.1. (After Borgia & Linneman 1990.)

Figure 2.22 Photograph of an outcrop of an accretionary levée of the prehistoric lower lava field (A2₁). The levée, which has cogenetic breccia on both sides, might be misidentified as a dike (compare with levées shown in Figure 2.16). Similarly, planar sections of levée walls, especially those with flow striations (see Figs. 2.8 & 11e), may be mistaken as fault surfaces.

Intersection of two cones
The drainage of a volcanic cone tends to be radial and tributary systems are rare. Consider the case of two cones growing next to each other (Fig. 2.27). A zig-zag pattern of the drainage between the two cones may suggest structural control (i.e. conjugate faults) of stream valleys. Such a pattern can be identified between the cones of Arenal and Chato (Fig. 2.1). This distribution of lineaments is simply produced by the intersection between the well developed radial drainage pattern of Chato and the younger Arenal whose flows "flood" the valleys of the older cone Chato (Fig. 2.1 & 27).

Lava flow fronts as fault scarps
Apparent structural control may also be the interpretation of features formed by erosion of lava flow fronts that are deposited on top of a thick pyroclastic sequence. Differential erosion of the two may lead to the formation of a cliff that resembles a fault scarp (Fig. 2.28). Such a scarp was evident on the south-west side of Arenal (Fig. 2.13a) before being covered by lavas of the present episode of effusion. This feature was initially interpreted as structural in nature.

Levées as fault scarps and dikes
The vertical walls formed by levées in collapsed channels and the striations on these walls may be incorrectly interpreted as fault planes (Fig. 2.11d & e). Some of these

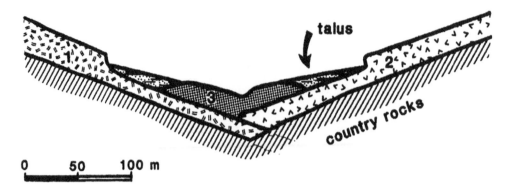

Figure 2.23   Sketch of apparent volcanic stratigraphic irregularities. Unit 1 is stratigraphically below units 2 and 3, but it may commonly be found at a higher topographic level.

vertical walls have also been erroneously interpreted as dikes, both on active and inactive flows (Fig. 2.16a & 22).

Monitoring future flows and hazard determination

The authors' experience at Arenal has convinced them that prudent volcanic hazard assessment requires a complete knowledge of lava flow emplacement mechanisms. Field geological constraints on both active flows and older flows are needed for complete characterization of emplacement mechanisms. A summary of the constraints deemed useful in the characterization of the Arenal system can be divided into morphometric and kinematic data.

Morphometry

Active flows
The geometry and dimensions of the lava flows in all stages of development and collapse should be acquired. For the frontal zone, this includes the slope and character of the underlying topography, the height and width of the snout and RFZ, the extent of talus slopes and the shape of the nucleus protrusion. Similarly for the channel zone, one would measure the width of the flow, the height of the levées and the level of the lava above or below the levées.

Dead flows
Some morphometric measurements may be made more safely on flows which are no longer active (i.e. "dead"). Since the frontal zone of flows changes little with death, the measurements suggested above may also be made on inactive flows. However, the geometry and dimensions of collapsed channels can only be determined after collapse

a

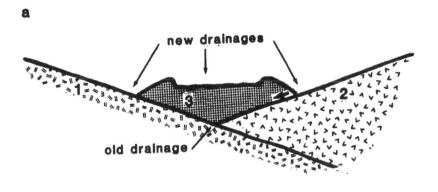

b

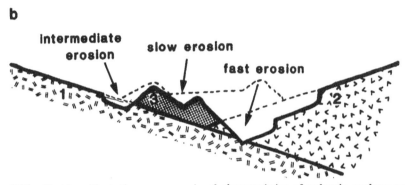

Figure 2.24 Sketches illustrating some erosional characteristics of volcanic geology common at Arenal. (**a**) This cross section shows how a stream valley may be split by the emplacement of a lava flow which followed that valley. Three new drainages may exist where only one did previously. Geological units decrease in age from unit 1 to unit 3. The heavy arrows mark the locations and directions of the most vigorous erosion. (**b**) The results of such selective erosion are portrayed in this cross-sectional sketch of the same drainage system. Notice that differential erosion produces valleys in which one, two or three different geological units are exposed.

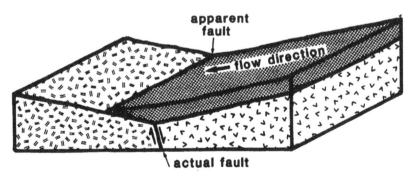

Figure 2.25 Sketch of fault misidentification due to lava flow emplacement. The correct location of structural features may be quite difficult to determine due to the rapid deposition of volcanic units such as lava flows and tephra. This is especially true where deposition is concentrated in valleys of a structural origin (see also bottom of Fig. 2.1).

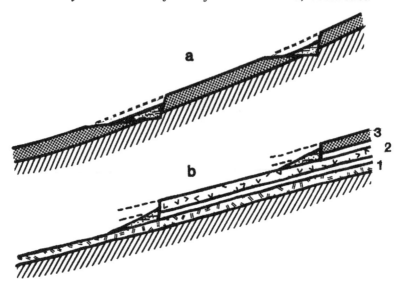

Figure 2.26 Sketch of the effect of differential erosion along a stream valley. These cross-sections show how similar morphological patterns may form by erosion through one lava field (illustrated in (**a**)) or several lava fields (illustrated in (**b**)). The cliffs and waterfalls are formed by backcutting due to the efficient erosion of basal debris relative to the massive interior of flows.

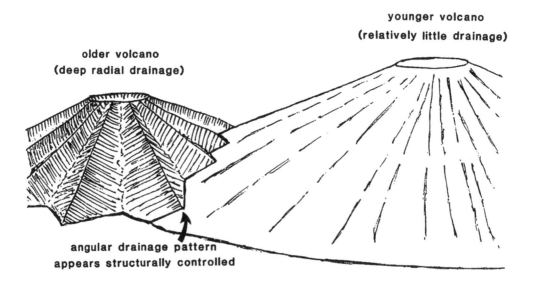

Figure 2.27 Sketch illustrating the pattern produced by the intersection of cones of different age. The unequal development of the characteristic radial drainage on the two cones produces an angular drainage pattern that may by mistaken as conjugate faults. Such an intersection occurs between Arenal and Chato volcanoes. The zig-zag pattern is evident in the south-east corner of the geological map (Fig. 2.1).

a

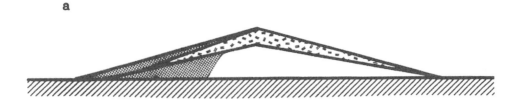

b

erosional,
not structural,
scarp

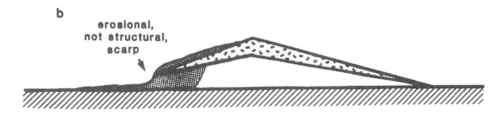

Figure 2.28   Sketch of the fault-like erosional scarp produced at a contact between lava and tephra units. The stippled pattern represents easily eroded tephra units. The distal edges of lava fields are the most likely sites for these scarps. Such a feature exists on the south-west side of Arenal and is obvious on the right side of the aerial photograph in Figure 2.13a.

has finished. These should include the width of the channel at the top and bottom of the levées, the depth of the channel and estimates of the collapsed crustal thickness. On a larger scale, the limits of individual lava fields must be known for any lava field and volcano growth model to be tested.

Kinematics

Unit flows
A variety of velocity measurements should be made at several different distances from the vent. These include the rate of lengthening of the flow, the rate of widening of the flow front, the rate of heightening of the flow front and the velocity distribution across the flow.

Composite flows
The complex nature of composite flows requires measurement of such factors as rates of change of size for individual lobes of a composite front and of the entire flow. In addition, the timing of unit flow formation at the crater must be known in order to quantify the growth of a composite flow. The timing of changes in outflow from the vent are critical in characterizing the birth and demise of composite flow systems. Similarly, the location of the principle vents must be known for the evaluation of lava field growth models.

Volcanic hazards

Complete hazard determination at Arenal must include pyroclastic eruptions that are not considered in our analysis of lava flow emplacement. Certainly, by far the greatest hazard is from pyroclastic surge eruptions and associated fall deposits. Sector collapse and catastrophic avalanching might be expected if the lava cone grows above its limit of stability

The hazards related to lava flow emplacement only include: (a) destruction of cultivated fields and pasture land; (b) damming of valleys forming small lakes (geological evidence exists for at least three such events); and (c) remotely threatening the San Gregado hydroelectric power dam. Local hot avalanches may originate from the lava flows, but these are usually small and confined to the summit area.

Summary of conclusions

The development of the two lava flow fields erupted from Arenal volcano during 20 years of continuous effusion have been studied in detail. By both monitoring active lava flows and studying prehistoric flows at various erosional levels, a complete model of the mechanics of lava flow emplacement has been assembled that extends from the most simple unit flow to the construction of the volcano. The observations of constructive volcanic features at Arenal are organized according to the same hierarchy of geological units which makes up the volcano: the *unit flow* is the most fundamental component; each *composite flow* includes a number of unit flows; a *lava field* is built of several composite flows; the *lava armor* (i.e. the volcanic edifice) consists of superposed lava fields. Both the unit flows and composite flows are elongate features generally possessing mirror symmetry and consist of distinct channel zones and frontal zones, each with unique form (geometry) and function (dynamics). The lava field consists of the flows accumulated from eruption from a single crater. Therefore different parts of the lava field lack fluid dynamic continuity with other parts. The lava armor is characterized by the parabolic intersections of the lava fields. The Arenal example illustrates the necessity of integrating all available information on active and ancient flows in order to achieve a valid lava flow model. This model forms the basis for the determination of the limited volcanic hazards from lava flows at Arenal.

Acknowledgements

Support for the field portion of this research was provided by the Latin American Program of the Associated Colleges of the Midwest and Centro Investigaciones Geofisicas, Escuela Centro Americana de Geologia of the Universidad de Costa Rica and Instituto Costarricence de Electricidad. Since that time, many colleagues have contributed to the authors' ideas via discussions and reviews of manuscripts. In particular the authors thank Steve Baloga, Stephen Blake, Michael Carr, Katherine Cashman, Jonathan Fink, Mike Kerwin, Henry Moore, Frank Spera, Robert Tilling and Ken Wohletz. The authors also thank Guillermo Alvardo for allowing them to publish his photo of Arenal's crater. Part of this work was completed while Borgia held an NRC associateship at the Jet Propulsion Laboratory, California Institute of Technology.

References

Bennett, F. D. & S. Raccichini 1977. Las erupciones del volcano Arenal, Costa Rica. *Revista Geographica de America Central, Costa Rica* **5**, 7–35.

Borgia, A. & S. R. Linneman 1990. On the evolution of lava flows and the growth of volcanoes. In *IAVCEI Proceedings in Volcanology*. Vol. 2, *Lava Flows and Domes: emplacement mechanisms and hazard implications*, J. H. Fink (ed.), 208–43. Berlin: Springer.

Borgia, A., S. R. Linneman, D. Spencer, L. D. Morales, J. A. Brenes 1983. Dynamics of lava flow fronts, Arenal volcano, Costa Rica. *Journal of Volcanology and Geothermal Research* **19**, 303–29.

Borgia, A., C. Poore, M. J. Carr, W. G. Melson, G. E. Alvarado 1988. Structural, stratigraphic, and petrologic aspects of the Arenal–Chato volcanic system, Costa Rica: evolution of a young stratovolcanic complex. *Bulletin of Volcanology* **50**, 86–105.

Cas, R. A. F. & J. V. Wright 1987. *Volcanic successions: modern and ancient*. London: Allen & Unwin.

Cigolini, C. & A. Borgia 1980. Consideraciones sobre la viscosidad de la lava y la estructura de las coladas del volcan Arenal. *Revista Geographica de America Central, Costa Rica* **11–12**, 131–40.

Cigolini, C., A. Borgia, L. Casertano L 1984. Intra-crater activity, aa-block lava, viscosity and flow dynamics: Arenal volcano, Costa Rica. *Journal of Volcanology and Geothermal Research* **20**, 155–76.

Kilburn, C. R. J. & R. M. C. Lopes 1991. General patterns of flow field growth: aa and blocky lavas. *Journal of Geophysical Research* **96**, 19,721–32.

Malavassi, E. 1979. *Geology and petrology of the Arenal volcano*, MS Thesis, University of Hawaii.

Melson, W. G. & R. Saenz 1968. *The 1968 eruption of volcano Arenal: preliminary summary of field and laboratory studies*. Smithsonian Center for Short-Lived Phenomena, Report 7.

Melson, W. G. & R. Saenz 1973. Volume, energy and cyclicity of eruptions of Arenal volcano, Costa Rica. *Bulletin Volcanologique* **37**, 416–37.

Reagan, M. K., J. B. Gill, E. Malavassi, M. O. Garcia 1987. Changes in magma composition at Arenal volcano, Costa Rica, 1968–1985: real time monitoring of open system differentiation. *Bulletin of Volcanology* **49**, 415–34.

Wadge, G. 1983. The magma budget of volcan Arenal, Costa Rica, from 1968 to 1980. *Journal of Volcanology and Geothermal Research* **19**, 281–302.

CHAPTER THREE

Aa lavas of Mount Etna, Sicily

Christopher R. J. Kilburn & John E. Guest

Abstract

The historical aa lavas on Mount Etna, Sicily, show repeatedly similar morphological associations, from the shapes of flow fields to textural variations among surface crusts. Fundamental structures are individual arterial flows, whose distribution governs the large-scale features of a flow field. At smaller scales, persistent trends along single flows involve channel and tube formation, modification of marginal levées, and changes in crustal texture from pahoehoe to aa. Flow front motion also changes from simple rolling forward to irregular, intermittent advance. In cross section, a flow can be divided into a massive core between upper and lower crusts. Crusts are especially important. They act as cool boundary layers for the hotter core and, though breaking locally to expose new lava at the surface, they can maintain net continuity across a flow, slowing widening at the front and favouring tube formation upstream. Autobrecciating basal crust may also insulate the core so that flow continues over snow without causing melting. The main structural changes of a flow are interpreted in terms of channel formation migrating downstream and surface crusting extending upstream with time. Together, these changes allow six basic facies to be recognized: poorly crusted, well crusted and roofed sheets, and poorly crusted, well crusted and roofed channels. These all relate to flow cooling and, since lava rheology is strongly temperature-dependent, may act as indicators of the local energy fluxes of a flow.

Introduction

The historical flow fields of Etna are hawaiitic in composition and most can be classified as aa (Chester et al. 1985). Although many have been covered by younger deposits, human construction or heavy vegetation, some 25–30 are well enough preserved for morphological studies. These cover areas of tens of square kilometres or less, and have maximum lengths and widths of 17 and 4 km (Lopes & Guest 1982). Although local

73

thicknesses may achieve several tens of metres, mean flow field thicknesses rarely exceed 10–15 m. Typical volumes and durations range from 10^6 to 10^8 m^3 and from hours to hundreds of days (Guest 1982, Lopes 1985), while mean effusion rates (volume/duration) normally lie between 1 and 100 m^3 s^{-1} (Lopes & Guest 1982, Romano & Sturiale 1982).

Despite the ranges of dimension and emplacement time involved, Etna's flow fields repeatedly show similar associations among morphological features. Such associations imply regularity in the overall growth of aa lavas, encouraging the view that their long-term behaviour can be forecast. Attention is here focused on the major morphological trends among individual flows on Etna. The trends involve changes with time and position, emphasizing the dynamic nature of flow evolution. The main features considered relate to surface crusting and channel development. These features are described and then drawn together in a simple facies model which helps identify flow regimes and provides a framework for more detailed studies in the future. To set flow development in context, however, the overall growth of a flow field is first briefly reviewed.

Planimetric shapes of aa flow fields

Flow fields from single eruptions may consist of one or more morphologically distinct flows. Following Walker (1971) and Borgia & Linneman (1990), a flow is defined as a circumscribed and continuously supplied body of lava which maintains fluid continuity along its length. It has static or slowly moving margins flanking a more mobile

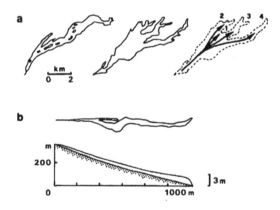

Figure 3.1 Shapes of aa flow fields. (a) Relatively narrow flow fields (left, 1981) often consist of only one arterial flow. Wider flow fields (middle, 1983) may have several arterial flows controlling their planimetric outlines (middle and right). Numbers (right) show order of emplacement of arterial flows, using data from Frazzetta & Romano (1984). (b) Plan and profile of simple aa flow (1979 Citelli). Note small variations in width and thickness for most of the flow length. The scale bar on the right refers to flow thickness.

central zone which, either roofed-over or with an open surface, feeds lava from the vent (main vent or secondary bocca) to the flow front.

When several flows form, the planimetric outline of a flow field is controlled by the shapes and distribution of a few major, or arterial flows (Fig. 3.1; Kilburn & Lopes 1991). Smaller streams fill in gaps between major flows and may locally modify the periphery of a flow field. A final flow field may thus appear as if emplaced as a continuous sheet, much larger than any of its constituent flows.

Most arterial flows on Etna are tabular in final form (Fig. 3.1), much longer than wide and much wider than deep: their ratios of length to mean width and of local width to local mean depth are commonly from 10 to 30 and from 10 to 100, respectively. Their lengths are generally less than 10 km. Times of emplacement are of the order of days, even when eruption continues for a much longer period (Fig. 3.2).

At the start of eruption, lava spreads out as a sheet, tens of centimetres deep. Because it is thin, the sheet is often confined by shallow depressions and may break up around local topography to form a braided network of small streams. Sheets are local features and rarely control the final planimetric form of an arterial flow. They are also usually preserved only among short-lived flows. Longer-duration lavas tend to bury initial sheets beneath overflows which, near the vent, may thicken a flow by several metres.

Lava advance is soon concentrated along a small number of directions. Ideally a single flow is established following the steepest slope. The flow margins come to rest, forming lateral levées and confining motion to a central channel; occasionally, parts of the channel may also stagnate to produce medial levées.

Extending from the vent, the channel structure dies out towards the flow snout, where lava spreads out laterally and fixes the initial flow width (Borgia et al. 1983, Lipman & Banks, 1987, Kilburn & Lopes 1991).

Crustal growth and crystallization of internal lava decelerate a flow front and cause it to thicken. Thickening, in turn, increases the driving force on the front, favouring

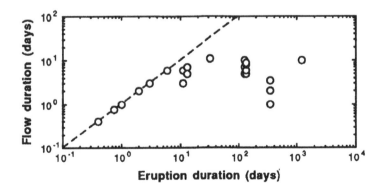

Figure 3.2 Arterial flows have durations of about 10 days or less, even when eruption continues for much longer periods. The broken line shows the trend for eruptions producing one flow. The circles indicate error.

renewed acceleration. The short-term interaction between thickening and cooling may result in near-steady advance or may induce alternating periods of faster and slower flow.

In the long term, cooling dominates and the flow widens, thickens and decelerates downstream. During the main phase of emplacement, mean velocities may drop from metres per second at the vent to centimetres per second at the front.

Mean thickening rates are typically less than 2 $m\,km^{-1}$, while flow widths can remain remarkably similar for much of the length of a flow (Fig. 3.1). Since downstream thickening induces an adverse pressure gradient against motion, bulk flow advance must be governed by the weight of lava.

In addition to frontal deceleration, lava thickening may result from (a) increasing discharge into a flow, (b) narrowing of the cross section of a channel due to inward cooling and crustal accretion, and (c) blocking of a channel by lava debris (e.g. broken crust or levée). However caused, thickening is first accommodated by flow swelling, local overflows or faster widening and lengthening. When these responses are insufficient, a new flow is formed, by marginal breaching or sustained overflow (Kilburn & Lopes 1988, 1991).

Common modes of breaching are the wholesale displacement of a margin (Guest et al. 1987) and local crustal puncturing to form secondary (or ephemeral, or rootless) boccas (Pinkerton & Sparks 1976, Cristofolini 1984). In the first case, the remobilized stretch of margin behaves as a flow front. In the second, lava from the hot channel interior escapes through the new bocca and repeats the earlier process of flow development. Frontal boccas are especially significant because they indicate that advance was initially halted by the crust and not by the flow interior; otherwise internal lava would not have been able to escape (Kilburn & Lopes 1988).

New flows may also form as an existing stream bifurcates about topographic highs. Cooling and topography thus encourage flow fields to evolve as a collection of distributary streams. One or more of these streams may be active at any moment, and so the supply rate to any particular front may be less than the eruption rate from the vent. Since Etnean flows tend to be longer when their supply rate is higher (Walker 1973, Wadge 1978, Lopes and Guest 1982, Ch. 10 this volume), the division of vent discharge among flows provides a self-imposed mechanism limiting flow field length. It also indicates that, at least for short-term forecasts of flow lengthening, knowledge of the flow supply rate is more useful than the eruption rate at the vent.

Lengthening, widening and thickening can occur throughout flow field growth. When many flows form, the underlying trend (Kilburn & Lopes 1988) is for the *dominant* growth mechanism to change with time (and, simultaneously, with decreasing effusion rate) from lengthening (of the initial arterial flow) to widening (by the lateral propagation of new flows) and, finally, to thickening (by flow superposition or by uplift of the flow surface as lava accumulates beneath). This trend, however, may contain several cycles of lengthening, widening and thickening. Two examples (the 1983 and 1989 flow fields) are summarized in Table 3.1 and show how initially superposed flows may extend significantly beyond earlier flows beneath.

Table 3.1 Lengthening cycles during aa flow field growth

---

1. *Main 1989 flow field* (Valle del Leone-Valle del Bove, 27 September–8 October 1989 (duration: 11 days); Barberi et al. 1990, R. Romano pers. comm.)

Days 1–5. Initial arterial flow halts at ~ 5.75 km from vent. The later stages are accompanied by lateral flow propagation and thickening.

Days 5–11. A superposed flow advances down the initial flow channel. By Day 8, the new lava reaches the end of the flow field and reactivates or overflows the original front. The second lengthening phase continues for 2 days (until day 10), extending the flow field by 1.25 km (~ 22% of its initial length). Eruption stops on Day 11.

2. *1983 flow field* (South flank, 28 March–6 August 1983 (duration: 131 days); Frazzetta & Romano 1984, Guest et al. 1987, Kilburn & Lopes 1988)

Days 1–16. Initial arterial flow (Capriolo flow) halts 4.5 km from vent on Day 7. The maximum flow field width (W_m) is ~ 0.3 km. Secondary lateral and superposed flows continue to be emplaced until Day 16.

Days 16–28. A superposed arterial flow (Sona flow) advances down the western edge of the flow field, increasing the flow field length to almost 7 km (an extension of ~ 55%) and W_m to ~ 1 km (an increase of ~ 230%) by Day 25. Secondary widening and thickening continue to Day 28.

Days 29–38. Another superposed arterial flow (Manfré flow) extends beyond the western edge of the Sona flow, increasing the flow field length and maximum width to respectively ~ 7.5 and 1.6 km by Day 38. Compared to the initial arterial flow, flow field length has increased by 67% and W_m by 430%.

Days 39-131. Flow field thickens by flow superposition and local accumulation of lava beneath existing surface. Between Days 58 and 62, a fourth arterial flow (Ardicazzi flow) lengthens the eastern side of the flow field, but does not increase the flow field's maximum width or length. Most other superposed flows are confined to upper reaches of flow field. Eruption ceases on Day 131.

For eruptions lasting about a month or less, the final flow field may appear as a collection of channel-fed flows. When eruptions are much longer lived, distributed tube systems may form beneath aa crusts. Once the flow field has fixed its planimetric shape, part or most of it may behave almost as if it were a lava-filled sponge (especially along the middle reaches of the flow). Increased lava pressure within tubes causes surface rupturing, the escape of small streams (often as pahoehoe tongues, including toothpaste pahoehoe) and the growth of tumuli. Much of the original structure of the early flows controlling flow field shape may thus be destroyed.

Despite variations in growth history, most of Etna's aa flow fields can be grouped into two planimetric types (Fig. 3.1; Kilburn & Lopes 1988, Hughes et al. 1990):

(a) Relatively narrow flow fields, associated with rapid emplacement within days. Frequently, only one or two arterial flows and few secondary streams are formed. Their maximum width-to-length ratios ($W_m:L$) are usually less than 0.14.

(b) Relatively wide flow-fields, containing several arterial flows and having

$W_m{:}L$ normally greater than 0.18. These are typically associated with longer-lived eruptions at lower mean effusion rates than in the case of narrower flow fields.

Such a planimetric grouping suggests that two preferred eruption regimes operate on Etna (Kilburn & Lopes 1988; a more detailed discussion is given in Hughes et al. (1990)). The key points are (a) that even elementary comparisons of overall flow field shape may identify constraints in a volcano's feeding system (see also Wadge 1978), and (b) that the existence of preferred final patterns among Etna's flow fields simplifies the task of forecasting their long-term behaviour. Importantly, though, similar *final* flow distributions may result from different time sequences of flow emplacement. The long-term behaviour of a flow field may thus be systematic, even if the propagation sequence of its arterial flows appears to be random (randomness perhaps resulting from local topographic variations or fluctuations in eruption rate).

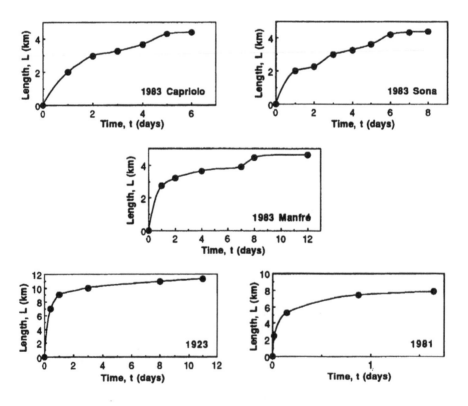

Figure 3.3 Flow lengthening. Mean rates of lengthening slow after opening stages of emplacement. The 1923, 1981 and 1983 Capriolo flows are initial arterial flows; the 1983 Sona and Manfré flows are successive arterial flows. Curves are simple interpolations. Estimated errors are ±10% on length and time. (Data from Ponte 1923, Frazzetta & Romano 1984 and Guest et al. 1987.)

Arterial flows

Rates of advance

Examples of flow advance rates (for both initial and successive arterial flows) are given in Figure 3.3. The lengthening curves tend to flatten some 10–50 hours after the start of emplacement; indeed, the flows reach 50% or more of their final length during their first 24 hours of activity. Implied velocities range from at least 0.023 m s^{-1} (1983) to perhaps 1–2 m s^{-1} (1923, 1981) at the beginning of eruption, dropping to 0.01 m s^{-1} or less by the second day.

Rapid slowing is favoured by shallowing terrain, decreasing discharge rate or increasing frontal resistance. Topographic profiles (e.g. Fig. 3.4) exclude shallowing slopes as a unique control, while available data cannot test the significance of the other two processes. Such a test is an important goal for future studies: if discharge rate dominates, frontal velocity may be limited by conditions at the vent; but, if frontal resistance is more important, flow advance may become independent of processes upstream.

Zones of flow

On the basis of field aspect alone, an arterial flow can be divided into several constituent zones. How the zones are defined depends on the features of interest. For aa lavas, two basic divisions can be made according to (a) the state of the crust, and (b) the degree of channel development.

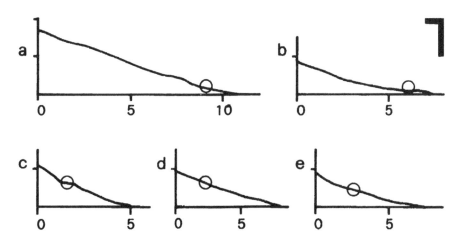

Figure 3.4 Underlying slopes for flows in Figure 3.3: (**a**) 1923; (**b**) 1981; (**c**) 1983 Capriolo; (**d**) 1983 Sona; (**e**) 1983 Manfré. The circles show where mean frontal velocities slowed significantly. These locations correspond to shallowing slopes in parts b and c, but not obviously so in parts a, d or e. The horizontal scale is in kilometres. The vertical scale is graduated every 1000 m. The scale bars at the top right represent 1 km.

(a) *Poorly crusted* and *well crusted* zones can be distinguished by the continuity of darkened crust across the flow width (the whole flow width or, for channel zones (see below), the channel width). Poorly crusted surfaces are dominated by incandescent lava. They normally occur close to a vent, where lava motion is strong enough to repeatedly disrupt chilled crust. Well crusted surfaces consist almost entirely of darkened crust. They develop downstream as a flow decelerates. Local incandescence may be seen (especially along any channel/levée boundaries) but accounts for only a few per cent or less of the surface area.

Surface colour, rheology and rates of thermal energy loss are all temperature-dependent properties (see Ch. 10 this volume). Visible by day and night, therefore, the poorly/well crusted boundary provides not only an identifiable surface isotherm, but also acts as a natural reference location for following the thermal and rheological (and, ultimately, structural) evolution of a flow (see also the experimental models of Fink & Griffiths (1990)). In particular, a well crusted surface may eventually create a static roof over a flow (whether or not a channel has formed; see later), initiating tube formation.

(b) *Channel* and *sheet* zones are distinguished by the presence and absence, respectively, of static marginal levées. Emplaced initially as a sheet, a flow soon develops marginal levées along its upper reaches, so defining its channel zone. Since channel lava continues moving, the levées are soon significantly older than the lava passing between them. The age difference between levée and channel material decreases downstream, becoming zero at the flow snout (Borgia & Linneman 1990, Kilburn & Lopes 1991). At the same time, distinct levée structures disappear and the channel zone gives way to the *frontal sheet zone* or, more simply, *frontal zone* of a flow (Kilburn & Lopes 1991).

The frontal zone, in turn, can be subdivided into the rear frontal zone (RFZ), which merges upstream with the channel zone, and the snout frontal zone (SFZ), which represents the leading edge of a flow (Kilburn & Lopes 1991). The RFZ is characterized by constant or weakly increasing depths in the downflow direction; the rates of downstream thickening are smaller than the nominal 2 m km^{-1} in the channel zone. Widening is very slow and, in the absence of significant levées, downstream motion may continue close to the flow edge.

The SFZ marks the sloping flow snout, where newly arriving lava spreads radially and fixes the front's initial width (Borgia et al. 1983). It is much smaller than the RFZ and, while the SFZ extends backwards from the flow tip by distances comparable to the mean RFZ thickness, the RFZ itself may be 10 times or more longer.

The two sets of lava zones described above provide a framework for an initial facies description of conditions along a flow and how these conditions change during emplacement. They emphasize crusting and channel formation as the fundamental processes governing aa flow growth. Clearly other factors can be added, such as changes in crustal morphology, or whether or not channels evolve into tubes. Such additional features are described below before the facies model is presented.

Channel lava and surface structures

Although classified as aa, an arterial flow commonly contains subordinate stretches of pahoehoe (Fig. 3.5). The typical downstream change in crustal morphology follows the sequence filamented pahoehoe, cauliflower aa then rubbly aa (Fig. 3.5). This section describes how the surfaces evolve. Their principal morphological features are summarized in Tables 3.2 & 3; more details can be found in Macdonald (1953, 1967), Wentworth & Macdonald (1953), Lipman & Banks (1987), Rowland & Walker (1987) and, specifically for Etnean lavas, Kilburn (1990).

Channel lava shows gross structural regularity in transverse section. Three main units with transitional contacts may be recognized (Figs 3.5 & 6, and Table 3.4): upper and lower units, composed of external crusts (continuous and broken) and fractured inner lava whose fractures reach the surface (top or bottom, respectively), separated by a larger middle unit of massive lava (see also Ch. 2 this volume). More detailed unit descriptions are given in Table 3.4.

Also in transverse section, channel surfaces may be nearly horizontal or convex upward. Surface velocity increases inwards from the levées, but at a rate which

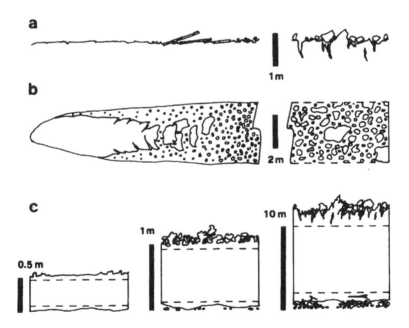

Figure 3.5 Morphological changes along a flow. (a) Longitudinal surface profile and (b) corresponding plan view show downstream change (left to right) from continuous and slab pahoehoe (both white) to cauliflower aa (small, irregular circles) and rubbly aa (detached diagrams on right). The full width of rubble aa flow is not shown. (c) Transverse profiles show textural changes and divisions into upper, middle and lower units (separated by broken lines). Cauliflower surfaces (middle) protrude upwards but rubbly surfaces (right) cleave downwards. Basal surfaces are smoother than top surfaces.

Table 3.2 Pahoehoe surface textures

Filamented and slab pahoehoe

Pahoehoe surfaces are often found at the head of a flow and may extend downstream for tens of metres. They are normally flat-lying and smooth at lengthscales of decimetres or more, although they may locally have a ropy appearance, ropes reaching tens of centimetres wide. However, at scales of centimetres or less, the crusts are highly irregular, commonly forming an intricate network of black, glassy ridges, or filaments.

When disrupted into slabs, the slab undersides often lack strong filament development. More common textures are (a) pasty, millimetre-sized spines, apparently caused by tearing from lava beneath and (b) rounded ridges and protrusions, centimetres across, resembling a collection of coalescing vesicles seen from below. The second texture suggests that some slabs were separated from underlying lava by a cushion of gases trapped beneath the chilled crust.

Subordinate pahoehoe types found on Etna are described in Kilburn (1990).

---

decreases towards the centre. Published measurements on crusted flows indicate parabolic (Oddone 1910; aa crust on 1910 lava) or shallower (Kilburn 1990; mixed pahoehoe–aa crust on 1983 and 1984 lava) velocity profiles. Because of crustal inhomogeneity and disruption, such profiles may not reflect lateral velocity changes beneath the surface; however, they demonstrate that local rates of surface shearing (i.e. local surface velocity gradients across the flow width) are smaller at the channel centre than at its margins. Secondary rolling may occasionally be evident, lava moving downwards against the levées and upwards near the channel centreline (Booth & Self 1973).

As lava moves down a channel, cooling promotes crustal growth, while stresses induced at least by the static margins favour crustal disruption. Interaction of lava cooling and deformation thus controls how a channel surface evolves with time. Deformation initially dominates near a vent. Chilled skins continually disrupt at the subcentimetre scale, yielding an incandescent (often orange-yellow) surface dotted with tiny patches of darker skin. At larger length scales, the surface may develop a weak lumpiness, forming domes centimetres to decimetres across and centimetres or less tall. Skin fragments at this stage do not accumulate at the surface. Either they are locally reincorporated into the flow, or the surface must be stretching and exposing new lava between separating patches of broken skin.

Downstream slowing is associated with decreasing surface shear rates. Chilling becomes more effective and unbroken skin extends radially from isolated centres, producing blackened clots, centimetres or more across; on lumpy surfaces, the clots tend to grow from dome crests. Near the middle of the channel (where shear rates are lowest) the clots merge to form a glassy pahoehoe crust, typically with a filamented texture (Table 3.4). At sufficiently low shear rates, pahoehoe crust covers the whole channel, producing a static roof. More often, deformation is too fast near the channel

Table 3.3 Aa surface textures

---

1. Cauliflower aa

Aa surfaces are highly uneven at decimetre to metre scales and normally support a layer of broken debris. Alongside and immediately downstream from filamented pahoehoe crusts, the lava surface develops protrusions, up to tens of centimetres tall and similar distances apart. These resemble deformed cauliflowers, having irregular outlines and a lumpiness at the centimetre scale. Their interiors are dense. When they break from their stems, they provide a loose fragmental cover over the continuous surface. The cover is rarely deeper than two or three times the size of the largest aa cauliflowers; it is normally poorly packed and may show a net reverse grading.

Essential textures of aa cauliflowers are millimetre-sized spines. Close to filamented pahoehoe crusts, the spines are usually black, glassy and dagger-like (stiletto spines), and may have associated patches of filamented surface. Further downstream, the spines become stubby, equant and less sharp; they also lose their glassy appearance, often acquiring an ochre or magenta-brown colour. The change from stiletto to stubby spines is associated with increases in groundmass crystallinity and rates of crystallization (Kilburn 1990).

2. Rubbly aa

Aa debris changes downflow from irregular cauliflowers to subrounded massive fragments. The transition is marked by an overall paling (typically to ochre-grey) in surface colour. Some fragments are abraded cauliflowers. The majority are local autobreccia which, reaching several tens of centimetres across, are larger on average than typical aa cauliflowers.

At the millimetre scale, aa rubble shows a wide range of textures. Most characteristic are subangular, equant protrusions resembling granulated sugar. These sugary surfaces are especially conspicuous when oxidized ochre-red and become more abundant with depth through the loose rubble deposit. Indeed, near the base of the deposit (typically tens of centimetres deep), the autobreccia may be tightly packed in a matrix of granules detached from sugary surfaces.

The main lava surface beneath the loose debris is fractured and highly irregular. Rubble tends to fill-in surface depressions and may leave exposed, by as much as metres, upstanding sections of underlying massive lava. Other tall pinnacles may result from upturned fragments (e.g. from a broken levée) carried downstream by the flow.

At scales of tens of metres, the mean surfaces of both cauliflower and rubbly aa tend to be even. Over shallowing slopes and near the toe of a flow, surface buckling may produce convex-downstream folds (or ogives). Crossing most of the surface, ogives may reach wavelengths and amplitudes of the order of 10 m and 1 m respectively.

---

margins and the central pahoehoe is flanked by isolated clots (Fig. 3.5). The clots rotate about near-vertical axes (their inward-facing sides moving downstream) and can twist upwards by tens of centimetres to form aa cauliflowers (Table 3.3). The cauliflowers may finally detach from their thinning stems, new cauliflowers growing in their place.

Shearing simultaneously tears the edges of the pahoehoe crust, fragments migrating with aa cauliflowers to the channel margins. The central pahoehoe narrows downstream until the whole surface is aa. Continuous pahoehoe crust may also disappear by wholesale failure to slabs (Fig. 3.5, Table 3.2). Breakage may be initiated by crustal extension, convex tears across the surface facing upstream. Subsequent deceleration (especially towards the end of emplacement) induces crustal buckling and either the overriding of

Table 3.4   Internal structure of channel lava

---

1. Upper unit.

The upper unit includes chilled lava crust and its autobreccia, together with any underlying zone of fractured massive lava in which the fractures extend to the surface. It tends to thicken downstream, from centimetres to metres, as the surface evolves from filamented pahoehoe to rubbly aa.

Close to boccas, final pahoehoe or cauliflower aa surfaces may also be cut to depths ~10 cm or less. Distinctive are clefts which curve, convex upstream, across the direction of flow. They are usually less than metres apart and support tear-spines or filamented surface textures. In longitudinal cross-section, sets of clefts tend to dip upstream, near-surface angles ranging from less than 25° to subvertical with respect to the mean flow surface. Occasionally, they may dip downstream, opening out to produce a series of wave crests.

2. Middle unit.

The middle unit consists of massive lava. Its thickness, typically decimetres to metres (although tens of metres can be attained), may be from 2 to over 10 times greater than that of massive lava in the other units. It may be homogeneous or divided into subunits (centimetres to decimetres thick for a unit thickness of metres) based on variations in vesicle content and shape- and size-frequency distributions, even though crystal content and distribution may be similar in adjacent subunits. Incipient internal rupturing may also occur, especially near the transitions to the upper and lower units.

3. Lower unit.

The lower unit comprises the uneven base of massive lava (typically thinner than its top-surface counterpart) and any fragmental deposit beneath. Basal unevenness is usually most pronounced at scales of centimetres to decimetres. Small fractures may penetrate upwards from the bottom by as much as tens of centimetres. Incipient tear-structures are frequent, often oriented at angles <45° with respect to the mean basal surface. The underlying fragments are typically contorted and may reach decimetres across. Although some fragments may derive from surface talus, deposited ahead of a flow front and then overridden, many appear to result from autobrecciation along the flow base.

---

already broken slabs, or the arching and rupture of continuous crust. Separation of the crust from underlying lava may occur by direct tearing or through formation of a gas layer beneath the solid crust (Table 3.2). The result is a collection of slabs, decimetres to metres across but centimetres thick, settling as imbricated or chaotic piles. Faster lava emerges from beneath the slabs and chills as pahoehoe or aa. If pahoehoe, the previous sequence is repeated until the surface is irreversibly cauliflower aa.

The change from cauliflower to rubbly aa is less abrupt than the pahoehoe–aa transition. It is marked by a general rounding of loose fragments (Fig. 3.7) and a net increase in their size (to several tens of centimetres across). Most of the aa rubble is locally derived autobreccia, abraded cauliflowers making a minor contribution (Table 3.2).

It is very difficult to observe rubble formation directly. By the time aa cauliflowers are abundant, the channel surface consists almost exclusively of loose debris (Fig. 3.8) and, apart from occasional protrusions of massive lava, the continuous underlying surface is rarely exposed. Disrupting processes are thus hidden from view.

Figure 3.6 Transverse section through rubbly aa. Note cleaved top surface and zone of autobrecciation to right at base. Upper debris cover has variable thickness. Massive middle unit is about 3 m thick on the left side.

Figure 3.7 Contrast between pale, abraded aa rubble and fresher, darker cauliflower aa, emplaced from secondary bocca in rubbly aa flow. The compass case is 10 cm long.

Figure 3.8 Typical rubbly aa flow surface.

Sections through solidified flows (Figs 3.5 & 6, and Tables 3.2 & 4), however, show rubbly aa surfaces cut by clefts, commonly subvertical and metres to tens of metres apart. They have irregular profiles and tend to subdivide upwards into branches of smaller-spaced cracks. As well as cutting across the flow, clefts also develop subparallel to the downstream direction, extending uninterrupted for tens of metres or less. They are lined with tear spines or sugary, friable-looking protrusions (Table 3.2; see also Fig. 3.16), and are often associated with internal cracks (not breaking the surface; Fig. 3.5). Cleft tips also tend to be surrounded by narrow, irregular tears, millimetres long (Fig.

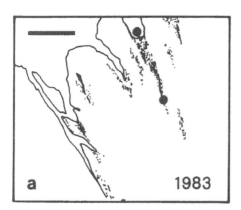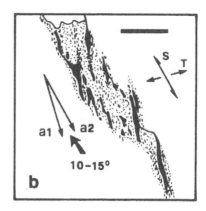

Figure 3.9 Surface clefts and internal cracks in the upper unit of a 1983 rubbly aa flow. (a) Looking downflow. The scale bar represents 10 cm. (b) Close-up of section between filled circles in part a. Internal cracks are surrounded by tiny tears. Mean apparent inclination of large crack axes (a1) is 10–15° steeper than the mean apparent trend of the whole torn unit (a2). Possibly a2 follows a shearing direction (S), while a1 runs perpendicular to the direction of tension (T). The scale bar represents 5 cm.

3.9). Notably absent are the upward protrusions characteristic of cauliflower aa. As with the pahoehoe–aa transition, therefore, the change in aa debris reflects a change in how the continuous lava surface deforms.

Lava levées

Single lateral levées
After the opening stages of emplacement, widening ceases along much of a flow and lateral motion is confined mostly to spreading at the flow front. Extending from the vent to some distance behind the front, static levées appear at the flow margins, producing the channel structure characteristic of Etnean flows.

These first levées may be preserved in lava sheets and along the lower-middle reaches of more evolved flows. They have massive interiors and show the tripartite structure seen in channel lava (Table 3.4). Termed *initial* levées (Sparks et al. 1976) when incandescent and without a layer of surface debris, they are more often already crusted upon formation. Particularly if the crust is aa, their outer margins may be flanked by fallen surface autobreccia, producing the *rubble* levées of Sparks et al. 1976; note that, in this context, the term "rubble" refers to aa debris in general, including both cauliflower and rubbly aa. In view of their massive interiors when solidified, these early margins are collectively labelled *massive* levées (Table 3.5).

Marginal solidification promotes discontinuities between channel and levée. The original massive levées are often modified by lateral overflow or accretion (Sparks et al. 1976), caused by channel thickening or migration of broken crust. Pahoehoe and aa fragments, for example, tend to rotate to the channel margins. If the channel is brimful, the fragments push on to and may cascade over the bounding levées. If the channel surface is lower, the fragments eventually accumulate along the channel–levée margin and, if they pile up sufficiently, may again roll on to the levées. The resulting *debris* levée (Table 3.5), with massive core and breccia coating, may thus resemble first-formed rubble levées. For a given type of autobreccia (e.g. cauliflower or rubbly aa), however, the loose coatings over rubble levées and channel material have comparable thicknesses while, because of fragment accumulation, the loose covers for debris levées tend to be much thicker (Table 3.5).

Debris levées may also form when the channel thickens, especially if thickening is slow enough for crustal growth to prevent overflow of continuous lava. If thickening is faster, and its amount exceeds a threshold value (equivalent to the depth of rigid channel crust and breccia cover, plus the minimum amount needed for fluid underlying lava to spread), then the levées may be coated by pahoehoe or cauliflower aa sheets, typically tens of centimetres thick. Successive thickening phases give a layered structure to such *overflow* levées (Sparks et al. 1976), levels of massive lava alternating with autobrecciated horizons (Figs 3.10 & 11, and Table 3.5).

Fragments (particularly early clots) migrating from near the channel margins may instead be hot enough to weld to the top and sides of a levée. Crustal overhangs grow

Table 3.5   Single Levées

1. Massive levées.

Massive levées have the internal tripartite structure of channel material (Table 2) and may have pahoehoe or aa surfaces. They are preserved usually among simple sheets or towards the distal zone of evolved flows. They have been termed initial levées when incandescent, but rubble levées when flanked by their own autobreccia (Sparks et al. 1976).

In rubbly aa flows, their external margins typically dip outwards at 40–45°. Their inner margins are normally abrupt, revealing a sheer face of massive lava. Exceptionally, channel drainage may cause a levée to tilt inwards and settle subvertically (Fig. 3.10).

2. Overflow levées.

Together with massive levées, overflow levées predominate on Etna's aa lavas. Common along a flow's middle and upper reaches, they have layered interiors, levels of massive lava alternating with autobrecciated horizons.

Each massive level represents a channel overflow. It normally has a pahoehoe or cauliflower aa upper surface and an autobrecciated base; the fragmental horizons thus consist mainly of basal autobreccia or aa cauliflowers, centimetres to decimetres across. Both massive and autobrecciated horizons are frequently decimetres thick and extend laterally for tens of metres or less.

The underlying massive cores may have pahoehoe, cauliflower aa or rubbly aa textures and so their thicknesses may range from tens of centimetres to metres. Successive overflows may build a levée to heights of several metres, its external surface dipping outwards by as much as 25° (Fig. 3.10). Inner slopes are steeper and normally dip inwards, the draping massive levels grading laterally into the channel material. As the slopes become subvertical, sagging overflows may break to reveal a massive surface. Occasionally, the drape remains intact, curving back on itself with depth and producing an outward-dipping inner margin.

3. Debris levées.

Typically narrower than the channels they enclose, debris levées are associated mainly with cauliflower aa and with accumulations of poorly-sorted lava fragments. With outward-dipping external slopes of about 30°–35°, the fragmental piles may reach metres in height and width, although tens of centimetres are more common. The massive lava core is rarely seen through the debris along the outer levée margins. Channel drainage may reveal the inner face of the core as a steep, inward-dipping or vertical massive unit. Above the core, the debris may again settle to slopes (dipping inwards) of 30°–35°. Debris levées generally have much thicker fragmental covers than rubble levées with similar surface autobreccia.

4. Accretionary levées.

Compared with other levée types, upstanding accretionary levées are less-frequently preserved on Etna's lavas. Associated with pahoehoe and cauliflower aa, they appear close to vents as narrow, upward-tapering walls, tens of centimetres wide and less than 1–2 m tall. Both inner and outer margins dip outwards at angles generally steeper than 50°. Their outer surfaces consist of welded lava debris which, on their inner surfaces, are usually coated by massive lava.

5. Swollen levées.

Lateral intrusion of channel lava may modify all the above levées, particularly their massive cores. As well as simple swelling, intrusion may lead to internal levée autobrecciation and to local extrusions through the levée exterior.

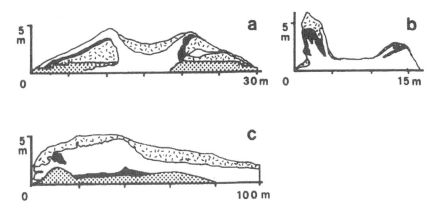

Figure 3.10 Channel/levée structures. White, massive lava. Irregular dashed stipple, loose surface rubble. Black, basal and interlimb autobreccia. Close stipple, wall in part a or ground and artificial scree due to road-works in part c. (**a**) Prehistoric flow, looking upstream. (**b**) 1971 Fornazzo flow, looking upstream. (**c**) A 1983 flow, looking downstream. Massive overflows (a & b) arch over levée cores or become upturned during channel drainage. In part c, thicker lava (left) lies at the head of a drained channel; no clear levées are apparent.

inwards over the channel (Sparks et al. 1976, Guest et al. 1987). If accompanied by slow lava thickening, distinctive, inward-sloping walls may be built upon the levée core. Such *accretionary* levées (Table 3.5; Sparks et al. 1976) occur near to vents and, because of their upward narrowing, may close over to form a channel roof.

Simple overhangs may also lead to channel roofing. Frequently, however, they remain as arches over the active channel. Overhangs are not always recognizable as such in the field and may give a misleading impression of relative channel/levée widths.

Figure 3.11 Overflow levée, looking upstream. Massive overflows arch over the levée and grade left into a massive unit of drained channel lava. Thickness of the massive unit by the channel (near "K" marked on outcrop) is about 1.5 m.

If the channel level subsides, an overhang remains unsupported and may start to crack away from the main levée core. Fluctuations in channel level may repeatedly shift the overhang upwards and then leave it to sag. Eventually the overhang detaches completely to produce a crustal raft as much as tens of metres across (Guest et al. 1987, Lipman & Banks 1987). Carried downstream, a raft may jam in a channel constriction, promoting first upstream overflows and then drops in channel level if the raft is dislodged. This process may be a significant cause of channel-level fluctuations along the upper reaches of a flow where (a) channels tend to be narrowest and (b) crustal fragments remain hot enough for welding to create the original overhangs.

Where fluid continuity persists between the channel and levée interior, flow thickening may also be accommodated by levée swelling, channel material intruding laterally rather than overflowing (Guest et al. 1987, Lipman & Banks 1987, Kilburn & Lopes 1988). Although it can occur along the whole length of a flow, lateral intrusion dominates overflow towards a flow front and when lava is rubbly aa. For low intrusion rates and small rheological gradients across the internal channel–levée contact, intrusion may maintain the massive interior of the first-formed levée. It may not be clear, therefore, whether massive levées in solidified flows reflect initial levée dimensions or include the effects of later swelling.

Higher intrusion rates promote more complex levée modifications. After a rise in channel level, static levées can be remobilized and begin pushing slowly outwards, even though well behind a flow front. While such *active* levées (Guest et al. 1987) swell and spread, intruding lava may break through their flanks, either as viscous protrusions or as small drapes over external debris. Extrusions may in turn be buried by debris avalanches down the levée exterior. At least locally, therefore, levées can acquire internal layering in the absence of channel overflows. Indeed, for higher intrusion rates and lava viscosities, it is conceivable that local layering may also result from internal autobrecciation, induced either by levée shearing or by rupture around the edges of a forced intrusion.

Cooling and modification by channel lava tend to increase the size of levées with time. The effects are most pronounced near a vent, where the levées form earliest and so are exposed for longest to lateral intrusion and overflow. Thus, while downstream levées may account for 5–10% of the width of a flow, near a vent the proportion may reach as much as 80%. Late-stage channel drainage, moreover, may leave levée tops standing several metres above the final channel surface.

Although some drained channels have upward-narrowing cross sections (Walker 1967, Sparks et al. 1976), the majority on Etna have rectangular or slightly upward-widening outlines. Whatever their geometry, the levée–channel junctions are often abrupt, indicating lateral discontinuity during at least the final stages of flow. Exposed inner levée walls frequently show massive lava surfaces (possibly only a thin veneer in some cases) crossed by subparallel lineations, marking the path of draining lava (Fig. 3.12).

Along distal parts of a flow, differences in final surface level between channel and levée are less evident. Levée–channel contacts must be inferred by weak variations in

Figure 3.12 Inner levée wall, looking upstream along a drained channel. The wall is subvertical, dipping slightly outwards. Lineations and flute-like protrusions (lower half) show the direction of channel lava (flow to lower right). Flutes may result from accretion of channel lava. The levée surface is about 2 m above the top row of flutes.

surface level – although this is often a subjective judgement – or, when possible, either by abrupt lateral changes in crustal morphology, or by the presence of longitudinal clefts as much as metres deep (Fig. 3.13). Frequently, however, even such indirect evidence is lacking and it is not clear from final structures that distinct levées develop towards the toe of a flow.

Multiple, medial and compound levées
Channel margins may also consist of two or more adjacent levées (Fig. 3.13). Such *multiple*, or nested, levées are associated with an overall waning in flow discharge rate (e.g. Oddone 1910, Naranjo et al. 1992). The resulting structures range from inward-descending steps (Fig. 3.13) to a collection of levées of comparable height, distinguishable only by angular depressions at their junctions or by differences in the character and size/frequency distribution of their debris coatings (Fig. 3.13). Similar structures also result when a drained channel is reoccupied by successively narrower streams, each draining to leave a pathway for a following flow.

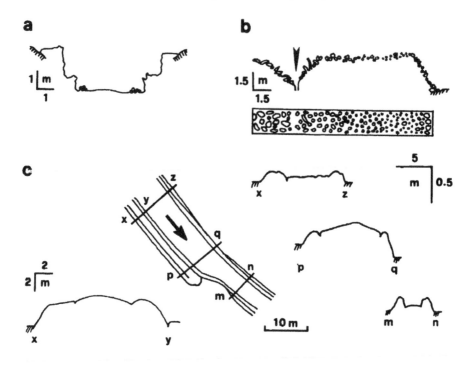

Figure 3.13 Multiple levées. (**a**) Inward-stepping profile looking downstream from a secondary bocca. The surface is mainly pahoehoe. Note topographic control. (**b**) Sharp contact (arrow) between a levée (right) and a partially drained channel, looking downflow. Note changes in sizes of aa debris over the levée surface. (**c**) Three transverse sections (diagonally right: xy, pq, mn) across a channel with multiple levées. Levées are asymmetric on either side of the flow due to stagnation of the original flow margin (xp) behind topographic high. xy shows a section across the stagnant zone.

Ideally, each member of a marginal levée has a corresponding member on the opposite flow margin. This symmetry may be broken by differences in local conditions, such as when one side of a flow banks against an obstacle while the other is unconstrained (Fig. 3.13). It may also be broken if *medial* levées form within an active channel. Associated with topographic irregularities, growth of medial levées is favoured by waning flow and may cause channels to split into braided networks or to evolve as separate streams.

When several flows are propagated, later streams may transform existing flows into *compound* levées. This occurs especially after the flow field has established its outline and new streams begin to fill the spaces between earlier flows. When the available gap is too narrow, the emplaced streams function as levées and, particularly if undrained, their original nature may be lost beneath lateral overflows from the new lava.

Lava tubes

Lava tubes are common in Etna's aa lavas (Ponte 1949, Greeley 1987), normally those with durations of weeks or more rather than days. They are most evident along the upper reaches of a flow, where channels are narrowest, and when drained show a range of cross-sectional shapes, from approximately circular and elliptical to triangular, the apex upwards, resembling a Gothic arch (Guest et al. 1980). Their vertical dimension is typically between 1 and 5 m.

Four processes have been recognized by which lava channels evolve into tubes (Peterson & Swanson 1974, Greeley 1987): inward growth of channel crust, rooted to flow margins; repeated overflow and accretion causing levées to arch over channels (Fig. 3.14); jamming of crustal fragments in channel restrictions; and wholesale attachment of a complete channel crust to the bounding levées. Importantly, all these mechanisms require crustal formation.

Inward-levée growth appears to have been the most frequently observed process on Etna (e.g. Guest et al. 1980). It may produce feeding tubes extending as much as kilometres from the vent and made visible by skylights in the channel roof, caused by incomplete arching or later roof collapse. Such tubes may merge into the channel or frontal zones of a flow, or they may terminate directly as secondary boccas. In the extreme, they may merge with and extend through the frontal zone to feed lava escaping from a secondary bocca near the snout. Since levées are poorly developed in the frontal

Figure 3.14 Roof of a lava tube revealed at the roadside. Arched limbs of massive lava meet over the tube, indicating roofing by inward levée growth due to repeated overflow.

Figure 3.15 Slowly moving 1983 flow front, covered mainly by loose aa rubble. Flow towards centre right. The massive protrusion (centre) from the lava core stands about 1.5 m tall. Note the heat haze to the right.

zone, such tube extension shows that rigid carapaces can also form directly from simple crusting over the whole width of a flow sheet.

As with pahoehoe lava (Peterson & Swanson 1974, Greeley 1987), tube formation favours increased flow lengthening by reducing rates of heat loss. This is especially the case when tubes are not completely full, allowing hot gases to collect beneath the tube roof and to maintain the flowing surface at near-eruption temperatures. Tubes may also remain active after burial by new streams propagated from elsewhere in the flow field. In such a situation, surface observations may significantly underestimate the total flux of lava being transported downstream (Guest et al. 1980).

If once a tube system has developed, the contained lava pressure increases through an increase in discharge rate, downstream tube blockages, or general downstream flow deceleration, overflows may occur through existing skylights. A greater increase in pressure may cause the roof to arch upwards and split open, producing either a linear tumulus or vertical ramparts at the flow surface. Lava may then exude through the tension cracks to form small flows and squeeze-outs over the deformed roof.

Flow fronts

An aa front advances by a mixture of continuous flow and rupture. When poorly crusted, it may roll forward as a caterpillar track. Surface skins of pahoehoe or, possibly, cauliflower aa are carried around the curved front, either peeling-away or being pulled

94

under the main lava. Such motion is common to early fronts of near-vent overflows, which may also begin to cleave vertically while rolling ahead. During the 1989 eruption of Etna, for example, aa overflows 50 cm thick developed frontal clefts (centimetres to decimetres apart) within an hour from the beginning of overflow.

As lava evolves towards rubbly aa, flow fronts thicken and acquire a bipartite syle of motion. Their lower portions continue to flow as a continuous fluid with little external fragmentation. Their upper portions advance as a single unit, sometimes steadily, sometimes lurching onwards in a series of hiccups. Irregular blocks calve or tear from the leading edge, preferentially eroding the surface layers and dumping debris into the path of the flow. Hence, although the flow top may move faster, the front maintains a backward-sloping profile (~40°) because of surface avalanching.

Eventually the front appears as a collection of chilled and incandescent aa rubble. Movement occurs by pushing the rubble ahead, the snout developing an almost linear profile with a 40–50° dip. Across the whole face, sections of internal lava irregularly extrude through the debris, emerging as oozing sheets or viscous protrusions Fig. 3.15).

Both types of extrusion occur at all levels. Sheets (often pahoehoe) drape the front and may advance ahead by a few metres. The main front catches up with them, extruding new sheets in the process. In the extreme, oozing extrusions may instead establish a sustained secondary bocca.

Viscous protrusions, in contrast, tear off as blocks, metres or less across. Tearing is accompanied by glowing cascades of sand-sized granules. Under a scanning electron microscope, the particles appear roughly cubical with angular edges and flat faces; even when fresh they show signs of intense chemical alteration.

Macdonald (1972) has suggested that granules result from attrition along fracture surfaces in a flow. Two other observed mechanisms, however, may be more significant. The first is rapid chilling of newly exposed surfaces. Cooling cracks break millimetre-sized irregularities in the surface, causing its skin to crumble suddenly into a granule cascade.

Figure 3.16 Friable or sugary textures on rubbly aa surfaces. Features resemble lightly welded agglutinate. Most, however, appear to result from primary autobrecciation. Note the abundance of granules. The lens cap is about 5 cm across.

The second mechanism is the tearing process itself. Tiny cracks appear along an incipient plane of failure, causing rupture as they join together. Detached blocks have the sugary, or friable, textures (Fig. 3.16) observed on cleaved rubbly aa surfaces (Table 3.2). However, when hammered open immediately after falling, the same blocks split smoothly to reveal incandescent massive interiors. The smooth fractures caused by sudden blows emphasize that granulation is produced during slower rupture.

Wide fronts often develop lobate leading edges. For rubbly aa, lobes metres thick are typically metres to tens of metres wide. Advancing simultaneously, adjacent lobes may fall behind and then overtake their neighbours such that they maintain similar long-term mean velocities. Otherwise, some lobes stagnate and flow is restricted to a smaller combined front.

Lobes may form after (a) amalgamation of independent streams, (b) marginal breaching or (c) bifurcation about topographic highs (Wadge & Lopes 1991). It is also possible that they appear spontaneously, without an external influence. Lobe formation may then reflect inherent frontal instability, perhaps analogous to the break-up of liquid films caused by surface tension effects (Huppert 1982). The lava supply rate may be insufficient to keep the whole front moving, due to a drop in discharge rate or to an increase in frontal resistance. In the first case, the lava attempts to reduce its cross-sectional area. In the second, the flow thickens to maintain advance and, because of higher outward pressures, starts to widen; once again, the supply rate cannot keep the whole front continuously active and the lava divides into narrower streams feeding individual lobes.

Ground heating

The ground beneath lower units usually represents the pre-eruption surface. No evidence has been found for downward lava erosion, even beneath drained channels. This might partly reflect a biased outcrop distribution. Ground erosion is favoured by high lava velocities and heating for long periods. Such conditions are most likely to be satisfied near the main vents of long-lived eruptions, precisely the locations where cross sections are rarest on Etna. If ground erosion does occur, it must be a near-vent phenomenon.

Lava heating, however, can produce baked horizons (of soil or ash-and-cinder deposits) immediately below basal units. Examples under some 1983 flows reach 50 cm in thickness. Baking gives the ground a characteristic reddish-brown, terracotta colour (literally, terracotta = cooked earth) and, indeed, thicker Etnean deposits have been mined in the past for use in pottery.

The most problematical effect of basal heating occurs when lavas pass over snow or wet terrain. Mudflows may be triggered by snow melting (Romano & Vaccaro 1986) and phreatic explosions by water vaporization (Cucuzza-Silvestri 1949, Cumin 1954, Romano & Vaccaro 1986). In 1843, 36 people were killed as an advancing front exploded (Rodwell 1878), while phreatic lava fragments set woodland alight during the 1950–51 eruption (Cumin, 1954). At the other extreme, flows can advance over snow without causing significant melting (Ponte 1923, Cucuzza-Silvestri 1949); Ponte (1923) even reported the weight of a flow turning underlying snow to ice.

The impact of basal heating is evidently sensitive to the thermal structure at the bottom of a flow. Young fronts have negligible basal breccia and thin thermal boundary layers, favouring high rates of downward heat transfer and, hence, snow melting and vaporization. More mature fronts, in contrast, may rest on substantial breccia deposits and have thicker thermal boundary layers in their massive interiors; basal heat transfer rates are small and allow only weak interactions with any wet surfaces beneath.

Dangers from aa flows

The main threat from Etna's lavas is invasion of land and property. Apart from the first few hours of emplacement, flows rarely advance faster than a kilometre per day and are slow enough to be avoided. Such generalizations are irrelevant, of course, if a vent opens within or just upstream from an inhabited area, particularly at night. Natural selection, however, has determined that most of Etna's towns lie more than a day's flowage away from common vent locations. More detailed analyses of the lava hazard from Etna are given by Guest & Murray (1979) and Chester et al. (1985); other summaries for lavas in general can be found in Bolt et al. (1984) and Blong (1984).

Most of the dangers during flow monitoring derive from high lava temperatures and transient flow processes (see Ch. 14 this volume for practical precautions). Although incandescent surfaces can be strong enough to support a person, radiant heat makes breathing difficult and can ignite unprotected clothing within minutes. Even at distances of several metres, repeated exposure to incandescent surfaces (e.g. during sample collection) can produce the irritation of bad sunburn. At the other extreme, degassing (normally of steam) may cause scalding and respiratory problems.

Heating of air over open channels frequently allows atmospheric vortices to form (Cumin 1954, Whitford-Stark & Wilson 1976). Usually exhausted within minutes, a vortex may whip up finer fragments from the lava surface and, especially if several occur in succession, may render observations difficult without eye protection.

Common transient phenomena are channel overflows, levée instability and avalanching of surface scree. These phenomena are hazardous because they can occur quickly and without warning. For example, fluctuations in effusion rate or the sudden unblocking of a crustal raft wedged between levées may send a lava wave, perhaps 2–3 m high, surging downstream and, at the same time, feeding lateral overflows. The overflows may move away from the channel at rates of several metres per second (Guest et al. 1987). They are also frequently silent.

Typical forms of levée instability are failure of levée overhangs, reactivation of levée spreading and secondary bocca formation. As mentioned previously, overhangs are not always recognizable as such, and it is wise to check for surface cracking before establishing an observation point on what appears to be a channel margin.

Levée reactivation is often slow and may take tens of minutes to become apparent.

It is most likely to occur during periods of lava thickening in a channel. Since it is slow, the major hazard is less related to shifting of the margin itself than to the triggering of debris avalanches down its sides. Rubble 30–50 cm across may weigh from about 70 to 300 kg; commonly arriving from heights of 3–10 m, these require more than hard hats to offer protection.

Finally, as with channel overflows, secondary boccas may form silently and within minutes. Again, the major threat is posed by high initial velocities of escaping lava.

Less frequent, but especially dangerous in compensation, are phreatic explosions triggered as flows move over snow or wet ground. They are most likely to occur when the lava base lacks a breccia layer and, in addition to the earlier discussion, have also been observed as lava invades residential areas, burying water wells and cisterns (Cucuzza-Silvestri 1949).

Crustal behaviour and flow emplacement

Many of the features described here are linked to crustal behaviour. This is hardly surprising since only the outer parts of active lavas can normally be observed. However, since crusts act as the boundaries between the lava interior and surrounding environment, they play a key role in flow emplacement (see Ch. 10 this volume). One example is provided by how lavas advance. As snouts mature towards rubbly aa, their styles of advance evolve from a simple rolling forwards to puncturing of outer layers by viscous protrusions or oozing sheets. Oozing sheets may cool with pahoehoe surface textures, suggesting that parts of the frontal core may have rheologies similar to that of lava upon eruption. One factor changing snout motion may thus be increased crustal restraint as lava becomes rubbly aa (lava crusting and flow advance is discussed further in Ch. 10).

More immediate effects of crustal growth can be seen as lava travels from vent to snout. Most obvious is tube formation by surface roofing. Other implications are as follows:

(a) Boundary layer growth. Growth of the upper and lower units is consistent with inward-migration of thermal boundary layers from the flow top and base (Archambault & Tanguy 1976, Pinkerton & Sparks 1976). The upper unit is invariably thicker than the lower unit (Table 3.4), indicating faster heat losses at the flow surface. The simplest interpretation is that radiation and crustal autobrecciation allow faster heat losses to the atmosphere than can occur to the ground. However, for free-surface flow, viscous shearing is greater near the ground than the free surface. Another significant factor slowing basal cooling may be higher rates of viscous heating (with respect to the upper unit) near the bottom of a flow (see also Ch. 7 this volume).

(b) Crustal transport. Downstream changes in surface texture occur throughout emplacement, textures near the vent being usually pahoehoe or cauliflower aa. The persistence of textural change indicates that, while some broken crusts are buoyed downflow, such a transport mechanism cannot dominate the whole length of a flow, otherwise crusts formed near vents should be common over the entire surface.

Textural change instead appears to be governed by surface rupture. Rupture continuously exposes underlying lava which may have had a cooling history completely different from earlier crusts. The longer lava remains in the cooler upper part of a flow, the more crystallized it is likely to be when revealed at the surface. Different crustal types are thus not related by direct transitions.

For textures to evolve downstream, rupture must allow underlying lava to escape ahead of existing crust. This implies that the surface is pulled downflow, either by the front or by underlying lava, as cooling retards the crust more strongly than the hotter lava core. In other words, the crust tries to decelerate more quickly than the core; hence, although a flow may slow downstream, its crust remains in tension. Supplementary rupture mechanisms are tension associated with shear deformation against marginal levées (the analogous case for glaciers is described by Nye (1952)), crustal cooling, and stretching of crust under its own weight.

(c) State of newly exposed lava surfaces. Before chilling effects are important, the main lava surface, beneath any debris layer, evolves from an even to lumpy (pahoehoe and cauliflower aa) to cleaved (rubbly aa) morphology. These features resemble the deformation patterns (Nye 1952, Hulme 1974, Blake 1990) expected for fluids whose viscosities decrease under increasing shear stress (including pseudoplastic and plastic fluids) as occurs for solidifying magma (e.g. Shaw et al. 1968, Shaw 1969, Pinkerton & Sparks 1978, Hardee & Dunn 1981). For such fluids, curved paths of maximum shearing (or slip lines), induced by levée friction, cross-cut each other over the channel surface. If viscosity drops abruptly above a critical stress, deformation concentrates along the cross-cutting directions, isolating zones (or lumps) of more slowly deforming lava. Chilled skin grows preferentially over the lumps, further concentrating flow around their peripheries. Upward twisting begins and aa cauliflowers are formed.

Slip line curvature produces lumps of different sizes across the flow width, larger mean sizes being favoured by larger critical stresses (see Blake (1990) for a discussion on lava domes). Lumps are thus bigger when lava is more crystallized upon exposure. Higher viscosities also slow slip line flow and inhibit cauliflower growth. The surface tends to advance as a plug. Accelerating downstream, clefting may develop when the normal stresses exceed the tensile strength of the lava (Nye 1952). Again guided by slip line directions, clefts cut convex-upstream paths across a channel and, away from the levées, may also run along its length (Nye 1952).

The change from even to cleaved morphologies thus suggests that, moving downstream, new surfaces acquire more pronounced non-Newtonian rheologies, higher viscosities and larger critical stresses. This trend is consistent with lava becoming more crystallized before exposure, as inferred above from downstream changes in surface texture (see also Kilburn 1990).

(d) Crustal continuity. Lack of clear levées in the frontal zone suggests that widening may initially be limited by crustal restraint and not by a yield strength of the lava core as suggested by Hulme (1974). On the other hand, aa development requires crustal autobrecciation. This apparent contradiction can be resolved if (i) autobrecciation is local, such that unbroken crust stretches across a stream, either surrounding or beneath the broken surface, or (ii) crustal discontinuities across the whole flow are sufficiently short lived (as a result of cooling and rehealing) to have a negligible influence on longer-term flow widening.

Lava flow facies

The underlying theme of this chapter is that lava flows change continuously during emplacement: initially isothermal sheets develop channels and tubes; massive levées are modified by intrusion, accretion and overflow; crustal textures evolve from pahoehoe to aa; and once-simple snout advance becomes more complex with time. The two major trends associated with these changes are the formation of channels and surface crusting; superimposed are modes of levée modification and textural changes from pahoehoe to aa. Essential changes in lava state can thus be described by combining the poorly/well crusted and sheet/channel zones (described earlier) to identify six basic associations or *active facies* (so described to distinguish them from more conventional "passive" facies which use associations after flow emplacement). These facies are (Fig. 3.17): poorly crusted, well crusted and roofed sheet flow, and poorly crusted, well crusted and roofed channel flow. Note that, although roofed flow denotes tube formation, if the tube is only partially filled, the flowing lava surface beneath may be poorly or well crusted.

The extent and location of these facies change during emplacement (Fig. 3.17). At the start of effusion, the whole flow belongs to the poorly crusted, sheet flow facies. With time, channels begin to develop close to the vent and extend downstream, while the well crusted zone migrates upstream from early formation near the flow front (Fig. 3.17). Because of channel growth, clear roofing is first favoured at some point nearer to the vent than the front, and later extends both up- and downstream (Fig. 3.17).

An alternative way of viewing such changes is to follow flow evolution at specified positions. Two distinct reference locations are by the vent (fixed) and at the front (moving). These are defined informally as the proximal and distal regions, respectively, intermediate positions along a flow lying in the medial region.

One set of evolutionary sequences in the three regions is shown in Figure 3.18. The proximal (P) region evolves from the poorly crusted to roofed channel facies (P1–P4); a distinct medial (M) region coincides with the well crusted channel facies; and the distal (D) region develops from the poorly crusted to well crusted sheet facies. Figure 3.18

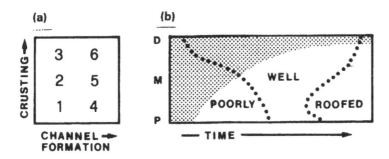

Figure 3.17 Active facies for aa lava flows. (a) Facies: 1, poorly crusted sheet; 2, well crusted sheet; 3, roofed sheet; 4, poorly crusted channel; 5, well crusted channel; 6, roofed channel. (b) Example of facies migration during emplacement. Sheet flow is stippled, channel flow clear. The dotted lines divide poorly crusted, well crusted and roofed zones. P, M and D correspond to proximal, medial and distal regions.

also illustrates how accessory changes can be used to subdivide the six basic facies, levée morphologies varying in the proximal and medial regions (note also the active levées in the medial region), while crusts evolve from pahoehoe to rubbly aa in the distal region.

All these changes occur because of cooling and reflect the sensitivity of lava rheology to temperature. The *form* of the changes, however, depends on how the mechanical energy of a flow can overcome cooling-induced energy losses. It should thus be possible eventually to describe facies changes in terms of prevailing energy fluxes (see also Fink & Griffiths 1990). Indeed, a major goal for the future is to seek links between facies and controlling parameters. Such links should help identify means of simplifying the governing flow equations and improve quantitative modelling of lava behaviour.

Conclusions

The evolution of aa lavas is a dynamic process. It promotes a range of structural changes during emplacement, from formation of channels and tubes to smaller-scale changes in surface texture (pahoehoe and aa). The final form of a flow therefore depends on its stage of evolution when the supply of lava is terminated.

Essential structural associations can be described in terms of the downstream progression of channel formation and the upstream migration of well developed surface crusts. Both features relate to flow cooling and, because lava rheology is strongly temperature-dependent, their associations may provide direct indicators of local energy fluxes.

Lava crusts are especially important. They grow inwards as thermal boundary layers but, through local failure, continually allow new material to be brought to the surface. Thus, while a lava core remains thermally insulated, new lava surfaces become more

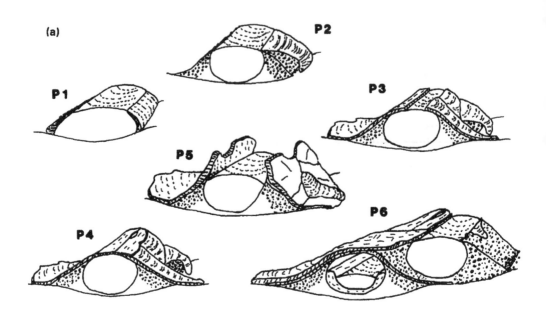

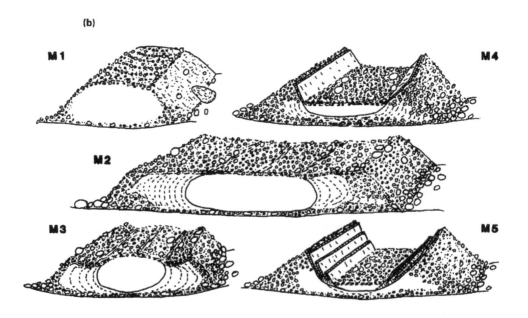

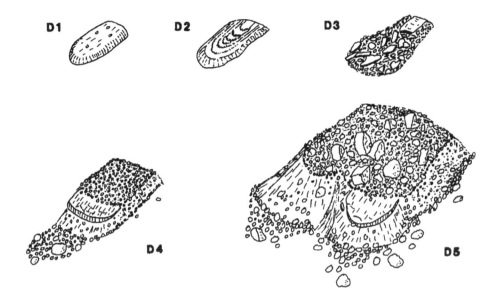

Figure 3.18 Examples of facies changes in proximal, medial and distal regions.
(a) Proximal region. P1, static levées start forming shortly after eruption. P2, channel overflows of aa cauliflowers, with or without continuous lava, build up static levée. P3, levées grow inwards, arching over channel. Overflows continue and parts of levées may break away. P4, channel roofs over, except for occasional skylights, from which lava may exude when the tube is full. P5, lava pressure in the tube may become high enough to arch the roof into nearly vertical ramparts. Tumuli (not shown) may also form. P6, large breakouts further upstream may produce new flow alongside or over the established tube. New flow may pass through stages P1–P5. Discharge may also continue unobserved through the established tube.
(b) Medial region. M1, the surface is well crusted. Bounding levées may be active and spreading outwards. Lateral intrusions may pierce levée to feed lava toes. Note possible granulation. M2, flow has widened, but levées are now static. Note slower-moving lava along the inner margins of levées. M3, cooling and reduced discharge causes motion to concentrate near the channel centre. Note the humped transverse profile of the surface. M4, continued drop in discharge rate causes lowering of the channel lava. Massive lava is veneered against inner levée walls. Previous overflows may also sink inwards. Hot, stagnant lava remains in levée cores. M5, fluctuating discharge along the channel may produce a series of levée veneers. Situations M4 and M5 may occur in place of P5 and P6 if a tube does not form in the proximal region.
(c) Distal region. D1, soon after eruption, the front is covered with patches of thin, darkened and glassy skin, which are rapidly broken and overrun. D2, ropy textures, convex downflow, develop over the whole front as the skin thickens (typical of pahoehoe). D3, crust breaks into slabs. D4, crust becomes aa. Surface debris forms scree, which may be overrun by oozes of internal lava. Granulation may begin. D5, crust is rubbly aa. Rubble is larger than surface debris in D4. Internal lava may extrude as oozes or upstanding protrusions (which may later crumble and fall). Cascades of granules are common.

103

solidified downflow before exposure, coinciding with the morphological trend from pahoehoe to rubbly aa.

Crusts also appear to maintain continuity across flow width in the frontal zone. Frontal widening can thus be inhibited even without obvious levée development. In the extreme, crustal restraint may favour flow propagation as new streams grow through punctures the carapace of an existing flow.

This chapter provides only an overview of aa flow processes. Much work still needs to be done to refine definitions of active facies and to integrate these into a more general description of flow field evolution when several flows are involved. This should provide better constraints for quantitative flow models and thereby improve methods of forecasting their behaviour.

Acknowledgements

Many people have discussed aa lava flows with the authors and, in particular thanks are due to Angus Duncan, Wyn Hughes, Rosaly Lopes-Gautier, Harry Pinkerton and Romolo Romano. Dave Rooks (University College London) also greatly helped with his patient photographic assistance.

References

Archambault, C. & J. C. Tanguy 1976. Comparative temperature measurements on Mount Etna lavas: problems and techniques. *Journal of Volcanology and Geothermal Research* **1**, 113–25.

Barberi, F. A. Bertagnini, P. Landi (eds) 1990. Mt Etna: the 1989 eruption. *Gruppo Nazionale per la Vulcanologia, Consiglio Nazionale delle Ricerche*. Pisa: Giardini.

Blake, S. 1990. Viscoplastic models of lava domes. In *IAVCEI Proceedings in Volcanology*. Vol. 2, *Lava flows and domes: emplacement mechanisms and hazard implications*, J. H. Fink (ed.), 88–126. Berlin: Springer.

Blong, R. J. 1984. *Volcanic hazards: a sourcebook on the effects of eruptions*. Orlando: Academic Press.

Booth, B. & S. Self 1973. Rheological features of the 1971 Mount Etna lavas. *Proceedings of the Royal Society, London* **A274**, 99–106.

Bolt, B. A., W. L. Horn, G. A. Macdonald, R. F. Scott (eds) 1984. *Geological hazards*. New York: Springer.

Borgia, A. & S. R. Linneman 1990. On the mechanisms of lava flow emplacement and volcano growth: Arenal, Costa Rica. In *IAVCEI Proceedings in Volcanology*. Vol. 2, *Lava flows and domes: emplacement mechanisms and hazard implications*, J. H. Fink (ed.), 208–43. Berlin: Springer.

Borgia, A., S. Linneman, D. Spencer, L. Morales, L. Andre 1983. Dynamics of the flow fronts, Arenal volcano, Costa Rica. *Journal of Volcanology and Geothermal Research* **19**, 303–29.

Chester, D. A., A. M. Duncan, J. E. Guest, C. R. J. Kilburn 1985. Mount Etna. The anatomy of a volcano. London: Chapman & Hall.

Cristofolini, R. 1984. L'eruzione etnea del 1983. *Atti della Accademia Gioenia di Catania* **160**, 39–78.

Cucuzza-Silvestri, S. 1949. L'eruzione dell'Etna del 1947. *Bulletin Volcanologique* **9**, 81–111.

Cumin, G. 1954. L'eruzione laterale etnea del novembre 1950–dicembre 1951. *Bulletin Volcanologique* **15**, 1–70.

Fink, J. H. & R. W. Griffiths 1990. Radial spreading of viscous gravity currents. *Journal of Fluid Mechanics* **221**, 485–501.

Frazzetta, G. & R. Romano 1984. The 1983 Etna eruption: event chronology and morphological evolution of the lava flow. *Bulletin Volcanologique* **47**, 1079–96.

Greeley, R. 1987. *The role of lava tubes in Hawaiian volcanoes*. US Geological Survey Professional Paper 1350, 1589–602.

Guest, J. E. 1982. Styles of eruption and flow morphology on Mt Etna. *Memorie della Società Geologica Italiana* **23**, 49–73.

Guest, J. E. & J. B. Murray 1979. An analysis of hazard from Mount Etna volcano, Sicily. *Journal of the Geological Society of London* **136**, 347–54.

Guest, J. E., J. R. Underwood, R. Greeley 1980. Role of lava tubes in flows from the Observatory Vent, 1971 eruption on Mount Etna. *Geological Magazine* **117**, 601–6.

Guest, J. E., C. R. J. Kilburn, H. Pinkerton, A. M. Duncan 1987. The evolution of lava flow fields: observations of the 1981 and 1983 eruptions of Mount Etna, Sicily. *Bulletin of Volcanology* **49**, 527–540.

Hardee, H. C. & J. C. Dunn 1981. Convective heat transfer in magmas near the liquidus. *Journal of Volcanology and Geothermal Research* **10**, 195–207.

Hughes, J. W., J. E. Guest, A. M. Duncan 1990. Changing styles of effusive eruption on Mount Etna since AD 1600. In *Magma transport and storage*, M. P. Ryan (ed.), 385–405. London: Wiley.

Hulme, G. 1974. The interpretation of lava flow morphology. *Geophysical Journal of the Royal Astronomical Society* **39**, 361–83.

Huppert, H. E. 1982. Flow and instability of a viscous current down a slope. *Nature* **300**, 427–9.

Kilburn, C. R. J. 1990. Surfaces of aa flow-fields on Mount Etna, Sicily: morphology, rheology, crystallization and scaling phenomena. In *IAVCEI Proceedings in Volcanology*. Vol. 2, *Lava flows and domes: emplacement mechanisms and hazard implications*, J. H. Fink (ed.), 129–156. Berlin: Springer.

Kilburn, C. R. J. & R. M. C. Lopes 1988. The growth of aa lava flow fields on Mount Etna, Sicily. *Journal of Geophysical Research* **93**, 14,759–72.

Kilburn, C. R. J. & R. M. C. Lopes 1991. General patterns of flow field growth: aa and blocky lavas. *Journal of Geophysical Research* **96**, 19,721–32.

Lipman, P. W. & N. G. Banks 1987. *Aa flow dynamics, Mauna Loa 1984*. US Geological Survey Professional Paper 1350, 1527–67.

Lopes, R. M. C. 1985. Morphological analysis of volcanic features on Earth and Mars. Ph.D. thesis, University of London.

Lopes, R. M. C. & J. E. Guest 1982. Lava flows on Etna, a morphometric study. In *The comparative study of the planets*, A. Coradini & M. Fulchignoni (eds). Dordtrecht: Reidel.

Macdonald, G. A. 1953. Pahoehoe, aa and block lava. *American Journal of Science* **251**, 169–91.

Macdonald, G. A. 1967. Forms and structures of extrusive basaltic rocks. In *The Poldervaart treatise on rocks of basaltic composition*. Vol. 1, *Basalts*, A. A. Poldervaart & H. H. Hess (eds), 1–61. New York: Wiley-Interscience.

Macdonald, G. A. 1972. *Volcanoes*. Englewood Cliffs, New Jersey: Prentice-Hall.

Naranjo, J. A., R. S. J. Sparks, M. V. Stasiuk, H. Moreno, G. J. Ablay 1992. Morphological, structural and textural variations in the 1988-1990 andesite lava of Lonquimay volcano, Chile. *Geological Magazine* **129**, 657–78.

Nye, J. F. 1952. The mechanics of glacier flow. *Journal of Glaciology* **2**, 82–93.

Oddone, E. 1910. L'eruzione Etnea del Marzo–Aprile 1910. *Bolletino della Società Sismologica Italiana* **14**, 141–203.

Peterson, D. W. & D. A. Swanson 1974. Observed formation of lava tubes during 1970–1971 at Kilauea volcano, Hawaii. *Studies in Speleology* **2**, 209–22.

Pinkerton, H. & R. S. J. Sparks 1976. The 1975 sub-terminal lavas, Mount Etna: a case history of the formation of a compound lava field. *Journal of Volcanology and Geothermal Research* **1**, 167–82.

Pinkerton, H. & R. S. J. Sparks 1978. Field measurements of the rheology of lava. *Nature* **276**, 383–4.

Ponte, G. 1923. The recent eruption of Etna. *Nature* **112**, 546–8.

Ponte, G. 1949. Riassunto delle principali osservazioni e ricerche fatte sull'Etna. *Bulletin Volcanologique* **9**, 65–80.

Rodwell, G. F. 1878. *Etna: a history of the mountain and its eruptions*. London: C. K. Paul.

Romano, R. & C. Sturiale 1982. The historical eruptions of Mt. Etna (volcanological data). *Memorie della Società Geologica Italiana* **23**, 75–97.

Romano, R. & C. Vaccaro 1986. The recent eruptive activity on Mt. Etna, Sicily: 1981–1985. *Periodico Mineralogico* **55**, 91–111.

Rowland S. K. & G. P. L. Walker 1987. Toothpaste lava: characteristics and origin of a lava structural type transitional between pahoehoe and aa. *Bulletin of Volcanology* **49**, 631–41.

Shaw, H. R. 1969. Rheology of basalt in the melting range. *Journal of Petrology* **10**, 510–35.

Shaw, H. R., T. L. Wright, D. L. Peck & R. Okamura 1968. The viscosity of basaltic magma: an analysis of field measurements in Makaopuhi lava lake, Hawaii. *American Journal of Science*, **261**, 255–64.

Sparks, R. S. J., H. Pinkerton, G. Hulme 1976. Classification and formation of lava levées on Mount Etna, Sicily. *Geology* **4**, 269–271.

Wadge, G. 1978. Effusion rate and the shape of aa lava flow fields on Mount Etna. *Geology* **6**, 503–6.

Wadge, G. & R. M. C. Lopes 1992. The lobes of lava flows on Earth and Olympus Mons, Mars. *Bulletin of Volcanology* **54**, 10–24.

Walker, G. P. L. 1967. Thickness and viscosity of Etnean lavas. *Nature* **213**, 484–5.

Walker, G. P. L. 1971. Compound and simple lava fields. *Bulletin Volcanologique* **35**, 579–590.

Walker, G. P. L. 1973. Lengths of lava flows. *Philosophical Transactions of the Royal Society, London* **A274**, 107–18.

Wentworth, C. K. & G. A. Macdonald 1953. Structures and forms of basaltic rocks in Hawaii. US *Geological Survey Bulletin* **994**, 1–98.

Whitford-Stark, J. & L. Wilson 1976. Atmospheric motions produced by hot lava. *Weather* **31**, 25–7.

CHAPTER FOUR

Extraterrestrial lava flows

Rosaly M. C. Lopes-Gautier

Rosaly M. C. Lopes-Gautier

Abstract

Volcanism has been one of the major processes shaping the surfaces of the terrestrial planets. Lava flows have been identified on the Moon, Mars, Venus and on Jupiter's moon Io. Due to the expense and complexity of sample-collecting missions, the study of extraterrestrial lavas has largely relied on the interpretation of remotely acquired imaging, topographic and, to a lesser degree, spectroscopic data. Models relating the final flow morphology to eruption characteristics and magma chemistry have been particularly important tools in the interpretation of these data. This chapter reviews what is known about lava flows on other planets, what data are currently available, and how lava flow models developed for terrestrial lavas have been applied to other planets. Some of the limitations and problems associated with these models are discussed, as well as what types of study and data are needed for further developing our understanding of extraterrestrial flows.

Introduction

Volcanism is a fundamental process shaping the surfaces of the terrestrial planets. During the last three decades diverse and often spectacular volcanic features have been revealed by spacecraft images. These features include not only lava flows and basaltic-type volcanoes, which are present on the Moon, Mars, Venus, Jupiter's moon Io and, possibly, Mercury, but also structures interpreted as pyroclastic flows, silicic domes, products from "ice volcanism" and, in the case of Io, possible sulfur flows. Active extraterrestrial volcanism has so far only been confirmed on Io, where it occurs on a vast scale, and on Neptune's moon Triton, where active geysers were observed by Voyager 2. Volcanoes on the Moon and Mars are thought to be extinct; in the case of Venus this question remains open.

One important difference between volcanism on Earth and on the other planets is

Figure 4.1 The nearside of the Moon, showing the dark areas (maria) which represent extensive, flood-like basaltic effusions, and the lighter-colored highlands. (Mount Wilson Observatory photograph.)

plate tectonics. Because plate tectonics does not appear to have operated on other planets, with the possible exception of Venus, extraterrestrial volcanic features can be quite different from those on Earth. For example, the much larger sizes of the volcanic edifices on Mars have been in part attributed to the lack of plate movement, allowing repeated eruptions in the same location to persist for a long time, building massive structures.

Most of the available data on extraterrestrial lava flows are morphological in nature, obtained by imaging systems aboard orbiting or fly-by spacecraft. Limited data on surface composition have been obtained by spectral analysis using both ground-based instruments and experiments aboard spacecraft. However, the most desirable way of studying the compositions of planetary surfaces is by means of analysis of samples brought to Earth by sample-collecting missions. The drawbacks of such missions, whether they are manned or unmanned, are their much greater complexity and considerably higher cost compared with remote sensing missions. Because rock samples have so far only been collected and brought back from selected sites on the Moon, only a few *in situ* measurements by lander craft have been performed on Venus and Mars, and spectral data on planetary surfaces in general have been obtained mainly at low resolutions, the study of extraterrestrial lava flows has relied largely on morphological data, using empirical and theoretical models which link conditions of emplacement to final morphology.

The following sections will review what is currently known about lava flows on the Moon, Mars, Venus and Io. Readers interested in a description of the types of remote sensing instruments deployed in spacecraft missions are referred to Greeley (1987). General reviews on the geology and volcanology of the terrestrial planets and satellites may be found in Taylor (1975, 1982), Guest & Greeley (1977), Johnson (1978), Carr (1980, 1981), Wilson & Head (1983), Carr et al. (1984), Greeley (1987), Saunders et al. (1991) and Head et al. (1991).

The Moon

Lava effusions have been extremely important processes on the Moon. The dark areas of the lunar surface, called maria (Fig. 4.1), are results of extensive, flood-like basaltic eruptions which spanned a long period in the Moon's history, peaking between 3.9×10^9 and 3.2×10^9 years ago. Mare basalts are mainly found on the lunar nearside, where they flooded older impact basins such as Imbrium. Studies of individual flows, however, are made difficult by the fact that few flow fronts can be distinguished on the maria, and their source areas are rarely identifiable. Most vents are thought to have been fissures which have subsequently been covered over by lavas. The scarcity of visible flow fronts and margins are possibly due to the lavas being too fluid to preserve the fronts, or because outlines have been eroded by numerous small meteorite impacts. Some flows on Mare Imbrium present clear outlines (Fig. 4.2) and can be traced back for several hundred kilometers, but even in those cases it is very difficult to identify their source areas, as the flows tend to overlap one another. Detailed mapping of the Imbrium flows by Schaber (1973) suggests that their source area was a fissure some 20 km long and that the late-stage lavas have travelled several hundred kilometers from their sources over gentle gradients.

Thickness estimates of mare flows have been diverse (Spudis et al. 1986), but experimental work on the mare samples (Brett 1975, Grove & Walker 1977) suggests that they were derived from relatively thin cooling units, only tens of meters thick. Such results are in good agreement with photogrammetric measurements and other photo-geological studies (Schaber 1973; Moore et al. 1978) which indicate that the thicknesses of young flows from Mare Imbrium are usually of the order of 10 m. Significantly thicker flow units were thought to have formed early in the history of the maria (Greeley

109

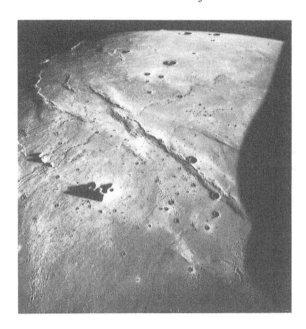

Figure 4.2 Oblique view across Mare Imbrium, one of the Moon's youngest maria, taken by Apollo 15. Note the lava flow running from the top right to the lower left of the image, and the mare ridges running from the top left to the lower right. (Apollo 15-1555.)

1976, Guest & Greeley 1977) and these may have required hundreds or even thousands of years to cool and solidify (Basaltic Volcanism Study Project 1981).

The difficulties in mapping individual lunar flows from source to toe and in obtaining reliable measurements of flow dimensions limits the application of both empirical and theoretical models to infer eruption parameters. Nevertheless, several attempts have been made to model the rheology of lunar lavas. Hulme (1973, 1974) and Hulme & Fielder (1977) used a Bingham plastic model to calculate yield strengths of lunar flows and, in some cases, also derived effusion rates from channel morphology. Yield strengths for Mare Imbrium lavas were found to be of the order of 400 Pa and flow rates to be of the order of 2×10^4 to 8×10^4 m^3 s^{-1} (assuming a Bingham viscosity of 10 Pa s). Impact-melt flows associated with impact craters which are non-volcanic in origin, having been produced by heat from the impact process, generally showed higher yield strengths, up to 2×10^4 Pa. Three impact-melt flows showed well defined channel morphology which was used to derive effusion rates. These effusion rates were found to be generally lower than those derived for Imbrium lavas, ranging from 260 to 1,000 m^3 s^{-1}.

Moore and Schaber (1975) also used a Bingham model, pointing out that the existence of a Mare Imbrium flow lobe whose thickness ranged from 7 to 20 m on a gentle gradient (average 0.13°C) implied that the flow might have stopped because of its yield strength. They approximated the critical flow depth required by Hulme's (1974) formula to the observed thickness of the flow and estimated its yield strength to be 100–200 Pa, slightly larger than values measured for Hawaiian lavas, but lower than the value of 400 Pa reported by Hulme (1974) for a nearby flow. Moore & Schaber also estimated the minimum times for the flow to cool by radiation from the liquidus to the solidus (13–40 hours). In a later work, Moore et al. (1978) calculated the yield strength of several lunar crater (impact-melt) flows. They used, whenever possible, the three Bingham model-derived equations to calculate the yield strengths of these flows and found yield strengths of the order of 2×10^4 Pa, consistent with the values reported by Hulme & Fielder (1977). However, the use of the different equations resulted in an average difference of about 1.7 for the values of yield strength. After applying the same method to terrestrial (and Martian) lavas, Moore et al. (1978) concluded that both lunar and Martian lavas are more akin to terrestrial basalts than to terrestrial andesites, trachytes and rhyolites.

Moore et al. (1978) did point out that application of Bingham fluid concepts to flows measured by remote methods must be done with caution, since it is still not clear whether the Bingham fluid model is valid for lava flows (see also Kilburn & Lopes 1991, Chs 7 & 10 this volume). Even if the model is valid, measurement problems remain; for example, Hulme's basic assumption that the final channel morphology is representative of the initial passage of the flow through it may be flawed, as channel morphology is subject to change by the variable dynamics of magma supply and flow field development, such as flow breaching (Wadge & Lopes 1991).

The high extrusion rates derived by studies such as that of Hulme & Fielder (1977) may imply a basic difference between terrestrial and lunar magmas and crustal environment. However, calculations by Wilson & Head (1981) comparing the ascent and eruption of basaltic magma on the Earth and Moon show that the differences between terrestrial and lunar magma rheologies and crustal environments do not lead to gross differences between the effusion rates expected on the two planetary bodies for similar-sized conduits or fissures. Therefore, Wilson & Head (1981) argued that the presence of very long lava flows on the Moon (and, by implication, possibly very high discharge rates) suggests only that tectonic and other forces associated with the onset of some lunar eruptions were such as to allow wide fissures or conduits to form. They calculated that the surface widths of elongate fissure vents need be no wider than 10 m to allow mass eruption rates up to 10 times larger than those proposed for terrestrial flood basalt eruptions (e.g. 3×10^4 kg s^{-1} m^{-1} proposed by Swanson et al. (1975) for the Columbia River basalts), and that 25 m widths for the fissure vents would allow rates 100 times larger. Fissures only 10 m wide would be difficult to locate in most of the currently available orbital images unless characteristic features such as constructional pyroclastic deposits or collapse depressions are present.

Sample analysis

Laboratory analysis of samples collected and returned by the Apollo missions have vastly improved our understanding of lunar volcanism in terms of chemistry and mineralogy. Remote spectroscopic studies from the Earth (Pieters 1978) and from the Apollo X ray and γ ray experiments (Davis & Spudis 1985) have also contributed to our understanding of the mineralogy of the lunar volcanic surfaces. The Moon's lack of an atmosphere enables measurements of the surface to be made directly, which aids the interpretation of remote measurements.

Lunar samples collected from the maria have made possible a variety of laboratory studies to determine chemical composition and magma rheology. However, such rheological studies refer to magma viscosity rather than yield strength as, so far, no reliable method of estimating yield strength as a function of temperature and composition is available (Wilson & Head 1981). Estimates of viscosity from laboratory simulations of lunar lavas have been made (Murase & McBirney 1970a) and values of about 1 Pa s (10 P) at 1,400°C were obtained. These low viscosity values, together with thermal conductivity studies of simulated lunar basalts (Murase & McBirney 1970b) which show that the heat loss from the active lunar lavas would have been very small, suggest that the lunar basalts may have been able to flow for long distances, even over the shallow slopes of the maria.

In terms of composition, analyses of mare lavas revealed that they are basaltic, but also that there are important differences between the lunar and terrestrial basalts (Taylor 1975), notably the absence of detectable H_2O in the lunar samples, and the higher abundance of iron, magnesium and titanium. This may imply a low silica content, perhaps 38–42% as was found for the Apollo 17 basalts (Lunar Sample Preliminary Examination Team (LSPET) 1973). More peculiar are some samples collected from the Oceanus Procellarum region by Apollos 12 and 14, which are characterized by enrichment in incompatible elements, including potassium (K), the rare-earth elements (REE) and phosphorus (P). These non-mare basalts have been designated KREEP basalts, but only a few of the KREEP samples are thought to be igneous rocks crystallized from internally generated melts. Most available KREEP samples are impact-melt rocks and impact breccias, formed as a result of the extremely high temperatures and pressures involved in a crater-forming meteorite impact. However, there is growing evidence that KREEP contaminated many, and maybe most, lunar magmas as they oozed towards the surface (Binder 1982, Warren & Wasson 1980).

Differences in texture between the terrestrial and lunar lavas are also evident from sample analysis, which has shown that the lunar lavas lack the alterations found in terrestrial basalts due to chemical weathering and hydrothermal activity. Sample analysis also indicates that most lunar lavas were derived from deep within the mantle, at depths ranging from about 150 km to as much as 450 km, and that the majority of lavas were erupted between 3.1 and 3.9×10^9 years ago. However, other studies suggest that lavas were erupted on the Moon before 4×10^9 years ago (Ryder & Spudis 1980, Taylor et al. 1983) and also as recently as 1×10^9 years ago (Schultz & Spudis 1983).

There is still some doubt as to the composition of possible volatiles in lunar magmas. Sample analysis and calculations by Sato (1976, 1977) suggest that an important source of volatiles in mare lavas was a chemical reaction between carbon and iron oxides at pressures less than 170 bars, which produced metallic iron, CO and CO_2. Thermodynamic calculations suggest that CO becomes dominant as the pressure decreases. Sample analysis by Housley (1978) suggested that between 250 and 750 ppm of CO was typically produced in the mare basalts as they erupted.

Small-scale features

The lunar surface lacks large-scale volcanic constructs, such as shield volcanoes. The absence of these features may be due to the common occurrence of high effusion rate flows on the Moon, as recognizable individual shields can only build up when the mean distance flowed by the lavas from one source area is substantially less than the mean spacing between sources (Head & Gifford 1980). The relatively few shield-like structures recognized on the Moon (Greeley 1976) are small (a few kilometers in diameter) in comparison with the lava flows, and are not considered to have been the sources for significant quantities of lavas. The same applies to other small-scale features such as domes and cones (Smith 1973). Some elongate fissure-like structures with widths of many tens to hundreds of meters have been proposed as fissure vents for the mare lavas (Head 1976, Schultz 1976) but, according to the calculations by Wilson & Head (1981), these features are far too wide to represent the true widths of the fissure vents, and are more likely to be the result of collapse around the vent after the eruption ceased.

Apart from lava flows, the most distinct features seen on the maria are sinuous rilles and mare ridges. Sinuous rilles (Fig. 4.3) consist of winding channels which may have a rimless pit at one end. They are found mainly around the outer edges of the maria and are interpreted as collapsed lava tubes or drained lava channels (Greeley 1971a, Taylor 1975, Guest & Greeley 1977). Sinuous rilles can be used to help map flow directions and general source areas. In terms of size, they are considerably larger than their terrestrial lava tube counterparts, as an example, Hadley Rille on Imbrium is over 130 km long and 5 km wide in places. Head & Wilson (1981) proposed that the size difference between sinuous rilles and terrestrial lava channels and tubes can be accounted for by the difference in gravity between the Earth and the Moon and by the higher discharge rates attributed to the lunar lavas.

Mare ridges (Fig. 4.2), also called wrinkle ridges, are prominent mare features which can be tens of kilometers long. The current consensus of opinion is that the ridges are compressional features (Muehlberger 1974) and that the period of major ridge production was synchronous with, or closely followed, the emplacement of major mare basalt sequences (Pieters et al. 1980). However, others (Strom 1971) have suggested that ridges were formed by lavas erupted along fractures, or by a combination of volcanic and tectonic processes (Guest & Greeley 1977). Recent identification and analysis of a number of terrestrial analogues (Plescia & Golombeck 1986) suggest that wrinkle ridges result from anticlinal folding above thrust faults that break the surface.

Lunar volcanism is not exclusively confined to the maria, and there is evidence that some highland units are also volcanic in origin. A prime candidate is the Apennine Bench formation, a light-coloured plains unit between the Imbrium and Serenitatis basins. Orbital geochemical data indicate that this unit is composed of KREEP basalt, and interpretation of the geology suggests that the unit was emplaced by extrusive igneous processes (Hawke & Head 1978, Spudis 1978, Spudis & Hawke 1985). However, no lava flows have so far been recognized in this unit.

Figure 4.3 Sinuous rilles are common on the lunar maria. Shown here is Hadley Rille, which is about 300 m deep and can be traced for tens of kilometers. The image is approximately 160 km across. (Apollo 15-0587.)

Future data acquisition

The Moon still offers many puzzles which may remain unresolved until new missions provide the data needed. Although ground-based work should not be neglected, as telescopic reflectance spectra of the Moon and laboratory spectral studies of lunar samples can still significantly advance our knowledge of the surface mineralogy, investigators are urging NASA to fund return missions to the Moon. Amongst those proposed is Lunar Observer (Nash 1991), an unmanned polar orbiting mission which will seek to extend our global knowledge of the Moon, particularly in terms of geochemical and geophysical studies. However, it must be stressed that great progress on the nature and emplacement of the lunar lavas can still be made with present data using comparative studies of lunar and terrestrial lavas. In particular, improved models of flood basalt eruptions could lead to a greater understanding of how the mare lavas were emplaced.

Mars

Mars has a richer variety of volcanic landforms and distinctive lava flows than the Moon, including some of the most spectacular volcanic edifices and flow fields in the Solar System. In general terms, the problems associated with the interpretation of Martian lavas are similar to those discussed for the Moon, such as the difficulty in obtaining reliable flow dimension measurements. In particular, flows can rarely be mapped back to their source areas, and thickness measurements are subject to large errors. No samples have yet been returned from Mars, though it has been proposed that the shergottite, nakhlite and chassignite meteorites (termed SNC meteorites) found in Antarctica are Martian in origin (Wood & Ashwal 1981). Topographic and spectroscopic coverage are more limited for Mars than for the Moon and studies using spectral reflectance are made difficult by the fine, iron-rich aeolian dust which blankets much of the Martian surface (Christensen 1982, Bell et al. 1989). However, the imaging data obtained by the two Viking orbiter spacecraft and its predecessor Mariner 9 are comprehensive and, at the time of writing, we look forward to additional coverage by Mars Observer. Some of the presently available Viking images have a resolution as high as 10 m per pixel, though commonly the resolution is about 100 m per pixel. The two Viking landers also obtained valuable data, performing *in situ* experiments on the northern hemisphere plains which included analysis of samples by X ray fluorescence techniques. It is generally thought that no crystalline rocks were sampled and that all the two dozen or so samples analyzed by the landers consisted of partly consolidated, weathered soils. Results (Table 4.1) showed that these samples appear to be derived from mafic to ultramafic source rocks and thus are grossly similar in composition to terrestrial and lunar basalts (Clark & Baird 1979). More recently, Burns (1988) proposed that the present-day Martian regolith is similar to terrestrial gossans, which are iron-rich oxidized cappings over sulfide-bearing rocks. Burns (1988) suggested that the Martian gossans may have been formed by reactions involving iron-rich ultramafic rocks similar to terrestrial komatiites.

Table 4.1 Chemical composition of Martian, lunar and terrestrial samples (wt%)

Component	Sample number (*)					
	1	2	3	4	5	6
SiO_2	44.7	42.8	44.48	45.03	37.79	49.34
Al_2O_3	5.7	NA	11.25	7.27	8.85	17.04
FeO	—	—	11.38	21.09	19.66	6.82
Fe_2O_3	18.2	20.3	3.00	—	—	1.99
MgO	3.3	NA	17.32	16.45	8.44	7.19
CaO	5.6	5.0	9.54	8.01	10.74	11.72
K_2O	< 0.3	< 0.3	0.40	0.06	0.05	0.16
TiO_2	0.9	1.0	NA	2.54	12.97	1.49
SO_3	7.7	6.5	—	—	—	—
Cl	0.7	0.6	—	—	—	—
Total	91.8	NA	98.37	100.45	98.50	95.75

From Greeley & Spudis (1981).
NA, not available.

*Sample identification: 1, Martian sample S1, Chryse Planitia (Toulmin et al. 1977); 2, Martian sample U1, Utopia Planitia (Toulmin et al. 1977); 3, model Martian lava, calculated composition (McGetchin & Smyth 1978); 4, lunar mare basalt, Apollo 12 olivine normative 12009 (Papike et al. 1976); 5, lunar mare basalt, Apollo 17 high-Ti 70215 (Papike et al. 1976); 6, terrestrial basalt, oceanic tholeiite (Engel et al. 1965).

Geological units

The global image coverage of Mars shows that there is a marked dichotomy between the northern and southern hemispheres (Carr et al. 1977, Carr 1980). The northern hemisphere is formed by relatively young lava plains dotted with volcanic structures, some of which are spectacularly large. The southern hemisphere appears to be much older, as shown by the heavily cratered terrain, and the volcanic structures located there also appear more degraded than those in the northern hemisphere. The dichotomy is accentuated by a difference in elevation between the two hemispheres; on average, the southern plains are 1–3 km higher than the Mars datum. Most of the northern hemisphere stands at elevations below the datum, the main exceptions being the volcanic provinces of Tharsis and Elysium. The Tharsis region forms a bulge some 8,000 km across, with a summit elevation 10 km above the datum, while the Elysium region shows a much smaller but still significant bulge. It is clear that the Tharsis bulge has played a major role in the tectonic evolution of Mars but its origin is still uncertain. It may be linked to mantle convection associated with the separation of the core (Carr 1981).

Most of the volcanoes on Mars are concentrated in the northern hemisphere in the Tharsis and Elysium regions. The most conspicuous volcanic structures are giant shield volcanoes, of which Olympus Mons is the largest known volcano in the Solar System (Fig. 4.4), being over 25 km high and some 600 km in diameter. Other giant shields are Ascreus, Pavonis and Arsia Montes which sit atop the Tharsis bulge aligned south-west to north-east. These are morphologically similar to Olympus Mons, having shallow flank slopes (4–6°), complex calderas and numerous lava flows discernible on the summit

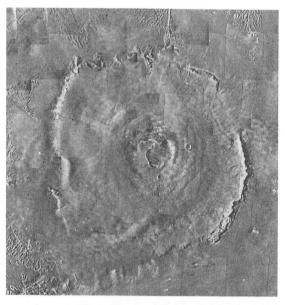

Figure 4.4   The largest volcano on Mars, Olympus Mons, which is about 600 km across and 25 km high. (Viking 211-5360.)

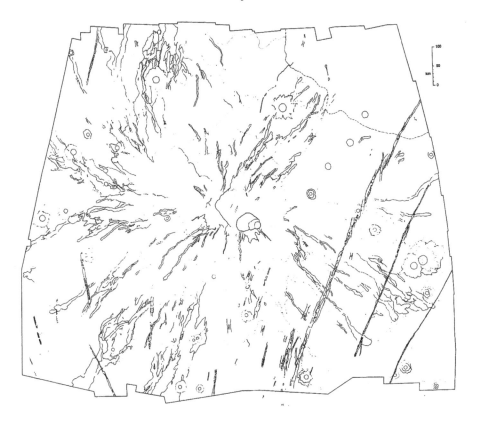

Figure 4.5 Map showing the distribution of lava flows on Alba Patera, Mars. Flows are predominantly oriented radially to the central caldera complex. (From Schneeberger & Pieri 1991, courtesy of D. Pieri.)

region and lower flanks. One significant difference, however, is that younger lavas from surrounding plains have buried the lower flanks of the three Tharsis shields, while the lower flanks of Olympus Mons stop abruptly at a scarp several kilometers high. The origin of the Olympus Mons scarp, along with that of the enigmatic corrugated terrain surrounding the volcano (the aureole), is still uncertain. It is possible that they both resulted from mass movement removing the outer flanks of the volcano (Lopes et al. 1982).

Also located in the northern hemisphere is Alba Patera (Fig. 4.5), a peculiar, shallow structure some 1,600 km across and 6 km high, topped by two nested calderas surrounded by graben (Carr et al. 1977, Mouginis-Mark et al. 1988). "Patera" is a collective term for a variety of unusual "saucer-shaped" features which often have a central caldera (Greeley & Spudis 1981). Some of the clearest and best defined lava flows on Mars are seen on images taken around Alba's summit caldera and on the lower flanks (Cattermole 1990, Lopes & Kilburn 1990, Schneeberger & Pieri 1991).

Other volcanic structures found on Mars are domes, also named tholii. These are relatively small volcanoes which may, in some cases, have experienced explosive activity (Mouginis-Mark et al. 1982a). Pyroclastic activity may also have formed several

117

breached cones which are morphologically similar to cinder cones on Earth (Plescia 1981, Tanaka & Davis 1988), though some of these appear to have lava flows emanating from them (Mouginis-Mark et al. 1992). In addition to all the above, Mars has thousands of subkilometer-sized hills on the northern plains (Frey & Jarosewich 1982) which are themselves volcanic in origin.

The volcanic plains on Mars are not unlike the lunar maria, having features such as wrinkle ridges (Sharpton & Head 1988), sinuous rilles (Schaber 1982), and overlapping flow lobes, which are mainly found around the periphery of shield volcanoes. Detailed geological maps of the flow lobes found on the plains have been prepared by Scott & Tanaka (1986). Many of these flows seem to originate several hundreds of kilometers from the large volcanic constructs (Schaber et al. 1978, Mouginis-Mark et al. 1982b), suggesting that vents, fissures and feeder dykes are quite numerous within much of Tharsis (Mouginis-Mark et al. 1992). Source areas for specific flows, however, have not been identified in the presently available images.

Numerous lava flows lobes of several different morphological types are seen on the flanks of the large shield volcanoes. The major flow types are well represented on Alba Patera and were originally described by Carr et al. (1977). More recently, Schneeberger & Pieri (1991) have produced a detailed map of the flows in this region (Fig. 4.5). The major flow types are: (a) sheet or tabular flows (Fig. 4.6), which can be several hundreds of kilometers long; (b) tube-channel flows or crested flows (Fig. 4.7), which have positive vertical relief with an axial apex which coincides with a valley and/or alignment of pits, and (c) tube-fed flows presenting wide "levéed" marginal structures (Fig. 4.6), possibly similar to terrestrial levéed flows. The source areas of some of these flows on Alba Patera can be inferred (for example, some can be traced to the edge of a caldera) and therefore it is possible to obtain reasonably reliable estimates of flow length, area and widths (Lopes & Kilburn 1990). However, even in those cases where the outline and lobate front of the flow can be seen clearly, reliable thickness measurements are a major problem. Measurements of ground slope are also hard to obtain accurately. Topographic maps are available for the whole planet, but the topographic resolution is seldom sufficient on the scale of the sizes of most lava flows to determine the underlying ground slope of the flow.

Lava flow studies
Several studies have attempted to relate the morphology of Martian lavas to their conditions of emplacement or chemical composition. As for lunar flows, yield stress models have been applied to Martian flows by Hulme (1976), Carr et al. (1977), Moore et al. (1978) and Zimbelman (1985). Hulme (1976) applied the same technique he used on the Moon (Hulme 1974) to a lava flow on Olympus Mons imaged by Mariner 9, relating the levée width and local slope of the flow to the yield stress and, in turn, the yield stress to the composition of the lava. Hulme's results indicated that the yield stress of the lava was in the range 3.9×10^3 to 2.3×10^4 Pa and its effusion rate between 380 and 470 m^3 s^{-1}. The silica content estimated suggested that the lava was more silicic than typical Hawaiian lavas.

(a)   (b)

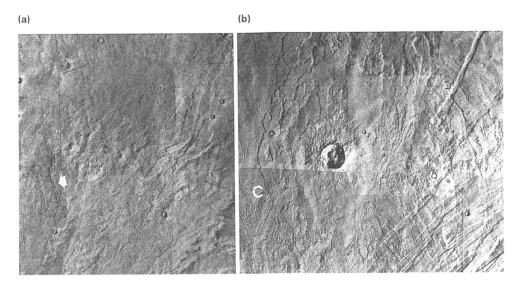

Figure 4.6 Various morphologically distinct types of lava flows on Alba Patera. (a) A tube-fed (levéed) flow is indicated by the arrow, other flows clearly seen in this image are of the tabular type. The image is about 100 km across. (From Viking rectified photomosaic MTM 45107.) (b) Sheet (tabular) flows dominate this massive flow field, but a less common tube-channel (crested) flow can be seen at the lower left (marked C). The image is approximately 180 km across. (From Viking MTM 45117.)

Using the higher resolution data acquired by Viking, Carr et al. (1977) used Hulme's technique on four levéed flows on the flanks of Arsia Mons and compared the results with those obtained using the alternative model proposed by Moore & Schaber (1975). Yield strengths obtained by both methods were found to be in reasonable agreement with one another, in the range 10^2–10^3 Pa. Effusion rates, assuming viscosities of 1–10 Pa s, were found to be in the range 10^3–10^4 m^3 s^{-1}. Moore et al. (1978) applied the Bingham model (using three distinct equations) to Martian flows and found the average yield strength for 11 Arsia Mons lavas to be of the order of 10^3 Pa, and those for three Olympus Mons lavas to be higher (of the order of 10^4 Pa), consistent with the results of Hulme (1976) for a different Olympus Mons flow. However, Moore et al. (1978) disputed Hulme's (1974) proposal that yield strength is simply related to silica content and argued that the apparent yield strength values were partly a function of topographic gradient. As discussed earlier, several authors (Moore et al. 1978, Crisp & Baloga 1990b) have pointed out the uncertainties involved in using the Bingham model and, in particular, the use of channel and levée morphology to infer lava properties. Nevertheless, the Bingham model has continued to be widely used in planetary volcanology (Zimbelman 1985, Cattermole 1987). Cattermole (1987) found yield strengths for nine Alba Patera lavas to be between 1.9×10^3 and 2.8×10^4 Pa, comparable with the ranges quoted above and with Zimbelman's (1985) results for Ascreus Mons lavas, which ranged from 1.2×10^4 to 3.8×10^4 Pa. The validity of these and other results mentioned above, however, can easily be disputed.

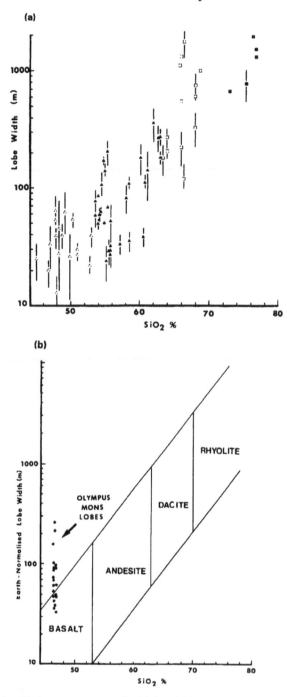

Figure 4.7 Relationship between average distal lobe width (*w*) and silica content. (a) Plot of *w* (units are log_{10} meters) with 1 standard deviation error bars against silica content for terrestrial lava flows, with compositions ranging from basalt (open triangles) to andesite (solid triangles), dacite (open squares) and rhyolite (solid squares). (b) Olympus Mons average distal lobe widths normalized to the Earth's gravitational field plotted on the same scale as part a. (From Wadge & Lopes 1991.)

A new approach using the Bingham model was taken by Wadge & Lopes (1991), who proposed that the widths of the distal lobes of lava flows are representative of the rheology of the lava, assuming that the lobes represent the arrest of free-flowing isothermal Bingham fluids on a slope. Lobe widths are a useful practical index because they are typically about an order of magnitude larger than lobe thicknesses and can be measured far more accurately on remote images. Moreover, lobes do not suffer from the changes in morphology that channels undergo during an eruption. Wadge & Lopes found a positive correlation between lobe width and silica content of the lava (Fig. 4.7) which is predictable from the isothermal Bingham model. This correlation was used to investigate 20 flows on the flanks of Olympus Mons. After lobe widths were measured and normalized to those that would be expected on Earth, they were found to be largely equivalent to those expected for terrestrial flows with andesitic/basaltic silica contents (Fig. 4.7).

A different method proposed for determining the rheological properties of Martian flows is the surface structure model of Fink & Fletcher (1978) and Fink (1980). This model relates the size and spacing of festoon-like ridges on flow surfaces to lava rheology, thickness of the thermal boundary of the flow, and applied stresses. Festoon-like ridges are seen on a variety of terrestrial lavas, ranging in size from centimeters, such as the ropes on pahoehoe lavas, to tens of meters, such as the ridges on rhyolitic flows. Because ridges on Martian flows are similar in size to those on terrestrial flows with a high silica content, some workers have compared them to rhyolitic, dacitic (Fink 1980) and trachytic (Zimbelman 1985) flows. In order to determine whether festoon ridges could be used to place constraints on the composition of flows, Theilig & Greeley (1986) used two Icelandic basaltic flows as analogues for Martian flows displaying festoon ridges. The Martian flows under investigation included those located west of Arsia Mons (Fig. 4.8). Theilig & Greeley found the viscosities determined for the Laki (Iceland) and the Arsia Mons flows to be comparable, mostly between 10^8 and 2×10^{10} Pa s. They pointed out that, even though these are high viscosity values for basaltic flows, they could be obtained by decreasing temperature, increasing solid content in the magma or decreasing gas content, all of which are related. They concluded, therefore, that since basaltic magma may have a high viscosity under specific conditions and large festoon ridges occur on some basaltic flows, ridge height and spacing may not represent compositional variations. Based on the morphological similarities between the Icelandic flows and the Martian flows under consideration, Theilig & Greeley concluded that the Martian flows they examined were emplaced as large sheet flows from basaltic flood-style eruptions. They proposed that the festoon ridges represented folding of the surface crust in the last stages of emplacement when viscosities were high, either due to cooling or to high-crystallinity lava being erupted under low temperatures.

More recently, radiation cooling models have become increasingly used to relate flow dimensions to eruption conditions. Pieri & Baloga (1986) proposed two models of radiative cooling, one assuming a thermally well mixed flow and a second assuming the lava flow to be "unmixed", that is, made up of a thermally homogeneous core covered by an infinitely thin crust. Radiation from the crust was characterized by a

N
△

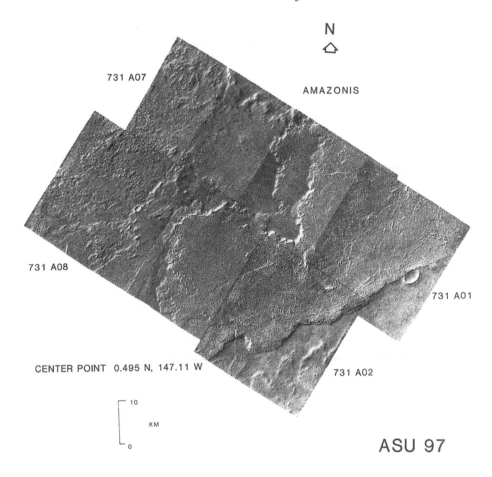

731 A07

AMAZONIS

731 A08

731 A01

CENTER POINT 0.495 N, 147.11 W

731 A02

10

KM

0

ASU 97

Figure 4.8   Flow lobes located west of the Martian volcano Arsia Mons. The distinctive arcuate festoon ridges, oriented normal to the inferred flow direction, are up to 50 km high and spaced 100–400 m apart. (From Theilig & Greeley 1986, courtesy of R. Greeley.)

constant "effective radiation temperature" which needs to be determined empirically. These two methods were used by Cattermole (1987) to derive the effusion rates of several Alba Patera lavas, assuming a range of initial temperatures. However, this approach has been criticized on several important points such as the assumption of an infinitely thin crust (Crisp & Baloga 1990a,b). Crisp & Baloga (1990a) have proposed a more refined model using a finite crust thickness and assuming partial core exposure at the surface which is more consistent with field observations. Crisp & Baloga (1990b) used this approach to calculate effusion rates of a flow on Ascreus Mons, previously mapped by Zimbelman (1985), to be in the range of 10 to 2×10^4 m^3 s^{-1}. Crisp & Baloga's model can only estimate the effusion rate with a minimum uncertainty of one order of magnitude due to the dependency of effusion rate on parameters which must be estimated from empirical studies of terrestrial lavas, such as the fractional area of the surface of the flow where the core of the flow is being exposed by cracking and overturning.

An approach consistent with Crisp & Baloga's method has been developed by Kilburn & Lopes (1991; see also Ch. 10 this volume) in the form of a flow field growth model which relates measurable parameters (maximum length, L_m; maximum width, W_m; average thickness, H; and average angle of underlying slope) to the duration of flow emplacement (T). The absence of gravity and viscosity terms in the model's equation (due to the presence of these factors in both the length and width terms which are ratioed) makes it particularly attractive for use on extraterrestrial lavas. Assuming that the extraterrestrial lavas in question are aa or blocky, and that similar conditions of flow field growth apply for the Earth and other planets (for more details see Kilburn & Lopes 1991), it is possible to use the flow growth model to calculate their duration of emplacement. Lopes & Kilburn (1990) did so for 18 well defined lavas on Alba Patera for which the vent locations could be inferred. Their results indicated typical average effusion rates to be of the order of 10^4–10^5 m^3 s^{-1} and durations which ranged from a few days for the single-type flow fields to over 200 days for the only multiple-type flow field. The high effusion rates could reflect a combination of larger source pressures, lower magma viscosities and larger fissure dimensions than on Earth (Wilson & Head 1983). Lopes & Kilburn (1990) estimated fissure dimensions for three Alba flows and calculated the average effusion rates per unit length of fissure (averaged over the duration of the eruption) to be between 5 and 15 m^2 s^{-1}, which are comparable to values for basaltic eruptions on Earth. Therefore, they concluded that the high effusion rate values could mainly reflect the large sizes of fissures on Alba and thus could not be used directly to infer lava composition by comparison with terrestrial examples.

Future data acquisition
The use of the lava growth model requires reliable measurements of flow length, width, thickness and underlying slope. At present it is difficult to find lavas on Mars for which all these measurements can be made with a satisfactory degree of accuracy (Lopes & Kilburn 1990). However, improved images and topographic coverage of the Martian surface will soon be acquired by the Mars Observer spacecraft. Apart from the improved imaging system, which will return data with a resolution as high as 1.4 m per pixel in selected areas (Komro & Hujber 1991), the spacecraft also carries the Mars Observer Laser Altimeter, or MOLA (Garvin & Bufton 1990). MOLA will have a vertical resolution as good as 1.5 m, which will allow the direct determination of heights of features such as lava flows and of ground slopes. Future missions are expected to address the question of the chemical composition of Martian rocks, performing further *in situ* experiments and, eventually, bringing samples back to Earth for analysis.

Venus

Venus is of particular interest to planetary geologists because it is the only planet in the Solar System of similar size and mass to the Earth and which, therefore, may possibly have had a similar geological history to the Earth's. The Magellan spacecraft has recently revealed in detail the remarkable range of volcanic and tectonic features which dominate the Venusian surface. Since these features are permanently obscured from view by the thick cloud cover which completely shrouds the planet, it was necessary to use radio waves to map the surface. Magellan used a synthetic aperture radar capable of mapping details as small as 120 m across, a resolution many times better than previously obtained by the earlier Soviet Venera spacecraft or from ground-based radar studies from Arecibo. Since Magellan started mapping in September 1990 an extraordinary variety of volcanic features has been revealed, including some very extensive lava flows. It is clear from the available images that Venus has undergone significant internal activity which produced features such as volcanic calderas, domes, folded mountain ranges, and extensive fault networks.

The overall density of impact craters on the surface of Venus indicates an average age for the surface of about 400 million years, young by planetary standards. It has been proposed that this young age is due to the resurfacing of large areas of the planet within the last 10 million years by relatively rare volcanic events which poured out large quantities of lava. In some areas active volcanism may have occurred as recently as the last few million years, as indicated by the total absence of impact craters on these areas. Although there have been suggestions that volcanism is still active on Venus (Robinson & Wood 1992), the evidence is still inconclusive.

Types of volcanic features

Over 80% of Venus' surface is composed of volcanic plains and edifices and the remainder, which is predominantly composed of tectonically deformed regions, is likely to be deformed volcanic deposits (Head et al. 1991). The young age of the surface is reflected in the pristine appearance of many of the volcanic features, which Magellan data show in remarkably sharp detail. Venusian volcanic edifices have a wide range of sizes and shapes but the most common are small shields generally less than 200 m high with diameters mostly between 2 and 8 km (Head et al. 1991). Also present on Venus are flat-topped, table-like features, and dome-shaped and cone-shaped edifices, all interpreted as volcanic in origin.

The fact that few discrete lava flows have been mapped on the shields, even though this may be due to insufficient resolution or radar contrast, has led investigators to doubt that the shields are a significant source for the extensive intershield lava plains, such as the dark intershield plains on Guinevere Planitia (Fig. 4.9), or indeed that shields and associated lavas have significantly contributed to resurfacing (Aubele & Slyuta 1990, Garvin & Williams 1990). More extensive lava flows are found in association with some of the larger (and less common) volcanic edifices, such as Sif Mons (Fig. 4.10), a 300 km diameter, 1.7 km high structure. Extensive lavas were emplaced from vents in and near Sif Mons, including a radar-bright flow some 300 km long and 15–30 km wide (Fig. 4.11). A similar flow (Fig. 4.11) appears to have emerged from a vent below the summit of Sif Mons and travelled some 400 km until it reached a topographically low region which caused it to turn (Head et al. 1991). Numerous other, smaller lava flows are associated with Sif Mons.

The question of whether pyroclastic volcanism has occurred on Venus is an important one since the high atmospheric pressure of Venus is expected to inhibit the exolution of volatiles from ascending magmas that leads to pyroclastic activity (unless the volatile content of the magma exceeds 4% by weight; Head & Wilson 1986). The identification of pyroclastic volcanism on Venus could, therefore, indicate that volatile-rich magmas have been erupted. Head et al. (1991) suggested one possible site in Guinevere Planitia where a 20 km diameter radar-dark unit appears to mantle the surrounding plains and may be the result of a Plinian-style eruption. Wenrich & Greeley (1992) identified five other sites which have radar-dark inferred ash deposits mantling radar-bright plains, and proposed that pyroclastic volcanism has occurred on these sites.

The presence of relatively steep-sided domes that resemble dacitic and rhyolitic domes on Earth may suggest the existence of more evolved lava compositions on Venus. The only data so far available on the surface composition of Venus has been obtained by five Soviet Venera landers. Geochemical data from all but one landing site indicated that the surface is similar in major-element composition to that of tholeiitic and alkali basalts (Surkov et al. 1984, 1987). However, data from one of the Venera 8 sites (Nikolaeva 1990) are more consistent with an intermediate to silicic composition. Magellan images show the location of the Venera 8 landing site (within about 1° of

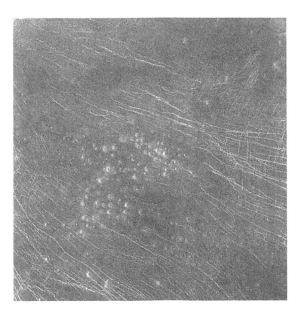

Figure 4.9 Magellan radar image showing a cluster of volcanic shields in Guinevere Planitia. The image is about 120 km across. (Magellan P38304.)

Figure 4.10 Magellan radar image showing Sif Mons. The image is about 450 km across. The two long, bright flows extending towards the top of the image are shown in greater detail in Figure 4.11. (Magellan P38054.)

uncertainty) to be near a pancake-like dome similar to andesitic, rhyolitic or dacitic domes on Earth (Head et al. 1991). Numerous Venusian domes, including those located south-east of Alpha Regio (Fig. 4.12), were examined by Head et al. (1991) and reported to be typically under 25 km in diameter, having heights of 100–600 m and volumes of 50–250 km^3, thus being larger than most rhyolite and dacite domes on Earth. Head et al. (1991) suggested that the large volumes may be due to the higher surface temperature

(b)

(a)

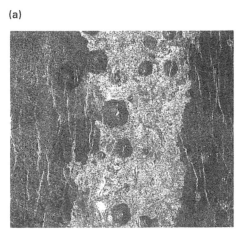

Figure 4.11 Magellan images of two flows from Sif Mons. (**a**)   Flow unit embaying small shields in the regional dark plains. (**b**)   Flow unit being locally diverted into small graben in dark plains near centre of image. (Magellan P38171, P38055.)

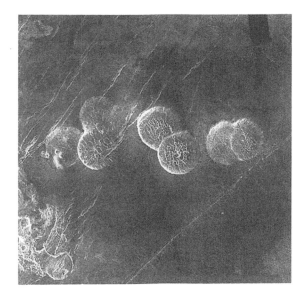

Figure 4.12 Magellan image of domes, each about 25 km in diameter, with maximum heights of 750 m, located south-east of Alpha Regio on Venus. (Magellan P37125.)

127

on Venus, which may permit magma to extrude more efficiently and to cool more slowly (and thus travel for greater distances) than similar magmas on Earth (see also Head & Wilson 1986).

Other volcanic constructs which characterize the Venusian surface are ring-like features called coronae which have been interpreted as products of local plume-like mantle upwelling (Solomon & Head 1990). Coronae are circular to elongate structures with diameters in the range 200–1,000 km, characterized by annuli of concentric ridges surrounding an elevated centre (Stofan & Saunders 1990; Head et al. 1991). Coronae are usually also characterized by interior flows, domes, and small (20–50 km across) edifices and exterior flows (Stofan & Saunders 1990). Some coronae show evidence that volcanism was linked to the uplift and radial fracturing and that volcanism continued in both the corona interior and exterior after the formation of the annulus (Head et al. 1991).

Among the spectacular volcanic features on Venus are large sinuous channels, some of which resemble lunar sinuous rilles. The Venusian channels typically have constant widths which range from 0.5 to 1.5 km, lack associated lava flow lobes or deposits and, in some cases, terminate in low-lying areas forming large plains deposits. They can be very long: for example, the sinuous channel in the south-west Guinevere Planitia (Fig. 4.13) is over 1,000 km long (Head et al. 1991). Venus also boasts the longest channel so far found in the Solar System, Hildr Fossa, which is 6,800 km long. These channels have been interpreted as resulting from low-viscosity lava (or lava emerging at high

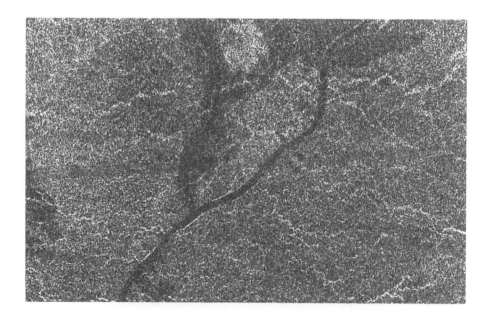

Figure 4.13 Magellan image showing sinuous channel in south-west Guinevere Planitia. The image is about 55 km across. The radar-bright margins have been interpreted by Head et al. (1991) as levées. (Magellan P36706.)

Figure 4.14 Magellan radar image of the massive flow field named Mylitta Fluctus. The image is approximately 800 km across. (Magellan P38289.)

effusion rates) becoming turbulent and thermally eroding a channel into pre-existing plains (Hulme 1973, Carr 1974, Huppert et al. 1987). It is possible that thermal erosion processes on Venus are helped by the high surface temperatures (Head & Wilson 1986).

Studies of lava flows

Venus has a great number of lava flows, ranging from those forming lava plains similar to the lunar maria and the Martian northern plains to well defined, thick lava flows. The great lengths and volumes of some of these flows have led to suggestions that they may be similar to terrestrial komatiites and flood basalts (Roberts et al. 1992). The most dramatic flow field, which has tentatively been named Mylitta Fluctus, covers an area of approximately 300,000 km^2 (Campbell et al. 1991, Head et al. 1991, Roberts et al. 1992). This massive flow field (Fig. 4.14) comprises flows several hundreds of kilometers long with widths ranging between 30 and 100 km in the medial and distal parts of the flow field (Roberts et al. 1992). Some flows show well developed channels and levées. The source area for most of the flows was shown by Magellan to be a caldera

129

some 40 by 20 km, the location of which had been previously predicted by Senske et al. (1991) on the basis of Arecibo data. Roberts et al. (1992) have mapped the stratigraphy of Mylitta Fluctus, identifying six major eruptive episodes and some individual flows within those episodes. They found that measurements of individual flow lengths and widths to be often impaired by flow superposition and that thicknesses of these flows could only be roughly estimated by examination of the regional topography. Although Magellan carries an altimeter, it is difficult to use its measurements to estimate flow thicknesses because of the large size of the altimeter's footprint (20 km in diameter in the Mylitta Fluctus region). The use of local topography by Roberts et al. to bracket flow thickness values yielded thicknesses of 30–90 m for the distal region of the flow field and a maximum estimate of 400 m for the thickness of the proximal region. Based on preliminary measurements, as well as general flow morphology and the presence of channels possibly formed by thermal erosion, Roberts et al. concluded that the lobate flows within Mylitta are basaltic in general composition, possibly emplaced at very high effusion rates or as high-temperature, low-viscosity basaltic komatiites.

Due to the great uncertainty in flow thicknesses, interpretation of eruption style and duration are subject to great uncertainties. The further uncertainty about whether the lavas are aa or pahoehoe in origin also poses problems for attempts at flow modelling since emplacement models are not necessarily applicable to both types (Kilburn & Lopes 1991). Pahoehoe textures have been suggested for many Venusian flows based on the polarization ratio (Campbell & Campbell 1991), Venera landing site morphology (Garvin et al. 1984) and on radar back scatter cross sections (Campbell & Campbell 1992). It is possible, however, that the most radar-bright flows interpreted as pahoehoe may in fact be aa, as moderately rough surfaces are thought to produce the brightest flow signatures (Campbell & Campbell 1992). Better means of correlating radar signatures to lava textures would greatly aid in the future application of flow models to Venusian flows; however, such studies must also wait until more reliable measurements of flow thicknesses and local topography become available.

Markedly different lava flows from those at Mylitta Fluctus are the unusually thick units investigated by Moore et al. (1992) which issue from a volcano in the plains between Artemis Chasma and Imdr Regio (Fig. 4.15). Moore et al. (1992) applied two techniques that rely on geometric image distortions to estimate flow thicknesses and obtained thicknesses ranging from 133 to 723 m for broad lobes with widths ranging from about 5,000 to over 40,000 m. These large thicknesses, together with the presence of regularly spaced ridges with large separations and with the large widths of the lobes, are indicative of evolved lavas with large silica contents.

Calculations of yield strengths by Moore et al. (1992) based on a model by Orowan (1949) which does not require topographic gradients to be known produced values ranging from 2×10^4 to 4×10^5 Pa, comparable to those for terrestrial rhyolite, andesite and basaltic andesites calculated in the same way. However, Moore et al. warned that compositional inferences should be viewed with caution because yield strengths of flow lobes calculated with Orowan's (1949) method and compared with silica content data by Wadge & Lopes (1991) show such scatter that a correlation between yield strength

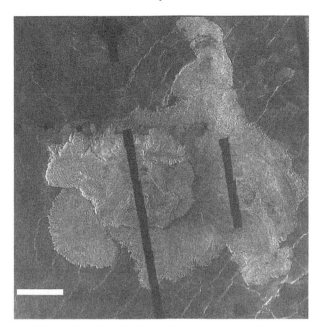

Figure 4.15 Magellan images of unusually thick lava flows on Venus from a volcano located on the plains between Artemis Chasma and Imdr Regio. The white bar is 50 km long. Black strips and rectangles are missing image data. (Magellan MIDRP 37S164, courtesy of H. Moore.)

and silica content is no longer present. Other rheological calculations by Moore et al. (1992) indicated Bingham viscosities to be quite large (ranging from 1×10^7 to 8×10^9 Pa s) and consistent with compositions more evolved than basalt. Although flows with these characteristics are rare on Venus, they indicate that silica-rich effusive volcanism (and possibly associated explosive volcanism) has played an important role in the evolution of the Venusian surface.

At the time of writing, the Magellan mission is still collecting data, and results are being published at a fast pace. The acquisition of stereo images should be of particular interest to studies of lava flows, as such images will allow more accurate determination of flow dimensions. Studies using radar properties are promising; for example, radar brightness variations among flows have been suggested as indicators of age (Kryuchkov & Basilevsky 1989), with bright, rough flows representing the youngest volcanic activity, though such variations are unlikely to be the result of ageing processes alone (Head et al. 1991). It is possible that a better understanding of the factors which influence radar brightness may allow relationships to be identified between flow brightness variations, flow texture, and variations in factors indicative of flow age.

Io

Io (Fig. 4.16), the innermost of Jupiter's Galilean satellites, is of particular interest to the study of extraterrestrial volcanism because it is the only body outside the Earth known to have active large-scale volcanism. Io's volcanic activity was first detected on images returned by the Voyager 1 spacecraft in 1979, which showed a spectacular plume rising about 200 km above the surface of the satellite (Morabito et al. 1979). Another Voyager instrument, the infrared interferometer spectrometer (IRIS), revealed enhanced thermal emission from parts of Io's surface (Hanel et al. 1979). Active volcanism was confirmed when the most prominent thermal emission detected by IRIS (about 17°C, in contrast to the surrounding surface at −146°C) was shown to coincide with one of the plumes. Higher-resolution camera images obtained later showed numerous lava flows and calderas (Carr et al. 1979), but no evidence that the lava flows were still active.

Io is a small body, similar in size (3,640 km in diameter) and density (3,500 kg m^{-3}) to our own moon, so it requires a heat source other than radioactive decay to drive its volcanic activity. An explanation for the activity was, in fact, put forward shortly before

Figure 4.16 Full-disk image of Jupiter's moon Io, taken by Voyager 1. Io is the first extraterrestrial body where active volcanism has been identified. (Voyager P-21457C.)

132

the Voyager encounters. Peale et al. (1979) proposed that tidal heating of Io generated by the gravitational interactions of Jupiter and the other Galilean satellites would cause melting of Io's interior and possibly lead to active volcanism.

Eighteen weeks after Voyager 1's dramatic confirmation of the prediction by Peale et al. (1979), new observations of Io's activity were made by Voyager 1's companion spacecraft, Voyager 2. Intense activity was still taking place but significant changes were detected between the two fly-bys, including the cessation of the largest plume, Pele (Smith et al. 1979a), and the change in the appearance of an area some 10×10^3 km^2 of associated surface deposits. Another plume, Loki, was found to have increased in size by about 50%, reaching nearly 200 km above the surface. A total of nine eruptive centres were identified by the Voyagers, of which Loki has remained the most active.

Since the Voyager fly-bys, Io's activity has been monitored from Earth (Johnson et al. 1984, Goguen et al. 1988) by means of infrared astronomical observations. These often take advantage of occultations of Io by Jupiter and by the other satellites, when timing the disappearance of each spot can give its surface position to a precision of up to 100 km in one dimension. These observations can also be used to determine the temperature of a hot spot and even to detect new spots (Goguen et al. 1988). Results obtained so far have shown that the Loki region is a dominant and persistent source of Io's thermal emission (Johnson et al. 1984, Goguen and Sinton 1985) though recent (1990) measurements (J. Spencer, personal communication) found Loki to have become fainter than during the Voyagers' encounters. Pele still seems to be dormant (Goguen et al. 1992) and it is now thought that Io has both persistent hot spots (such as Loki) and "Pele-class" activity, which is short lived.

Sulfur or silicates?

The relative importance of silicates and sulfur in the composition of Io's upper crust and volcanic melts has been a much debated issue since the Voyager fly-bys. Arguments favouring a composition predominantly of sulfur have been based on spectral and temperature data and on the surface colours shown by Voyager. On the other hand, a predominantly silicate composition is supported by the topography of some of Io's surface features and by other temperature measurements. Io's size and density, similar to the Earth's moon, are also indicative of a predominantly silicate composition.

The strongest evidence for the presence of at least a thin layer of sulfur on Io's surface comes from spectral observations (both by Voyager and from the Earth) and from laboratory studies (Fanale et al. 1979, Nelson et al. 1980, Nelson et al. 1987, Smythe et al. 1979). Voyager's IRIS experiment obtained a spectrum of sulfur dioxide gas over the erupting Loki volcano (Pearl et al. 1979) and ionized sulfur has been detected in the Io torus (Broadfoot et al. 1979), the trail of neutral and ionized particles that Io leaves behind in its orbit. Based on Voyager data, Smith et al. (1979b) and Kieffer (1982) interpreted the eruptive plumes as sulfur and SO_2 geysers. Spectral data still provides strong evidence for the presence of sulfur on Io, although Hapke (1989) has recently argued that the only unambiguously identified sulfur species on Io is SO_2, and that all the observed spectra can be modelled by a combination of SO_2 condensates and basalt.

Io's surface colours, which were first revealed in the Voyager images as predominantly red, yellow and orange, were cited as evidence for the presence of sulfur. Sagan (1979) attributed these colours to different anhydrous mixtures of sulfur allotropes, plus SO_2 frost and sulfurous salts of sodium and potassium. Pieri et al. (1984) argued that the sequence of colours displayed by the flows around the Ra Patera complex (Fig 4.17) was consistent with Sagan's interpretation, as supposedly cooler materials were found to be further away from the vent. Interpretations based on colour have, however, been contested on several grounds. Firstly, it was pointed out that the exact colours of sulfur can be drastically altered by the presence of even small amounts of other materials and by the temperature of heating, rate of cooling, and age of the sulfur compounds (Gradie & Moses 1983). Secondly, laboratory studies of sulfur in a vacuum carried out by Nash (1987) showed that the rapid quenching of high-temperature (and thus highly coloured) sulfur flow required by the model of Pieri et al. (1984) is not possible and that solid sulfur in a vacuum cannot preserve its original post-solidification colour. In addition, more precise calibration of the Voyager data indicated that Io's colours are yellowish-green, rather than orange-red (Young 1984). Although the new calibration does not rule out the presence of sulfur, it casts further doubt on interpretations based on colour. A further complication is introduced by the fact that it is not known whether Io's surface colours are also those of the crustal materials underneath, or whether they merely represent a thin surface layer perhaps only a few millimeters thick.

Initial measurements by the IRIS instrument on the Voyagers indicated maximum

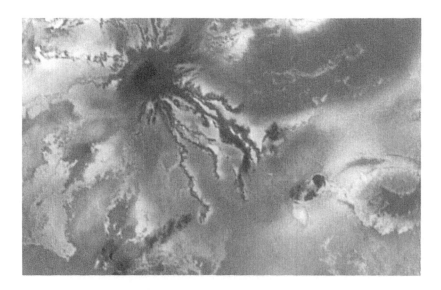

Figure 4.17 Voyager 1 image of Ra Patera on Io, showing the dark central caldera and surrounding flows. The total absence of impact craters suggests a young surface. The image is about 900 km across. (Voyager P-21277C.)

temperatures of about 600 K, consistent with the melting temperatures of sulfur. However, since lava bodies typically develop a cool crust as well as glowing cracks (Crisp & Baloga 1990a), the IRIS-derived temperatures do not rule out a silicate composition. Indeed, Carr (1986) developed a model of silicate volcanism on Io with a range of temperatures (matching the infrared observations) which could be attributed to different degrees of cooling of silicate materials.

Recent ground-based observations have tended to support a predominantly silicate composition. Johnson et al. (1988) reported a temperature measurement of about 900 K for a large eruptive event on Io and the measurements by Goguen et al. (1992) of a particular hot spot yielded temperatures of over 1,150 K. These temperatures are significantly above the boiling point of sulfur in a vacuum (715 K) and are consistent with those of silicate magma. Johnson et al. (1988) pointed out that although these measurements rule out pure molten sulfur as the major constituent of magma in the eruptions observed, possibilities other than silicate volcanism remain, as there are other sulfur compounds with boiling points higher than 900 K such as sodium polysulfides (Lunine & Stevenson 1985).

Further evidence in favour of a predominantly silicate composition for Io's crust is provided by the presence of high mountains, steep scarps, and caldera walls. Clow & Carr (1980) argued that, due to the low thermal conductivity and low melting point of sulfur, ductile behaviour would occur at depths of at most a few hundred meters for all reasonable values of heat flow in a dominantly sulfur upper crust. The occurrence of ductile behaviour at such shallow depths would prevent the formation of observed topographic features on Io, such as calderas with depths greater than a few hundred meters and scarps higher than 1,000 m, as these would not be self-supporting if they were composed primarily of sulfur. However, they added that a silicate crust with several per cent (but not tens of per cent) of sulfur included could satisfy both the mechanical constraints and the observed presence of sulfur on Io. They also pointed out that injection of silicate melts into sulfur deposits could be the cause of some of the plumes observed, as proposed by Reynolds et al. (1980). Eruptions of silicate magma into near-surface sulfur could also result in remobilization of sulfur to produce flows. The implication of Clow & Carr's analysis for the flows is that, while the observed relief for the sheetlike flows is compatible with the critical heights for sulfur, the presence of calderas (which appear to be their source) makes a dominantly sulfur composition for Io's upper crust unlikely.

Most planetologists are now tending towards the conclusion that both silicate and sulfur volcanism are taking place on Io. A likely scenario is one proposed by Carr (1985) in which injections of hot silicates from beneath a silicate and sulfur crust can remobilize the sulfur, causing flows and lakes within calderas which are maintained in a liquid state by the underlying hot silicates. This idea has been extended by Lunine & Stevenson (1985) to explain the temperatures and thermal fluxes in the Loki Patera region in terms of a convective sulfur lake heated by an underlying magma chamber. This model, which takes into account the physical and chemical processes in convective sulfur lakes, provides a way to relate the results from ground-based telescopic infrared observations of Io's surface heat flow to changes in the temperature of a sulfur lake and to variations in the output of silicate magma chambers in the crust.

Surface features and lava flows

Over 300 vent areas have now been identified on the Voyager images, most appearing as dark spots a few tens of kilometers across. In some cases higher resolution images of these areas show nearly black volcanic calderas which reflect less than 5% of the sunlight. Extensive lava flows have been mapped originating from several of these dark volcanic centres. Other geological features identified on Io's surface include long, curvilinear cliffs and narrow, straight-walled valleys a few hundred meters deep, as well as mountains several kilometers high and regions of layered terrain with extensive plateaus and mesas. Io has, however, a remarkable absence of impact craters, suggesting very high resurfacing rates (Johnson et al. 1979).

Schaber (1980) identified three major types of flow materials on Io which are associated with different vent morphologies: pit crater flows, shield crater flows and fissure flows. Pit crater flows are seen in the immediate vicinity of pit crater vents, generally extending from one side of the crater as massive coalescing flows. The flows generally present extreme colour and albedo variations and can be traced as far as 700 km from the individual vents. Shield crater flows are associated with shield constructs (such as Ra Patera, Fig. 4.17) which are concentrated in the equatorial region of Io. These flows are typically narrow and have sinuous paths, which may indicate that they flowed over significantly steeper slopes than the pit crater flows. Absolute values for the slopes, however, are not known. Individual shield flow lobes can be traced as far as 300 km, and colour and albedo can vary along their lengths. The last and rarest type of flow is that associated with elongate fissure vents. Schaber (1980) identified only four occurrences of fissure flows and suggested that some of these flows might be significantly thicker than those from pit and shield craters. Fissure flows may, therefore, have had considerably higher viscosities or yield strengths than the other types.

The uncertainties in the composition of surface materials, together with the poor morphological and topographic data currently available on all three types of Io flows prevents the application to Io of the empirical and theoretical flow models discussed in earlier sections. If the flows on Io are primarily composed of sulfur, an additional difficulty in modelling them is presented as relatively little is known about the morphology, emplacement mechanism, and physical and rheological properties of sulfur flows. Terrestrial sulfur flows are rare and only one example, the Siretoko-Iosan sulfur flow in Japan, has been observed while active (Wanatabe 1940). Among the few other examples described are those on Lastarria volcano in Chile (Naranjo 1985) and the 1950 flow on Mauna Loa which was studied by Greeley et al. (1984) as a possible analogue to the flows on Io. The Mauna Loa flow is thought to have been caused by the heat from basaltic flows of the 1950 eruption remobilizing secondary sulfur deposits which had accumulated on the flank of a basaltic cinder-and-spatter cone (Skinner 1970). The remobilization of sulfur by hot silicates may have occurred on Io; however, Io's flows are orders of magnitude larger in size than the Mauna Loa flow. Greeley et al. (1984)

suggested that molten sulfur could have flowed long distances on Io as a result of (a) relatively low viscosities in the melting range, (b) sustained effusion resulting from continued heating of the source area, (c) relatively low heat loss in the Ionian environment, even for thin sulfur flows (see also Fink et al. 1983), and (d) formation of flow tubes which effectively extend the source vent to the flow front. As a result of their fluidity and low melting temperature, sulfur flows on Io may form relatively thin veneers over other flows (possibly silicate in origin) and surface features.

The discovery of flows on Io which are possibly composed of sulfur has motivated research into terrestrial sulfur flows, including studies of industrial sulfur flows (Greeley et al. 1990) which are produced under well constrained conditions and provide a much needed alternative to field observations. The studies by Greeley et al. of the rheological properties of industrial sulfur flows showed that the flows remained mobile at temperatures below those at which sulfur has been observed to solidify. They attributed this mobility to small quantities of impurities that can change the physical properties of molten sulfur and proposed that similar changes on Io may allow flows to be emplaced over larger areas than laboratory studies might predict. The observations of industrial sulfur flows also showed that crusts were maintained on the surface of many of the flows for much of their emplacement history and that they possessed some buoyancy, probably due to trapped gases. If this is true for Io, where the trapped gas could be SO_2 vapour, development of such "unpredictably stable" crusts (Greeley et al. 1990) might allow flows to become efficiently insulated and, consequently, travel further.

An interesting part of the rheological studies of sulfur flows by Greeley et al. (1990) consisted of applying Hulme's (1974) model to obtain viscosities and flow rates which were compared to measured flow rates and published viscosity values. The results of Greeley et al. (1990) suggested that the Bingham model is only applicable in the latter stages of a sulfur flow when both the viscosity and yield strength increase in magnitude and the flow rate is low. It is possible that turbulence in the early stages of emplacement would make Hulme's model invalid, but Greeley et al. (1990) also pointed out that since the cooling history strongly affects the behaviour of these industrial sulfur flows, the use of a Bingham model may not be appropriate for them. This conclusion has also been supported by studies of a naturally occurring sulfur flow, as Naranjo's (1985) application of Hulme's (1974) model to the Lastarria flows predicted unreasonably high effusion rates. Other models relating morphology to flow emplacement described in earlier sections have not yet been tested on sulfur flows but it is unlikely that they would be applicable without at least some modifications. Specific models relating the morphology of sulfur flows to emplacement characteristics are clearly needed.

Future studies
The Galileo spacecraft is at present on its way to Jupiter and due to arrive in the Jovian system in December 1995. It is likely that Galileo will find that there have been significant changes in the morphology of the calderas, mountains and flows, and that new features will have been formed since the Voyager observations. Galileo carries several instruments which will greatly further our knowledge of Io's surface and

volcanism, including the Near-Infrared Mapping Spectrometer, which will obtain high spectral resolution data at a spatial resolution as high as 50 m. Such data should play a key role in determining the role of sulfur versus silicates. Other instruments aboard Galileo include the Solid State Imaging System, which will obtain images of selected areas on Io's surface with a resolution as high as 20 m per picture element, and the Ultraviolet Spectrometer which will map the erupting plumes. Together these instruments will provide a comprehensive data set from which many of the present questions regarding Io's volcanic behaviour may be answered. Meanwhile, further studies of sulfur flows on Earth would greatly aid in the interpretation of Io's flows, particularly if some morphological criteria could be established allowing the distinction between silica and sulfur compositions.

Summary

Studies of extraterrestrial lavas can be categorized as follows: (a) theoretical studies relating effusive volcanism to planetary parameters (gravity, atmospheric density), (b) interpretation of lavas mapped on spacecraft images and application of theoretical models which relate measurable morphological parameters to composition, rheology, duration and emplacement history, (c) spectral studies using ground-based or spacecraft acquired data, and (d) petrological studies based on available samples or on lander spacecraft measurements. This review has concentrated on category (b) studies because, first, the predominant type of data available on extraterrestrial lavas are imaging data and, second, the theoretical and empirical studies of terrestrial lavas, which are the core subjects of this book, are of key importance in the understanding of extraterrestrial lavas. It is important, however, to take into account the limitations of such models, the inherent assumptions within them, and the complex combination of physical, chemical and geological variables which can make results of theoretical and empirical studies uncertain. Nevertheless, the use of compatible models relating morphological parameters to different aspects of flow emplacement (eruption properties and magma composition) could greatly advance the understanding of the effusive volcanism of a planet, at least on a local scale. The further development and verification of models of flow behaviour using terrestrial data would represent a major step towards the proper interpretation of extraterrestrial lavas.

References

Aubele, J. C. & E. N. Slyuta 1990. Small domes on Venus: characteristics and origin. *Earth Moon and Planets* **50–51**, 493–532.

Basaltic Volcanism Study Project 1981. *Basaltic volcanism on the terrestrial planets.* New York: Pergamon Press.

Bell, J. F., T. B. McCord, P. G. Lucey 1989. High spectral resolution 0.3–1.0 microns spectroscopy and imaging of Mars during the 1988 opposition: characterization of Fe mineralogies. *Fourth International Conference Mars, Tucson, Arizona*, 67–8.

Binder, A. B. 1982. The mare basalt magma source region and mare basalt magma genesis. *Journal of Geophysical Research* **87**, A37–53.

Brett, R. 1975. Thickness of some lunar mare basalt flows and ejecta blankets based on chemical kinetic data. *Geochimica et Cosmochimica Acta* **39**, 1135–41.

Broadfoot, A. L., M. J. S. Belton, P. Z. Takacs, B. R. Sandel, D. E. Shemansky, J. B. Hoberg, J. M. Ajello, S. K. Atreya, T. M. Donahue, H. W. Moos, J. L. Bertaux, J. E. Blamont, D. F. Strobel, J. C. McConnell, A. Dalgarno, R. Goody, M. B. McElroy 1979. Extreme ultraviolet observations from Voyager encounter with Jupiter. *Science* **204**, 979–82.

Burns, R. G. 1988. Gossans on Mars. *Proceedings of the Lunar Planetary Science Conference* **18**, 713–21.

Campbell, B. A. & D. B. Campbell 1991. Comparison of 1988 Arecibo radar images of western Eistla Regio, Venus, and multipolarization airborne radar images of terrestrial terrains. *Lunar and Planetary Science Conference* **22**, 175–6.

Campbell, B. A. & D. B. Campbell 1992. Analysis of volcanic surface morphology on Venus from comparison of Arecibo, Magellan, and terrestrial airborne radar data. *Journal of Geophysical Research* **97**, 16,293–314.

Campbell, D. B., D. A. Senske, J. W. Head, A. A. Hine, P. C. Fisher 1991. Venus southern hemisphere: character and age of terrains in the Themis-Alpha-Lada region. *Science* **251**, 180–3.

Carr, M. H. 1974. The role of lava erosion in the formation of lunar rilles and Martian channels. *Icarus* **22**, 1–23.

Carr, M. H. 1980. The morphology of the Martian surface. *Space Science Reviews* **25**, 231–84.

Carr, M. H. 1981. *The surface of Mars.* New Haven: Yale University Press.

Carr, M. H. 1985. Volcanic sulphur flows on Io. *Nature* **313**, 735–6.

Carr, M. H. 1986. Silicate volcanism on Io. *Journal of Geophysical Research* **91**, 3521–2.

Carr, M. H., R. Greeley, K. R. Blasius, J. E. Guest, J. B. Murray 1977. Some Martian volcanic features as viewed from the Viking Orbiters. *Journal of Geophysical Research* **82**, 3985–4015.

Carr, M. H., H. Masursky, R. G. Strom, R. E. Terrile 1979. Volcanic features on Io. *Nature* **280**, 729–33.

Carr, M. H., R. S. Saunders, R. G. Strom, D. E. Wilhelms 1984. *The geology of the terrestrial Planets.* NASA Special Publication 469.

Cattermole, P. 1987. Sequence, rheological properties, and effusion rates of volcanic flows at Alba Patera, Mars. *Journal of Geophysical Research* **92**, E553–60.

Cattermole, P. 1990. Volcanic flow development at Alba Patera, Mars. *Icarus* **83**, 453–93.

Christensen, P. R. 1982. Martian dust mantling and surface composition: interpretation of thermophysical properties. *Journal of Geophysical Research* **87**, 9985–98.

Clark, B. C. & A. K. Baird 1979. Chemical analysis of Martian surface materials: status report. *Lunar and Planetary Science Conference* **10**, 215–7.

Clow, G. D. & M. H. Carr 1980. Stability of sulfur slopes on Io. *Icarus* **44**, 268–79.

Crisp, J. & S. Baloga 1990a. A model for lava flows with two thermal components. *Journal of Geophysical Research* **95**, 1255–70.

Crisp, J. & S. Baloga 1990b. A method for estimating eruption rates of planetary lava flows. *Icarus* **85**, 512–5.

Davis, P. A. & A. S. McEwen 1984. Photoclinometry: analysis of inherent errors and implications for topographic measurements. *Lunar and Planetary Science Conference* **15**, 194–5.

Davis, P. A. & P. D. Spudis 1985. Petrologic province maps of the lunar highlands derived from orbital geochemical data. *Journal of Geophysical Research* **90**, D61–74.

Engel, A. E. J., C. G. Engel, R. G. Havens 1965. Chemical characteristics of oceanic basalts and the upper mantle. *Bulletin of the Geological Society of America* **76**, 719–34.

Fanale, F. P., R. H. Brown, D. P. Cruikshank, R. N. Clark 1979. Significance of absorption features in Io's IR reflectance spectrum. *Nature* **280**, 761–3.

Fink, J. 1980. Surface folding and viscosity of rhyolite flows. *Geology* **8**, 250–4.

Fink, J. H. & R. C. Fletcher 1978. Ropy pahoehoe: surface folding of a viscous fluid. *Journal of Volcanology and Geothermal Research* **4**, 151–70.

Fink, J. H., S. O. Park, R. Greeley 1983. Cooling and deformation of sulfur flows. *Icarus* **56**, 38–50.

Frey, H. & M. Jarosewich 1982. Subkilometer Martian volcanoes: properties and possible terrestrial analogs. *Journal of Geophysical Research* **87**, 9867–79.

Garvin, J. B., J. W. Head, M. T. Zuber & P. Helfenstein 1984. Venus: the nature of the surface from Venera panoramas. *Journal of Geophysical Research* **89**, 3381–99.

Garvin, J. B. & J. L. Bufton 1990. Lunar Observer Laser Altimeter observations for lunar base site selections. *Proceedings of the Second Symposium on Lunar Bases and Space Activities in the 21st Century*. Houston: Lunar and Planetary Science Institute.

Garvin, J. B. & R. S. Williams 1990. Small domes on Venus: probable analogs of Icelandic lava shields. *Geophysical Research Letters* **17**, 1381–4.

Goguen, J. D. & W. M. Sinton 1985. Characterization of Io's volcanic activity by infrared polarimetry. *Science* **230**, 65–9.

Goguen, J. D., W. M. Sinton, D. L. Matson, R. R. Howell, H. M. Dyck, T. V. Johnson, R. H. Brown, G. J Veeder, A. L. Lane, R. M. Nelson, R. A. Mclaren 1988. Io hot spots: infrared photometry of satellite occultations. *Icarus* **76**, 465–84.

Goguen, J. D., W. M. Sinton, D. L. Matson et al. 1992. Io hot spots, satellite occultations of sources. *Icarus* (in press).

Gradie, J. & J. Moses 1983. Spectral reflectance of unquenched sulfur. *Lunar and Planetary Science Conference* **14**, 255–6.

Greeley, R. 1971a. Lava tubes and channels in the lunar Marius Hills. *The Moon* **3**, 289–314.

Greeley, R. 1971b. Modes of emplacement of basalt terrains and an analysis of mare volcanism in the Orientale Basin. *Proceedings of the Lunar Science Conference* **7**, 2747–59.

Greeley, R. 1987. *Planetary landscapes*. Boston: Allen & Unwin.

Greeley, R. & P. D. Spudis 1981. Volcanism on Mars. *Reviews of Geophysics and Space Physics* **19**, 13–41.

Greeley, R., E. Theilig, P. Christensen 1984. The Mauna Loa sulfur flow as an analog to secondary sulfur flows (?) on Io. *Icarus* **60**, 189–99.

Greeley, R., S. W. Lee, D. A. Crown, N. Lancaster 1990. Observations of industrial sulfur flows: implications for Io. *Icarus* **84**, 374–402.

Grove, T. & D. Walker 1977. Cooling histories of Apollo 15 quartz-normative basalts. *Proceedings of the Lunar and Planetary Science Conference* **8**, 1501–20.

Guest, J. E. & R. Greeley 1977. *Geology on the Moon*. London: Wykeham.

Hapke, B. 1989. The surface of Io: a new model. *Icarus* **79**, 56–74.

Hawke, B. R. & J. W. Head 1978. Lunar KREEP volcanism: geologic evidence for history and mode of emplacement. *Proceedings of the Lunar and Planetary Science Conference* **9**, 3285–309.

Head, J. W. 1976. Lunar volcanism in space and time. *Reviews of Geophysics and Space Physics* **14**, 265–300.

Head, J. W. & A. Gifford 1980. Lunar mare domes: classification and mode of origin. *The Moon and Planets* **22**, 235–58.

Head, J. W. & L. Wilson 1981. Lunar sinuous rille formation by thermal erosion: eruption conditions, rates and durations. *Lunar and Planetary Science* **XII**, 427–9.

Head, J. W. & L. Wilson 1986. Volcanic processes and landforms on Venus: theory, predictions, and observations. *Journal of Geophysical Research* **91**, 9407–46.

Head, J. W., D. B. Campbell, C. Elachi, J. E. Guest, D. P. McKenzie, R. S. Saunders, G. G. Schaber, G. Schubert 1991. Venus volcanism: initial analysis from Magellan data. *Nature* **252**, 276–87.

Housley, R. M. 1978. Modelling lunar eruptions. *Proceedings of the Lunar and Planetary Science Conference* **9**, 1473–84.

Hulme, G. 1973. Turbulent lava flow and the formation of lunar sinuous rilles. *Modern Geology* **4**, 107–17.

Hulme, G. 1974. The interpretation of lava flow morphology, *Geophysical Journal of the Royal Astronomical Society* **39**, 361–83.

Hulme, G. 1976. The determination of the rheological properties and effusion rate of an Olympus Mons lava. *Icarus* **27**, 207–13.

Hulme, G. & G. Fielder 1977. Effusion rates and rheology of lunar lavas. *Philosophical Transactions of the Royal Society, London* **A285**, 227–34.

Huppert, H. E., R. S. J. Sparks, J. S. Turner, N. T. Arndt 1987. Emplacement and cooling of komatiite lavas. *Nature* **309**, 19–22.

Johnson, T. V. 1978. The Galilean satellites of Jupiter: four worlds. *Annual Review of Earth and Planetary Sciences* **6**, 93–125.

Johnson, T. V., A. F. Cook II, C. Sagan, L. A. Soderblom 1979. Volcanic resurfacing rates and implications for volatiles on Io. *Nature* **280**, 746–50.

Johnson, T. V., D. Morrison, D. L. Matson, G. J. Veeder, R. H. Brown, R. M. Nelson 1984. Volcanic hotspots on Io: stability and longitudinal distribution. *Science* **226**, 134–7.

Johnson, T. V., G. J. Veeder, D. L. Matson, R. H. Brown, R. M. Nelson, D. Morrison 1988. Io: evidence for silicate volcanism in 1986. *Science* **242**, 1280–3.

Kieffer, S. W. 1982. In *Satellites of Jupiter*, D. Morrison (ed.), 647–723. Tucson: University of Arizona Press.

Kilburn, C. R. J. & R. M. C. Lopes 1991. General patterns of flow field growth: aa and blocky lavas. *Journal of Geophysical Research* **96**, 19,721–32.

Komro, F. G. & F. N. Hujber 1991. Mars Observer instrument complement. *Journal of Spacecraft and Rockets* **28**, 501–6.

Kryuchkov, V. P. & A. T. Basilevsky 1989. Radar-bright flow-like features as possible traces of the latest volcanic activity on Venus. *Lunar and Planetary Science Conference* **20**, 548–9.

Lopes, R. M. C. & C. R. J. Kilburn 1990. Emplacement of lava flow fields: application of terrestrial studies to Alba Patera, Mars. *Journal of Geophysical Research* **95**, 14,383–97.

Lopes, R. M. C., J. E. Guest, K. Hiller, G. Neukum 1982. Further evidence for a mass movement origin of the Olympus Mons aureole. *Journal of Geophysical Research* **87**, 9917–28.

Lunar Sample Preliminary Examination Team (LSPET) 1973. *Preliminary examination of lunar samples, Apollo 17*. NASA Special Publication, **330**, 7-1–46.

Lunine J. I. & D. J. Stevenson 1985. Physics and chemistry of sulfur lakes on Io. *Icarus* **64**, 345–67.

McGetchin, T. R. & J. R. Smyth 1978. The mantle of Mars: some possible geological implications of its high density. *Icarus* **34**, 512–36.

Moore, H. J. & G. G. Schaber 1975. An estimate of yield strength of the Imbrium lava flows. *Proceedings of the Lunar Science Conference* **6**, 101–18.

Moore, H. J., D. W. G. Arthur, G. G. Schaber 1978. Yield strengths of flows on the Earth, Mars, and Moon. *Proceedings of the Lunar and Planetary Science Conference* **9**, 3351–78.

Moore, H. J., J. J. Plaut, P. M. Schenk, J. W. Head 1992. An unusual volcano on Venus. *Journal of Geophysical Research* **97**, 13,479–93.

Morabito, L. A., S. P. Synnott, P. N. Kupferman, S. A. Collins 1979. Discovery of currently active extraterrestrial volcanism. *Science* **204**, 972.

Mouginis-Mark, P. J., L. Wilson, J. W. Head 1982a. Explosive volcanism on Hecates Tholus, Mars: investigation of eruption conditions. *Journal of Geophysical Research* **87**, 9890–904.

Mouginis-Mark, P. J., S. H. Zisk, G. S. Downs 1982b. Ancient and modern slopes in the Tharsis region of

Mars. *Nature* **297**, 546–50.

Mouginis-Mark, P. J., L. Wilson, J. R. Zimbelman 1988. Polygenetic eruptions on Alba Patera, Mars. *Bulletin of Volcanology* **50**, 361–79.

Mouginis-Mark, P. J., L. Wilson, M. T. Zuber 1992. The Physical volcanology of Mars. In *Mars*, H. H. Kieffer, B. M. Jakosky, C. W. Snyder, M. S. Matthews (eds), 424–52. Tucson: University of Arizona Press.

Muehlberger, W. R. 1974. Structural history of southwestern Mare Serenitatis and adjacent highlands. *Proceedings of the Lunar Science Conference* **5**, 101–10.

Murase, T. & A. R. McBirney 1970a. Viscosity of lunar lavas. *Science* **167**, 1491–3.

Murase, T. & A. R. McBirney 1970b. Thermal conductivity of lunar and terrestrial igneous rocks in their melting range. *Science* **170**, 165–7.

Naranjo, J. A. 1985. Sulphur flows at Lastarria volcano in the North Chilean Andes. *Nature* **313**, 778–80.

Nash, D. B. 1987. Sulfur in vacuum: Sublimation effects on frozen melts, and applications to Io's surface and torus. *Icarus* **72**, 1–34.

Nash, D. B. (ed.) 1991. *Lunar Observer: an essential robotic mission for planning human exploration of the Moon.* SJI Document 9. California: San Juan Capistrano Research Institute.

Nelson, R. M., A. L. Lane, D. L. Matson, F. P. Fanale, D. B. Nash, T. V. Johnson 1980. Io: Longitudinal distribution of SO_2 frost. *Science* **210**, 784–6.

Nelson, R. M., A. L. Lane, D. L. Matson, G. J. Veeder, B. J. Buratti, E. F. Tedesco 1987. Spectral geometric albedos of the Galilean satellites from 0.24 to 0.34 micrometres: observations with the International Ultraviolet Explorer. *Icarus* **72**, 358–80.

Nikolaeva, O. V. 1990. Geochemistry of the Venera 8 material demonstrates the presence of continental crust on Venus. *Earth, Moon and Planets* **50–1**, 329–41.

O'Reilly, T. C. & G. F. Davies 1981. Magma transport of heat on Io: A mechanism allowing a thick lithosphere. *Geophysical Research Letters* **8**, 313–6.

Orowan, E. 1949. Remark from the joint meeting of British Glaciology Society, British Rheologists' Club, and Institute of Metals. *Journal of Glaciology* **1**, 231.

Papike, J. J., F. N. Hodges, A. E. Bence, M. Cameron, J. M. Rhodes 1976. Mare basalts: Crystal chemistry, mineralogy and petrology. *Reviews of Geophysics and Space Physics* **14**, 475–540.

Peale, S., P. Cassen, R. Reynolds 1979. Melting of Io by tidal dissipation. *Science* **203**, 892–4.

Pearl, J., R. Hanel, V. Kunde et al. 1979. Identification of gaseous SO_2 and new upper limits for other gases on Io. *Nature* **280**, 755–8.

Pieri, D. C. & S. M. Baloga 1986. Eruption rates, areas and length relationships for some Hawaiian lava flows. *Journal of Volcanology and Geothermal Research* **30**, 29–45.

Pieri, D. C., S. M. Baloga, R. M. Nelson, C. Sagan 1984. Sulfur flows at Ra Patera, Io. *Icarus* **60**, 685–700.

Pieters, C. M. 1978. Mare basalt types on the front side of the moon: a summary of spectral reflectance data. *Proceedings of the Lunar and Planetary Science Conference* **9**, 2825–49.

Pieters, C. M., J. W. Head, J. B. Adams, T. B. McCord, S. H. Zisk, J. L. Whitford-Stark 1980. Late high titanium basalts of the western maria: geology of the Flamsteed region of Oceanus Procellarum. *Journal of Geophysical Research* **85**, 3913–38.

Plescia, J. B. 1981. The Tempe volcanic province of Mars and comparisons with the Snake River plains of Idaho. *Icarus* **45**, 586–601.

Plescia, J. B. & M. P. Golombeck 1986. Origin of planetary wrinkle ridges based on the study of terrestrial analogues. *Bulletin of the Geological Society of America* **97**, 1289–99.

Reynolds, R. T., S. J. Peale, P. M. Cassen 1980. Sulfur vapor and sulfur dioxide models of Io's plumes. *Proceedings of the IAU Colloquium 57, Kailua-Kona, Hawaii.*

Roberts, K. M., J. E. Guest, J. W. Head, M. G. Lancaster 1992. Mylitta Fluctus, Venus: rift-related, centralized volcanism and the emplacement of large-volume flow units. *Journal of Geophysical Research* **97**, 15,991–16,016.

Robinson, C. & J. Wood 1992. Recent volcanic activity on Venus: evidence from emissivity measurements. *Lunar and Planetary Science Conference* **23**, 1163–4.

Ryder, G. & P. D. Spudis 1980. Volcanic rocks in the lunar highlands. In *Proceedings of the Conference on Lunar Highlands Crust*, J. J. Papike & R. B. Merrill (eds), 353–75. New York: Pergamon Press.

Sagan, C. 1979. Sulfur flows on Io. *Nature* **280**, 750–3.

Sato, M. 1976. Oxygen fugacity and other thermochemical parameters of Apollo 17 high Ti basalts and their implications on the reduction mechanism. *Proceedings of the Lunar Science Conference* **7**, 1323–44.

Sato, M. 1977. The driving mechanism of lunar pyroclastic eruptions inferred from the oxygen fugacity behavior of Apollo 17 orange glass. *Eos, Transactions of the American Geophysical Union* **58**, 425.

Saunders, R. S., R. E. Arvidson, J. W. Head III, G. G. Schaber, E. R. Stofan, S. C. Solomon 1991. An overview of Venus geology. *Nature* **252**, 249–52.

Schaber, G. G. 1973. Lava flows in Mare Imbrium: geologic evaluation from Apollo orbital photography. *Proceedings of the Lunar and Planetary Science Conference* **4**, 73–92.

Schaber, G. G. 1980. The surface of Io: geologic units, morphology and tectonics. *Icarus* **43**, 302–33.

Schaber, G. G. 1982. Syrtis Major: a low relief volcanic shield. *Journal of Geophysical Research* **87**, 9852–66.

Schaber, G. G., K. C. Horstman, A. L. Dial 1978. Lava flows in the Tharsis region of Mars. *Proceedings of the Lunar and Planetary Science Conference* **9**, 3433–458.

Schneeberger, D. M. & D. C. Pieri 1991. Geomorphology and stratigraphy of Alba Patera, Mars. *Journal of Geophysical Research* **96**, 1907–30.

Schultz, P. H. 1976. *Moon morphology*. Austin: University of Texas Press.

Schultz, P. H. & P. D. Spudis 1983. The beginning and end of lunar mare volcanism. *Nature* **302**, 233–6.

Scott, D. H. & K. L. Tanaka 1986. *Geologic map of the western equatorial region of Mars, scale 1:15,000,000*. US Geological Survey Miscellaneous Map Series I-1802-A.

Senske, D. A., D. B. Campbell, E. R. Stofan, P. C. Fisher, J. W. Head, N. Stacy, J. C. Aubele, A. A. Hine, J. K. Harmon 1991. Geology and tectonics of Beta Regio, Guinevere Planitia, Sedna Planitia, and western Eistla Regio, Venus: results from Arecibo image data. *Earth, Moon and Planets* **55**, 163–214.

Sharpton, V. L. & J. W. Head 1988. Lunar mare ridges: analysis of ridge–crater intersections and implications for the tectonic origin of mare ridges. *Proceedings of the Lunar and Planetary Science Conference* **18**, 307–17.

Skinner, B. J. 1970. A sulfur flow on Mauna Loa. *Pacific Science* **24**, 144–5.

Smith, B. A., L. A. Soderblom, R. Beebe 1979a. The Galilean satellites and Jupiter: Voyager 2 imaging science results. *Science* **206**, 927–50.

Smith, B. A., E. M. Shoemaker, S. W. Kieffer, A. F. Cook II 1979b. The role of SO_2 in volcanism on Io. *Nature* **280**, 738.

Smith, E. I. 1973. Identification, distribution and significance of lunar volcanic domes. *The Moon* **6**, 3–31.

Smythe, W. D., R. M. Nelson, D. B. Nash 1979. Spectral evidence for SO_2 frost or adsorbate on Io's surface. *Nature* **280**, 766.

Solomon, S. C. & J. W. Head 1990. Fundamental issues in the geology and geophysics of Venus. *Science* **252**, 252–60.

Spudis, P. D. 1978. Composition and origin of the Apennine Bench formation. *Proceedings of the Lunar and Planetary Science Conference* **9**, 3379–94.

Spudis, P. D. & B. R. Hawke 1985. The Apennine Bench revisited. *Workshop on the Geology and Petrology of the Apollo 15 Landing Site*. LPI Contribution **581**, 57–9.

Spudis, P. D. (Chairman) and the Lunar Geoscience Working Group 1986. Status and Future of Lunar Geoscience. NASA Special Publication 484.

Stofan, E. R. & R. S. Saunders 1990. Geologic evidence of hotspot activity on Venus: predictions for Magellan. *Geophysical Research Letters* **17**, 1377–80.

Strom, R. G. 1971. Lunar mare ridges, rings, and volcanic ring complexes. *Modern Geology* **2**, 133–58.

Surkov, Yu. A., V. L. Barsukov, L. P. Moskalyeva, V. P. Kharyukova, A. L. Kemurdzhian 1984. New data on the composition, structure, and properties of Venus rock obtained by Venera 13 and Venera 14. *Journal of Geophysical Research* **89**, B393–402.

143

Surkov, Yu. A. et al. 1987. Uranium, thorium, and potassium in the Venusian rocks at the landing sites of Vega 1 and 2. *Journal of Geophysical Research* **92**, E537–40.

Swanson, D. A., T. L. Wright, R. T. Helz 1975. Linear vent systems (and estimated rates of magma production and eruption) for the Yakima basalt of the Columbia Plateau. *American Journal of Science* **275**, 877–905.

Tanaka, K. L. & P. A. Davis 1988. Tectonic history of the Syria Planum province of Mars. *Journal of Geophysical Research* **93**, 14,893–917.

Theilig, E. & R. Greeley 1986. Lava flows on Mars: analysis of small surface features and comparisons with terrestrial analogs. *Journal of Geophysical Research* **91**, E193–206.

Taylor, L. A., J. W. Shervais, R. H. Hunter, C.-Y. Shih, J. Wooden, L. E. Nyquist, J. C. Laul 1983. Pre-4.2 AE mare basalt volcanism in the lunar highlands. *Earth and Planetary Science Letters* **66**, 33–47.

Taylor, S. R. 1975. *Lunar science: a post-Apollo view.* New York: Pergamon Press.

Taylor, S. R. 1982. *Planetary science, a lunar perspective.* Houston: Lunar and Planetary Institute.

Toulmin, P., A. K. Baird, B. C. Clark, K. Keil, H. J. Rose, R. P. Christian, P. H. Evans, W. C. Kelliher 1977. Geochemical and mineralogical interpretation of the Viking inorganic chemical results. *Journal of Geophysical Research* **82**, 4625–34.

Wadge, G. & R. M. C. Lopes 1991. The lobes of lava flows on Earth and Olympus Mons, Mars. *Bulletin of Volcanology* **54**, 10–24.

Warren, P. H. & J. T. Wasson 1979. The origin of KREEP. *Reviews of Geophysics and Space Physics* **17**, 73–88.

Warren, P. H. & J. T. Wasson 1980. Early lunar petrogenesis, oceanic and extraoceanic. In *Proceedings of the Conference on Lunar Highlands Crust*, J. J. Papike & R. B. Merrill (eds), 81–99. New York: Pergamon Press.

Watanabe, T. 1940. Eruptions of molten sulphur from the Siretoko-Iosan volcano, Hokkaido, Japan. *Japanese Journal of Geology and Geography* **17**, 289–310.

Wenrich, M. L. & R. Greeley 1992. Investigation of Venusian pyroclastic volcanism. *Lunar and Planetary Science Conference Abstracts* **23**, 1515–6.

Wilson, L. & J. W. Head 1981. Ascent and eruption of basaltic magma on the Earth and Moon. *Journal of Geophysical Research* **86**, 2971–3001.

Wilson, L. & J. W. Head 1983. A comparison of volcanic eruption processes on Earth, Moon, Mars, Io and Venus. *Nature* **302**, 663–9.

Wood, C. A. & L. D. Ashwal 1981. SNC meteorites: Igneous rocks from Mars? *Proceedings of the Lunar and Planetary Science Conference* **12B**, 1359.

Young, A. T. 1984. No sulfur flows on Io. *Icarus* **58**, 197-226.

Zimbelman, J. R. 1985. Estimates of rheologic properties for flows on the Martian volcano Ascreus Mons. *Journal of Geophysical Research* **90**, D157–62.

Part II

MONITORING

Preface

One of the challenges (and frustrations) of lava studies is that monitoring active flows is plagued with practical difficulties. Most arise because lavas are hot and very viscous. As a result, it is often not possible to approach flows to make surface measurements (let alone take internal measurements using penetrometers) at desired locations and time intervals. Part II is thus devoted to monitoring strategies, the capabilities of current measuring techniques, and prospects for their improvement. Using Hawaiian flows as examples, Chapter 5 illustrates how to optimize ground-based observations and personal safety. Chapter 6 discusses monitoring objectives and instrumentation, while Chapter 7 focuses on measurements of convective heat flow. Chapter 8 concludes this Part by looking at the promise of remote-sensing techniques, on the ground, in the air and from space. Concerned mostly with thermal observations, it also explains how to procure satellite images for research purposes.

CHAPTER FIVE

Field observation of active lava in Hawaii: some practical considerations

Robert I. Tilling & Donald W. Peterson

Abstract

Kilauea and Mauna Loa volcanoes, on the Island of Hawaii, erupt frequently, producing basaltic lava that can exhibit a wide variety of flow regimes, including lava fountains, lava lakes and long-lived lava flows. Because the Hawaiian eruptions are generally non-explosive and relatively accessible, close-up observations of active lava can be made safely and easily, if reasonable precautions are taken. Among the most important observations that can be made are real-time measurements or estimates of lava discharge at the vent(s), of velocities of moving lava on the surface or in tube systems, and of the style, nature, and behavior of the flowing lava. Such observations are needed to complement and constrain the increasing number of theoretical and modelling investigations of the dynamics of active lavas.

The observations and the collection of molten lava samples can be made with the simple, inexpensive equipment (clinometer, tape rule, geological hammer, etc.) commonly used in geological fieldwork; some effective techniques for measuring lava movement and collecting fresh (molten) lava in Hawaii are given as examples. To minimize occupational hazards in working around active lava, observers must take some common-sense precautions, briefly reviewed in this chapter. Perhaps the two most important axioms to follow in observing active lava safely are: (a) if at all possible, do not work alone; and (b) always be alert, think ahead, and expect the unexpected.

Introduction

Over a span of more than 75 million years, countless outpourings of fluid basaltic lava created the Hawaiian ridge/Emperor seamounts chain as the Pacific Plate moved over the Hawaiian "hot spot" (Clague & Dalrymple 1987, 1989). The Hawaiian islands are the above-sea peaks of the south-eastern-most, and youngest, segment of this immense, 6,000 km long submarine volcanic mountain range.

The earliest observers of active lava in Hawaii were the Polynesians, who began to inhabit the islands in the 6th century. As a lasting legacy from these ancient observers, the Hawaiian words *pahoehoe* and *aa*, first introduced into the scientific literature by Dana (1849) and Dutton (1884), are now used worldwide to describe the two most common contrasting types of basaltic lava (Macdonald 1953, Peterson & Tilling 1980, Kilburn 1981, 1990). Since the discovery of Hawaii (Sandwich Islands) in January 1778 by Captain James Cook, thousands of popular accounts and scientific studies of Hawaiian eruptive activity and products have been published (e.g. see Macdonald 1947, Decker et al. 1987, Bevens et al. 1988, Wright & Takahashi 1989).

The other chapters in this book present diverse facets of recent studies – empirical, theoretical and socio-economic – pertinent to the monitoring of active lavas and their attendant hazards. The purpose of this chapter is two-fold: (a) to review some practical aspects of, and lessons learned from, making field observations of active lava; and (b) to emphasize the need for more and better field observations and measurements of moving lava *during* eruptions. In so doing, we draw heavily from our experiences gained as staff members at the US Geological Survey's Hawaiian Volcano Observatory (HVO) in the 1970s, when Kilauea volcano was almost continuously active (Peterson et al. 1976, Swanson et al. 1979, Duffield et al. 1982, Tilling et al. 1987). This chapter restricts itself to some aspects of working in the field; it attempts no analysis or interpretation of the observed phenomena.

Field observation of moving lava

Eruptions at Kilauea and Mauna Loa volcanoes, two of the world's most active volcanoes, typically take place either in the summit caldera or along one of the rift zones (Fig. 5.1). Frequent, generally non-explosive activity, combined with good accessibility, affords volcanologists many opportunities to make close-up observations of eruptive phenomena and to collect well documented lava samples with relative safety, if some common-sense precautions are taken (see p. 158).

Most eruptions begin with lava fountaining from one or more fissure vents ("curtain of fire"), and eruptive activity lasting more than a few days tends to become localized at a primary central vent. Depending on the nature of the activity and topographic controls, the erupted lava may move and/or accumulate in several modes: (a) ponding within pre-existing craters to form lava lakes (e.g. Wright et al. 1976, Peck et al. 1979, Helz 1987); (b) advancing as many small, short flows and piling up around the vent to

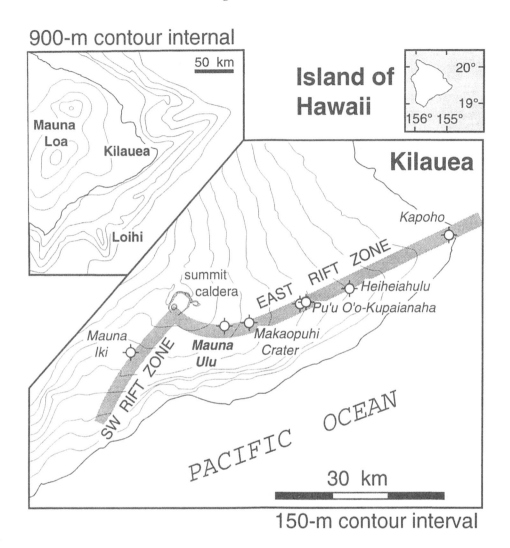

Figure 5.1   Index map of the island of Hawaii showing Mauna Loa, Kilauea and (submarine) Loihi volcanoes; Kilauea's east rift and south-west rift zones are indicated by the dark stippled swathe. The field observations discussed in this chapter were made mostly during the 1972–1974 eruption at Mauna Ulu on the upper east-rift zone; the continuous activity since 1983 has been localized farther downrift, in the vicinity of Pu'u O'o/Kupaianaha. (Modified from Tilling and Dvorak 1993.)

build volcanic shields (e.g. Tilling et al. 1973, Holcomb et al. 1974, Swanson et al. 1979, Tilling et al. 1987); (c) advancing rapidly as large, long flows fed by high-volume eruptions (e.g. Finch & Macdonald 1953, Lockwood et al. 1985, 1987, Lipman & Banks 1987); and (d) developing well defined, sustained flows that travel great distances (>10 km) – either via surface flows or lava tubes – to construct large flow fields, some of which extend to the ocean to form lava deltas, when the volume-rate of eruption is low

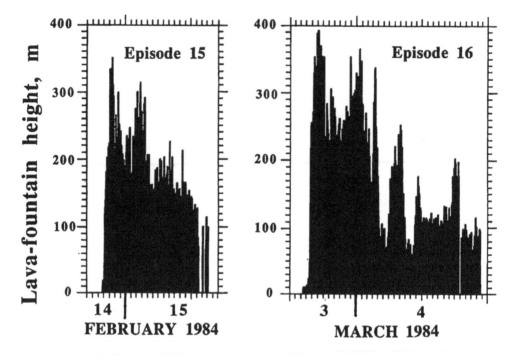

Figure 5.2 Example of the variation in lava-fountain height for two eruptive episodes of the Puu Oo eruption in 1984; each small subdivision on the abscissa is one hour. The heights were determined from photographs taken by a time lapse camera at Puu Halulu, about 1.5 km east-north-east of the vent. (After Wolfe et al., 1988, Fig. 1.24.)

to moderate and continuous (e.g. Swanson 1973, Moore et al. 1973, Peterson & Swanson 1974, Peterson 1976).

We make no attempt to discuss all the *possible* types of observations or measurements that can be made in the above-listed flow regimes typical of Hawaiian eruptions. Instead, we focus only on some general considerations that, from our own experience, seem to be most commonly encountered, if not most important. In practice, of course, what is or can be *actually* observed or measured is determined by local circumstances – including accessibility, weather conditions, and personal safety – as well as by the interests, available time, patience, background, resourcefulness and scientific equipment of the observers.

Eruptive vents

Some diagnostic observations to be made at eruptive vents include the beginning, duration and cessation of activity, as well as the lava discharge and its variation during the course of an eruption. Such information is important because it can be integrated with seismic, geodetic and other volcano-monitoring data to develop a model of the

150

volcanic "plumbing system", thereby leading to an improved understanding of how a volcano works. During a volcanic crisis, having such timely information is essential for civil defense officials in devising and implementing emergency response measures, including possible evacuation.

Unfortunately, quantitative determinations of lava discharge are impossible to make in real time, given the logistical difficulties inherent in working close to an active eruptive vent. Flow meters or other devices for measuring the flux of molten lava do not exist; thus, lava discharges at vents generally cannot be gauged precisely while an eruption is in progress. Under favorable conditions, however, such rates can be estimated – qualitatively but in real time – from field observations and measurements. Depending on eruption vigor, lava discharge can vary widely. For example, during low-level activity at Kilauea in January 1974, rates of 2–5 $m^3 s^{-1}$ were estimated (Tilling et al. 1987), whereas during the 1984 eruption of Mauna Loa, rates of 1×10^6 to 2×10^6 $m^3 h^{-1}$ were common although a few, brief peak rates as high as 3×10^6 to 5×10^6 $m^3 h^{-1}$ were estimated (Lockwood et al. 1985, Lipman & Banks 1987).

Some important variables to know in making the estimates include: height of lava fountain(s), size and geometry of the vent, dimensions and number of lava flows streaming away from the vent, and other aspects of lava movement. Lava fountain height can be measured with a theodolite, transit or inclinometer such as that in a Brunton compass, if the distance from the observation point to the fountain is known; it can also be visually estimated by reference to some nearby features of known height (e.g. crater wall, tree, cinder cone). An innovative method of systematically measuring lava fountain heights (Fig. 5.2) was employed recently by Wolfe et al. (1988), who projected images from time-lapse photography on to a computerized digitizing tablet; the measurements were scaled by periodic theodolite or transit measurements from points of known distance to the lava fountain, or by calibration of photographic frames by references to objects of known size and distance. With some or all of these variables determined, depending on eruptive and viewing conditions, the flow rate (lava volume per unit time) at the vent can be estimated. If the vent and the lava fed by it are entirely confined within a crater or depression of known approximate size and configuration, then the lava discharge can be obtained from the rise in the level of the ponded lava. Ideally, if there is no significant drainback or other complicating factors, estimates derived within a given time interval from several different sets of variables should agree within the necessarily large uncertainties inherent in the crude techniques used.

The conventional, and most widely used, method to determine lava discharge is simply to divide the volume of lava erupted by the duration of vigorous vent activity. This method yields more precise estimates, but the information can only be obtained after the cessation of eruptive activity, when field mapping of the erupted lava can be done. Even these estimates, however, are subject to uncertainties. While the area covered by a new flow can be measured accurately, the thickness, which generally is highly variable (e.g. Wolfe 1988 (Plates 2–5)), is more difficult to estimate. Another source of uncertainty in the determination of erupted volume (adjusted to dense-rock equivalent) is lava porosity, which can also be highly variable but is rarely measured systematically.

Figure 5.3 Aerial view of the active lava lake at the summit of Kupaianaha shield formed since July 1986; the lava lake surface is about 4–5 m below the crater rim. (Photograph taken on 18 February 1987 by J. D. Griggs, us Geological Survey's Hawaiian Volcano Observatory.)

While precise eruptive-volume data are important to volcanologists in reconstructing and analyzing the eruption *ex post facto*, they are of limited use to emergency response authorities confronting a volcanic crisis in progress. Much more useful is the real-time determination of lava discharges, even if crude, while observing vent activity. Significant variation in lava discharge at the vent, especially if accompanied by a measurable difference in lava temperature, may signal changes in eruptive mode. However, from experience we know that such observations of vent activity can be frustratingly difficult to make, even under the most favorable conditions.

Lava lakes

Lakes (ponds) of molten lava are commonly formed at Kilauea volcano, and two types of lava lakes have been distinguished (Swanson et al. 1979, Tilling 1987):

(a) *inactive* – results from the passive ponding of molten lava within pre-existing pit craters;

(b) *active* – gradually develops around one or more active vents by the construction of lava levées; can also form instantly if active vent(s) is located within a pre-existing crater.

Inactive lava lakes generally are formed during short-lived activity, lasting from less than a day (e.g. Hiiaka, 5 May 1973) to about a month (e.g. Kilauea Iki, 14 November to 20 December 1959). In contrast, *active* lava lakes, which may be considered as the surface expressions of the active magma column, are long lived, lasting from many months (e.g. Halemaumau, 1967–8) to several years (e.g. Mauna Ulu, 1969–71 and 1972–4) and even many decades (e.g. the pre-1924 activity at Halemaumau). More recent examples of long-lived lava lakes are those associated with the Kupaianaha and Puu Oo vents (Fig. 5.3) on Kilauea's east rift zone (Ulrich et al. 1987, Heliker 1988, Heliker & Wright 1991); both these lakes have been nearly continuously active from July 1986 to the present (November 1992).

Field observations of the formation and activity of lava lakes constitute a substantial part of the chronological narratives of many Kilauean eruptions. All aspects of lava lake activity – such as state of the surface (molten or crusted, smooth or rough, etc.), circulation pattern, gas-piston action, areas of crustal overturn and degassing, etc. – should be described as thoroughly as field time and conditions permit (e.g. see Wright et al. 1968, Richter et al. 1970, Swanson et al. 1972, 1979, Tilling et al. 1987). Such information, while often difficult to understand at the time of observation, may help constrain subsequent interpretive models of the eruption. Especially important is the careful documentation of the variations in the surface height of a lava lake. For an "inactive" lake, the rise of its surface can be related to the volume of lava filling the pre-existing crater, which in turn contributes to the estimation of lava discharge at the vent. For "active" lakes, the surface-height fluctuation directly reflects changes in the feeding magma column beneath and, hence, variations in vent activity.

The measurement of surface height relative to some datum (usually the rim of the crater or depression holding the lava) can be done easily without need for cumbersome surveying equipment. Measurements can be made by using an optical range-finder and inclinometer and sighting to one or more references on the lake surface (Swanson et al. 1979 (Fig. 49), Tilling 1987). The lake surface height to the datum is simply the product of the sine of the vertical angle and the range finder distance (Fig. 5.4a). This method, though extremely simple, gives satisfactorily reproducible results with different observers, with an uncertainty of ⩽ 5 % for determinations less than 100 m, and ⩽ 10% for greater distances (Fig. 5.4b,c). Periodic direct measurements made by human observers can serve as calibration for time-lapse photographs for the systematic remote monitoring of fluctuation in lake surface height in the intervals between field party visits to the lava lake.

In addition to the observations of lava lake activity that can be made while the eruption is in progress, various types of post-eruption field studies and *in situ* measurements can be made, once the lake surface has cooled sufficiently to form a solid crust capable of supporting people and equipment, including drill rigs. Such studies, which are beyond the scope of this chapter, include the analysis of developing cooling cracks, viscosity and temperature measurements, petrological and geochemical investigations, crystallization processes, numerical modelling, levelling surveys, seismic and other geophysical research, and other specialized topics. The enormous literature on Kilauea lava lakes, especially Kilauea Iki, amply attests to their serving as natural laboratories for

(a)

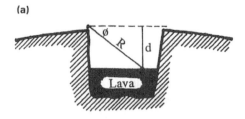

ø = Inclinometer angle

R = Rangefinder distance

d = depth to height of   = R sin ø
        lava-lake surface

Figure 5.4a Schematic of a simple method to measure the height of the surface of a lava lake using the crater rim as the datum. In general, several readings are taken on one or more surface features and the results averaged.

(b)

Figure 5.4b Observer using a range-finder to determine the distance to a reference point on the surface of Mauna Ulu lava lake; great care must be exercised in approaching the crater rim. (Photograph courtesy of the us Geological Survey's Hawaiian Volcano Observatory)

the study of the dynamics, cooling, crystallization, and differentiation of basaltic lava/magma; for a representative sample, the interested reader is referred to Peck (1966), Sato & Wright (1966), Shaw et al. (1968), Peck & Minakami (1968), Wright et al. (1976; see also references cited therein), Wright & Okamura (1977), Chouet (1979), Helz

(c)

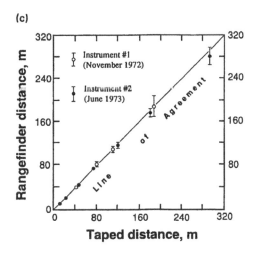

Figure 5.4c Results of two tests of range-finder measurements compared with measurements made by steel tape, involving two different instruments and several observers, some quite experienced and others complete novices; see Tilling et al. 1986 for details. Unless otherwise shown, ±1 SD is within the plot symbol.

(1980, 1987), Hermance & Colp (1982), Anderson (1987), Helz & Thornber (1987) and Helz et al. (1989).

Lava flows

Basaltic lava flows are the predominant products of Hawaiian volcanism and have been observed and documented with intense interest. Direct observations of active lava flows at Kilauea and Mauna Loa have greatly improved our general understanding of the processes and products of basaltic volcanism, not only in Hawaii but elsewhere in the world. Because Hawaiian lava flows are often more accessible – and less hazardous – to observe than are their active eruptive vents, there are many good descriptive accounts of lava flow phenomena, even by non-specialists, such as the Christian missionaries who first arrived in Hawaii in 1823 (e.g. Ellis 1825). In this section, we wish to re-emphasize the continuing need to describe lava flow activity as completely as possible, particularly with regard to measurement or estimation of flow rates.

Surface flows
Under favorable conditions, the determination of lava surface velocity is straightforward. Simply select some distinctive feature on the surface of the lava (a rafted slab or fragment of crust, a clot of more viscous material, long-lived gas bubble, etc.) and track the movement of this feature for a fixed distance and increment of time. This procedure is well suited to observations of flowing lava in well defined channels; Lipman & Banks (1987, see especially Appendix 57.2) give a detailed description of the methods used to

Figure 5.5   Volcanologists using a thermocouple to take the temperature of a flowing lava river during the March–April 1984 eruption of Mauna Loa. The aluminum shield provides protection against the intense radiant heat of the lava. (Photograph courtesy of the US Geological Survey's Hawaiian Volcano Observatory.)

observe the predominantly aa flows of the 1984 Mauna Loa eruption.

The surfaces of some active pahoehoe lava flows, however, lack distinctive markers; in such cases, the observer can simply throw in some foreign object (e.g. a rock or a branch) to serve as the marker. We have found that even very low-viscosity pahoehoe is capable of supporting such foreign markers for the observation time required, typically a few to tens of seconds, before they are consumed by, or sink into, the lava. These on-site determinations of flow velocity can be augmented by more precise, but not real-time, measurements based on analysis of photographs taken in rapid succession (at known time intervals) from the same camera station, or by frame-by-frame analysis of cine film or time lapse photography. In recent years, increasing use has been made of lightweight, high-resolution video cameras to record and analyze flow behavior, and to complement visual and photographic observations.

Because the flow rate differs from the margin to the center of the flow, and along the length of the flow according to local changes in gradient, several determinations should be made to fully characterize the range of rates. In addition, the observer should describe, and photographically document, channel dimension and configuration, as well as other attributes of lava behavior that might contribute to a quantitative analysis of the flow regime. If conditions permit, a sample of the active lava should be collected, and a temperature measurement made (Fig. 5.5). Unfortunately, except for extremely fortuitous circumstances (see p. 153), the depth of a channel generally cannot be measured during active flow of lava. Thus, an accurate knowledge of the cross-sectional area of an *active* lava flow – a key parameter in the calculation of the volumetric flow rate and in rheological analysis – can rarely be obtained; Lipman & Banks (1987, Appendix 57.2) discuss the constraints on estimation of channel depth. In some situations, the cross-sectional area of a lava channel can be reconstructed from post-eruption field observations.

With good accessibility, the determination of the rate of advance of the lava flow front is readily made by visual monitoring of the changing position of the front relative to landmarks at periodic intervals and noting the distances traversed per unit time. Flow fronts, especially for aa, typically advance at rates much slower than flow rates in upstream segments of the lava channel closer to the vent. Thus, the flow front generally can be adequately monitored by periodic rather than continuous observation. If the flow is advancing through developed areas, the road network serves as a handy reference grid for convenient measurement of distances covered between periodic observations. By prior placement of visually conspicuous markers (e.g. surveying ribbon, rock cairns, paint stripes on pavement) at 5 or 10 m intervals along roads in the path(s) of the advancing flow(s), two observers can easily and quite precisely track the movement of multiple lava flow fronts by periodic observations ("making the rounds") of the areas of interest. During some episodes of the current activity at Kilauea volcano (which began in January 1983), this simple but effective method was used to monitor lava flow advance through the Royal Gardens, Kalapana Gardens, and other subdivisions, providing timely and accurate information to the officials of Hawaiian Civil Defense for implementing emergency measures, including the orderly evacuation of residents. In July 1983, during episode 5 of the Puu Oo eruption, the measured average rates of flow front advance through the Royal Gardens varied from 0.4 to 7.4 m min^{-1}; however, some short-lived "surges" involving breakouts of thin, more fluid lobes travelled as fast as 25 m min^{-1} (Wolfe et al. 1988 (Fig. 1.7)).

The monitoring of the advance of lava flows spreading sluggishly on nearly flat ground can, on occasion, pose a special challenge to observers. At any particular time, much or all of the flow may appear to be immobile and crusted over, but careful inspection may reveal one or more active points around the flow perimeter where pahoehoe toes are slowly protruding, fed by intricately branching lava tubes. Typically, the points of activity shift from one locality to another, and the progress of the flow is sporadic, irregular and extremely difficult to monitor in detail. New lobes may also break out through the crust above the lava tubes but sometimes, instead of breaking through, pressure exerted by the confined lava beneath the crust bows the surface upward to form ovoidal swellings or tumuli. By a similar process, broad areas of the crust may be lifted by quasi-horizontal, sill-like injections of tube-fed lava. Such "inflated" flows (after Holcomb 1980 pp. 93–98)) tend to develop when a strong crust has formed and spreading of the perimeter has become very slow or has ceased, not allowing free egress of the still-molten lava. If the "inflated" area is small relative to the entire expanse of a flow that appears to be essentially static, the inflation process may be difficult to detect. More commonly, however, flow inflation occurs at a scale and rate to be recognizable visually. The process can be confirmed and measured by periodic level surveys from established points beyond the flow margins, marking the course to distinguish already solidified crust from newer flows during successive surveys; land-levelling surveys, calibrated by instrumental precise levelling, can readily provide the confirmatory evidence.

If a breakout should occur along the crusted perimeter of an inflated flow, the abrupt relief of pressure may cause localized subsidence or "deflation" by as much as a few meters. Such subsidences sometimes provide to observers the first clue that flow inflation processes have been operative. In some situations, especially in the interior parts of a flow, the observed "subsidence" is apparent and only reflects the more rapid, differential inflation of adjacent areas. Flow inflation phenomena, which have been common during the east rift activity at Kilauea since mid-1986, are being investigated systematically by visual and geophysical observations (e.g. Kauahikaua et al. 1990, K. Hon & J. Kauahikaua 1991, unpublished data, HVO).

Lava tube systems

With sustained eruptive activity at Kilauea, well integrated open channels commonly evolve into long-lived tube systems, which can transport molten lava great distances from the vent area with little loss of temperature and fluidity (see Peterson & Swanson 1974, Greeley 1987).

Lava tube systems permit another and quite unique means of observing moving lava. Portions of the roofs of the lava tubes commonly collapse and form openings or "skylights" (typically a few to tens of meters across), through which observations can be made of the active underground lava stream. The segments of the tube system between skylights can be studied by field geoelectrical techniques; for example, since 1970, VLF (very low frequency) tilt angle and EM31 loop/loop induction techniques have been used at Kilauea to complement the surface geological mapping of active lava tubes (Anderson et al. 1971, Zablocki, 1978, Kauahikaua et al. 1990). Because the roof of a lava tube is a good insulator, it provides an efficient shield against thermal radiation of the hot, molten lava, often allowing a much closer and safer approach to the lava stream than would be possible for an open channel.

In addition to the usual kinds of field observations that can be made at open lava channels, a long-lived tube system can offer a unique method for making measurements of *average* flow velocity over a long stretch of the lava river. A tree branch is tossed through a lava tube skylight, and the passage of the branch floating on the lava stream is timed at successive downstream skylights by observers with stop-watches who communicate by radio. Measuring the distances between the skylights and the passage time, lava flow velocity can be computed for each segment between a pair of skylights, and the results can then be averaged for all the segments measured. Because the lava tube contains oxygen sufficient to support combustion only in the vicinity of the skylights, a branch, though floating on lava with temperatures exceeding 1130°C, can last for hundreds of meters before being totally burned (Fig. 5.6). The measured velocities are minimum values, because they do not take into account the following

Figure 5.6   A tree branch floating on an underground lava stream bursts into flame as it encounters oxygen as a lava tube skylight (about 3 m across) is approached. Timing the passage of the branch at several skylights successively downstream makes possible an accurate determination of the *average* flow rate (see text). (From Tilling et al., 1987, Fig. 16.16.)

factors: (a) the linear distance between skylights measured on the surface is generally less than the actual length of a subsurface meandering lava stream; and (b) ponding (at the base of lava cascades) and eddies along the stream slow the passage of the branch. Average minimum flow rates measured in this manner in 1972 during the Mauna Ulu eruption varied between 0.2 and 3.0 m s^{-1}, depending on vent activity and the level of the feeding Alae lava lake (Tilling et al. 1987). For the current Kilauea east rift eruption (Heliker & Wright, 1991), flow rates of 1–2 m s^{-1} have been commonly observed during times of maximum flow (T. L. Wright 1990, personal written communication).

Lava tube skylights tend to provide better opportunities than at either lava lakes or open channels for measurement of lava temperature by means of optical pyrometers, because of the reduced atmospheric interference and the nearly isothermal conditions. Nonetheless, even the thinnest surface crust and associated fume of lava moving rapidly in the tube can apparently mask its true temperature, thus resulting in the minimum temperatures generally obtained by optical pyrometers. One favorably situated skylight permitted in 1972 the only direct measurement of the depth of an active lava channel made during the 1969–74 Mauna Ulu eruptions; an observer was able to plunge a stainless steel pipe through a small hole in the lava-tube roof to probe the bottom of the channel (Fig. 5.7). A similar "depth probe" also has been used to measure lava channel depth at Etna (Pinkerton & Sparks, 1976, Ch. 6 this volume). Analogous to the

Figure 5.7 An observer, using a stainless steel pipe, probes the depth of an active lava stream flowing in a lava tube. (From Tilling et al. 1987, Fig. 16.15.)

determination of the surface height of a lava lake, similar measurements or estimates can be made of the height of the lava stream (below the roof = ground surface) flowing within the tube. Although the variation in lava-stream height is generally not amenable to simple interpretation, sometimes it is possible to correlate the observed fluctuations with change in the lava discharge at the vent and/or blockages or constrictions downstream along the tube system.

Beneath the sea
Observation of moving lava in Hawaii is not restricted to volcanic activity on land. Under favorable circumstances and with some special equipment it has also been possible to make observations under the sea. However, the life-threatening hazards of diving on flowing lava cannot be overemphasized, even though, fortunately, no fatal accidents have occurred to date.

During the 1969–74 Mauna Ulu eruptions, as well as the current (1983 to the present) Kilauea activity, lava from vents on the east-rift zone entered the Pacific Ocean on several occasions, after flowing more than 10 km through well integrated tube systems on land (Swanson et al. 1979, Tilling et al. 1987, K. M. Kelly et al. 1989, unpublished data, HVO). Such lava entry into the sea added new land by building lava deltas (Moore et al. 1973, Peterson 1976, Kelly et al. 1989). In some instances, the tube system extended beyond the shoreline and fed lava directly to the submerged flow front offshore. For the first time in history, scuba diving scientists were able to observe and document submarine lava movement and the formation of pillow lava (e.g. Grigg 1973, Moore 1975, Tepley 1975). Since the early 1970s, a number of excellent films and videocassettes containing dramatic footage of submarine lava movement have been produced (e.g. *Fire Under the Sea – The Origin of Pillow Lava*, Lee Tepley 1975, *Eruption at the Sea*, Ka 'Io Productions 1988 and *Inside Hawaiian Volcanoes*, US Geological Survey and the Smithsonian Institution 1989).

Bathymetric, photographic and seismic data suggest that Loihi seamount, located about 30 km offshore of the south coast of of Hawaii, may be an active submarine volcano (Klein 1982, Malahoff 1987). Fresh, glassy samples dredged from the summit area of Loihi provide additional evidence of geologically recent volcanism (Moore et al. 1982). To date, however, none of the studies conducted at Loihi has obtained unambiguous evidence for historical eruptive activity. When the volcano is more thoroughly instrumented, it might be possible some day to detect definitive precursors and to recognize eruption onset sufficiently early to allow the timely mobilization and dispatch of investigators and equipment (including research submersibles) to Loihi, so that submarine eruptions can be directly observed.

Collection of lava samples

Many techniques can be used to collect fresh lava samples, depending on field conditions, available equipment, and the use to be made of the samples. Here we briefly mention some sample collection techniques that have been used in Hawaii with good success. It is important to collect, if at all possible, lava samples while still molten, so that they can be rapidly quenched. Quenched samples are most representative of the melt as it issues from the subsurface, before slow cooling and crystallization can modify the texture, vesicularity, and perhaps composition. In addition, rapidly quenched samples of Hawaiian lava provide the best means to derive meaningful temperature estimates using a well calibrated geothermometer based on the MgO or CaO content of the glass (R. T. Helz & C. R. Thornber 1987, unpublished data, HVO). Indeed, the lava temperatures so obtained are proving to be so reliable and superior that the HVO has abandoned all forms of field pyrometry in favor of the Helz–Thornber glass geothermometer (T. L. Wright 1990, written communication).

Fortunately, it is generally possible to collect molten lava under most eruption conditions in Hawaii. Some volcanologists have insisted – only half in jest – that any lava sample not collected "live" must be considered "weathered" at best, and "meta-morphosed" at worst. In the vicinity of an active vent, a convenient and generally the best means to obtain quenched material is to collect ejecta (e.g. spatter, pumice, bombs) from lava fountains. Such material is quickly quenched from the molten state during its passage in air. Ideally, the collector should choose material that he or she has observed during its fall. Depending on its size, the ejecta material can be too hot to handle, and may possibly remain incandescent for a short while, after striking the ground. Metal tongs can be used to carry hot samples to a safe place for further cooling. For obvious safety reasons, samples of lava fountain ejecta should be collected on the upwind side of the vent.

Experience in Hawaii has shown that the best samples are: (a) spatter collected via a cable hung over active fountaining areas (e.g. vent, edge of a lava lake); (b) spatter or pumice collected after falling from lava fountains; (c) glassy overflows from lava tube skylights; and (d) littoral spatter or surf zone samples where the lava tube empties directly into the ocean. When it is impossible to collect these types of samples, another widely used method is to extract molten material from the moving lava flow itself. This can generally be accomplished by approaching the side or the terminus of an active flow and inserting some implement to which lava will adhere when withdrawn. A geological hammer usually serves quite conveniently for this purpose (Fig. 5.8), on land as well as beneath the sea. Under conditions that preclude the safe approach to molten lava within arm's reach, longer steel pipes or ladles can be used to scoop a sample (Fig. 5.8), but these are awkward to carry long distances in rugged terrain. However, almost any implement can be used, so long as it does not melt in, or is otherwise destroyed by, the lava stream, or does not contaminate the sample. To minimize possible steel contamination of samples to be used for specialized geochemical studies, some investigators have collected material using ceramic tubes, wooden sticks or sturdy branches.

Rapid quenching of the lava sample minimizes groundmass crystallization and attendant oxidation. Vent ejecta samples are generally preferred to those taken from a lava stream, which undergo oxidation and, possibly, other effects during flowage. Moreover, with spatter samples, the time of ejection from the vent can be established exactly whereas, with flow samples, especially if collected far from the vent, the time of eruption is indeterminate. To accelerate the cooling of samples collected from lava flows, some workers (e.g. Neal et al. 1988) quench the sample by immersion in water. While this procedure may obtain a slightly better sample, it is too awkward under most conditions; it requires the collector to carry a supply of water and some suitable container (e.g. an empty coffee can) – adding to the logistical burden of doing fieldwork. Other factors being equal, air quench of small samples is probably at least as effective as water quench of large samples. To date, experience has shown that, for general petrological and major-element studies, any possible anomalies in analytical results that might be introduced by not water-quenching samples appear to be negligible, i.e. less than other sources of error. However, no systematic comparisons have been made of geochemical data obtained on sample pairs (air quench versus water quench) collected at the same time and place under identical field conditions.

Figure 5.8a Perhaps the most common method to collect active lava in Hawaii, a volcanologist uses a geological hammer to scoop lava from a pahoehoe lobe; the hand is protected from radiant heat by means of a canvas sample bag.

Figure 5.8b A long-handled sampler being used to ladle molten material from a pahoehoe lava stream.

Molten material below the ground surface (e.g. in a lava lake, flowing through a lava tube) is more difficult to collect than surface samples. The simple method of throwing in and the hauling up of a sampler attached to a flexible stainless steel cable – like drawing water from a well – has occasionally been successful. However, the logistics are difficult; the density and viscosity of the lava offer strong resistance to immersing any but stiff-handled samplers. A better method is to suspend the sampler via a cable above a fountaining area above the lava lake surface. Drilling studies of lava lakes also afford unique opportunities to collect the still-molten lava below the solidifying crust, in some cases many years after the eruption. For example, drilling studies conducted at Kilauea Iki lava lake in the 1970s were still able to sample molten lava erupted in 1959 (Helz 1980, 1987). The sampling technique involves drilling through the crust to penetrate the crust/melt interface. Melt is then collected by pushing in stainless steel or ceramic probes through the bottom of the drill hole; another collection method is to allow molten material to ooze into the hole and then to redrill through the the solidified ooze material (Wright et al. 1976, Wright & Okamura 1977, Helz 1987).

Safety precautions when working around active lava

Even with the relatively benign, generally non-explosive, Hawaiian eruptions, observers must be constantly vigilant about the volcano and be fully safety conscious when studying active lava. We review here some important safety precautions, which, while perhaps obvious to seasoned observers, may provide some basic guidelines to those less experienced. However, we have been surprised, sometimes dismayed, by the lack of common sense shown on occasion by even experienced geologists while observing eruptions in Hawaii. Thus, the points discussed below (with no order of priority implied) are in addition to the "normal" logistics and occupational hazards associated with any geological fieldwork.

    (a) *Wear protective clothing.* In working around active lava, it is important to wear appropriate field clothes (stout boots, long pants, long-sleeves, etc.). Most Hawaiian eruptive products contain abundant basaltic glass with only a small percentage of phenocrysts. The ejecta from the vent and the broken surfaces or pieces of solidified lava flows typically have jagged spines and sharp edges. Moreover, fresh lava near fractures can remain hot long after solidification because of continued streaming of hot volcanic gases. Sturdy gloves are necessary to protect the hands from the sharp lava, because stumbling and falling are hard to avoid despite careful footwork during fieldwork. It has been our experience that, without protective gloves, cuts, lacerations and, occasionally, burns of the hands often constitute the most serious injuries incurred during a fall. The wearing of proper footwear is essential to minimize potential injury in traversing the rugged terrain of active flow fields. The use of soft-material footwear (e.g. running shoes, tennis shoes) is especially inappropriate. Not only do they cut and melt easily, but they also do not provide sufficient ankle support and protection against the sharp edges and surface roughness of lava flows.

It is also advisable to have available some additional protective gear, especially if working near the active vent(s). Where falling vent ejecta or other debris poses a potential hazard, hard hats should be worn. In making observations in areas subject to volcanic fumes, the use of one or more lightweight paper (painter's particle) masks, previously soaked in a baking soda solution and then dried, provides quite effective protection for light to moderate sulphurous fume conditions. In a pinch, placing a wet handkerchief over one's mouth and nose offers some relief. In areas of heavy fuming, regular gas masks should be used. For protection from the sometimes intense radiant heat of active aa, shields made of flexed metal with wooden or canvas handles (e.g. Fig. 5.5) and foil-faced masks (e.g. welder's mask) and foil-lined suits or aprons can be used to good effect. However, unless the heat protection clothing is flexible and lightweight, its advantage can be negated by its bulk and awkwardness, making the wearer less agile and, hence, more susceptible to falling.

(b) *Learn to recognize potentially unstable ground.* The ground in the vicinity of active vents and lava flows can be quite unstable, and the sudden shifts or collapses of unstable ground represent a significant hazard to persons observing the eruptive activity. Examples include the edges of craters, especially those containing lava lakes (e.g. Tilling et al. 1975), routes of active lava tubes along which skylights may enlarge or new ones suddenly open, and crusted surfaces of actively inflating flows, where collapse (deflation) may occur abruptly with draining of molten lava and attendant pressure loss. Ground collapses on a larger scale can also happen during the development of active lava deltas with sustained entry of lava into the sea. The newly formed and growing coastal lava benches and littoral cones are highly unstable, and collapses involving blocks a few hundred meters long and tens of meters wide are not uncommon. For example, on 12 July 1988, a 100 × 25 m lava bench that had developed during the current Kilauea east rift activity dropped into the ocean within 20–30 seconds, accompanied by powerful explosions as water surged into the feeding lava tubes (SEAN 1988). Staff members of the HVO have witnessed each type of collapse event and realized the importance of learning to sense the potential hazards of unstable ground and to avoid it. All agree that suddenly shifting ground is likely the most dangerous, and least predictable, hazard around active vents and lava.

(c) *Work upwind of active lava.* Following this simple rule will avoid the common hazards in working around active vents or lava flows. Conditions downwind from active lava – falling ejecta from lava fountains, extreme temperatures, choking fumes, etc. – can be unpleasant and potentially injurious to health. For example, the air temperature immediately downwind of a lava tube skylight can be in excess of 800°C; areas of high temperature are generally indicated by thermal plumes marked by refraction and shimmering of air above the skylight, but such plumes are not always visible under fumy conditions. It is also prudent to avoid, to the extent possible, closed depressions where heavy gases may collect. Making observations when there is no wind or no well defined prevailing wind direction can be particularly hazardous, because erratic and sudden wind shifts can occur. Keep a keen nose; the smell of burning hair provides an immediate and unmistakable warning signal to *move away* from the problem!

(d) *Beware of new flows in vegetated areas.* When flows advance into areas of thick vegetation, organic matter is sometimes buried before being completely burned. Hydrocarbon gases are distilled from the unburned but thermally decomposing material and may be unable to escape upwards through the new lava, but instead migrate laterally through highly porous old rock. In this regard, fields of older pahoehoe flows are more treacherous, because pahoehoe more efficiently traps gases beneath the surface; aa flows allow easy release of gases through their rubble. The blending of the distilled organic gases with air can produce a highly volatile, flammable mixture, which may be ignited explosively by heat from the lava. Violent explosions are common along and well beyond the margins of the advancing flows. The explosions, some of which throw boulder-sized fragments and small trees for tens of meters, are random and no warnings precede them. Localities near such flows should be approached or traversed with extreme caution or, better still, avoided entirely.

(e) *Walk in the rough.* Pahoehoe lava flows can occur in a great variety of forms (Swanson 1973). In general, tube-fed pahoehoe tends to form smooth, stable flow surfaces, which can be walked on with great ease and minimal worry. However, fountain-fed lava in close proximity to the vent can form gas-rich, *shelly* pahoehoe flows, characterized by smooth but more bulbous surfaces (Fig. 5.9). While the smooth parts of these flows present invitingly easy paths for walking, they are in fact quite treacherous. As the name implies, shelly pahoehoe has a shell-like surface covering a honeycomb of large cavities formed by gas expansion during cooling and flow (Fig. 5.9). Some of the smoothly arched surfaces, which can be as thin as 1 cm, cover cavities as large as several meters across and more than a meter deep. These surface shells appear deceptively solid, but they are actually fragile and may collapse when stepped on. It does not take many falls (and scars from cuts) for the observer to learn that, in working on shelly pahoehoe surfaces, the least treacherous walking is along the roughest parts (e.g. creases, crenulations) of the flow. These lower parts of the flow tend to form where two or more cavities join and thus provide more structural support than the cavity roof (Fig. 5.9). Nonetheless, even with careful placement of the feet while traversing shelly pahoehoe, some collapse and falls are inevitable. During a fall, however, a person can minimize injury from the jagged, sharp edges of

Figure 5.9a Shelly pahoehoe lava freshly erupted during the 1972–74 Mauna Ulu activity at Kilauea volcano; the geological hammer on top of a large billow (upper center) gives scale. Its smooth, billowy surface – while appearing solid – is actually very fragile and often collapses when stepped on, posing considerable hazard to the unwary. (From Swanson, 1973, Fig. 2.)

Figure 5.9b Sketch showing the cavernous interior of shelly pahoehoe; the dots indicate places along the flow surface that tend to be structurally stronger and less likely to collapse when walked on. Traversing a field of shelly pahoehoe is more treacherous than crossing a field of aa lava.

the broken shells by arresting his or her forward motion, to fall as vertically as possible. This manner of controlled falling becomes almost second nature after a few painful lessons. Also, a walking stick or an ice axe can be helpful in controlling falls. As mentioned previously, sturdy gloves are mandatory to protect from cuts sustained in falls.

(f) *Look for escape routes.* The unexpected can and often does happen during eruptive activity – such as the opening of a new vent, a spurt in lava output, a sudden shift in the wind direction, a change in the trajectory of a lava fountain by an obstruction in the vent, "breakouts" of lava through a levée of a channel or lake, and various other unanticipated events. Observers of active lava should always try to have some back-up or escape routes in mind to evacuate the observation site if necessary; this important precaution is analogous to a prudent, well trained airplane pilot who continuously watches for emergency landing sites should unanticipated trouble strike. Common sense should govern all actions and decisions. Never become careless or complacent even under the most favorable of observing conditions, and always be aware of what nearby vents and flows are doing. Things can change quickly, often without warning.

(g) *Avoid prolonged exposure to conditions of high ambient temperature.* The simplest solution to this problem is to avoid or skirt, to the extent practicable, areas of high ambient temperature while approaching, or working in, places of interest. However, sometimes in order to make needed observations of active lava, it is necessary to work in hot areas. During some of our observations of the 1972–74 Mauna Ulu activity it was not uncommon to encounter air temperatures in excess of 60°C over wide expanses of fresh, still-hot flows. Prolonged exposure to such conditions can lead to severe fluid loss and resulting detrimental effects (heat cramps, heat exhaustion), which can be extremely serious if the condition strikes when the observer is in a remote area, far from relief. We speak from personal experience. While having salt tablets and an abundant supply of water help to compensate fluid loss, we also recommend that observers periodically retreat to an area of older, colder lava flows to cool off.

(h) *Try to avoid working alone.* Although applicable to doing geological field-work in any rugged or remote terrane, this axiom is especially important in observing eruptive activity. Because of the high risks in working around active lava, minor mishaps or medical problems can become serious if a colleague is not present to render assistance or summon help.

In the many decades of observing Hawaiian eruptions, only four serious accidents have occurred. In May 1924, during a major explosive eruption at Kilauea summit, a photographer, who ventured too close to Halemaumau crater, was killed when struck by a falling volcanic block. In 1970, 1983 and 1985 staff members of the HVO broke through ledges of seemingly solid lava crust and stepped into molten lava. Fortunately, each time, close-by colleagues instantly rendered assistance and summoned immediate transport. Experience gained from these exceedingly rare mishaps indicates that the observer – if he or she remains sufficiently cool-headed to remember under such stressful circumstances – can be more easily extricated by falling backwards on to solid ground, rather than trying to walk out. Although these observers all suffered burns – quite severe in two cases – each completely recovered to resume full and active lives. The outcome of these three accidents could have been disastrous, possibly fatal, if either of the observers had been working alone.

We acknowledge, however, that, under certain conditions and/or with insufficient personnel, working alone is sometimes unavoidable. In such situations, the lone observer must maintain frequent communication with colleagues by radio and exercise even greater caution while conducting the fieldwork.

Concluding remarks

Recent studies of active lava, including those in this volume, have become increasingly more sophisticated with advances in monitoring equipment, measurement of flow parameters, theoretical fluid dynamics, computer modelling, and numerical simulations. We sense that close-up field observations of active lava sometimes have been de-emphasized in favor of theoretical and experimental investigations. Field observations, however, are critical to guide and integrate all forms of instrumental monitoring to furnish realistic boundary conditions and constraints in computer modelling of flow dynamics and other eruptive phenomena, and to provide "ground truth" for volcano-monitoring data acquired by automated, unmanned systems (satellite or ground based). Wright & Swanson (1987) present a succinct, cogent rationale for the importance of "real-time study of volcanic activity", including systematic field observations, in furnishing a broad scientific framework for investigations of ancient volcanic terrains and for topical studies of volcanic/magmatic phenomena in general.

Given the substantial progress in volcanology in the 20th century, the reasons for the growing gap between field observations and laboratory studies are unclear. The personnel-intensive nature and logistical difficulties inherent in making on-site quantitative measurements of volcanic phenomena constitute one major factor. Another factor, we suspect, is that some people may feel that descriptive field observations are "old fashioned" and not fully scientific (i.e., not sufficiently quantitative). In addition, there may be a perception (misconception?) that it is easier to justify the time and cost of fieldwork if elaborate new equipment is used. In this paper, however, we have stressed that real-time field observations must form an integral component of multidisciplinary investigations of eruptive phenomena. Fortunately, despite the obvious difficulties, a number of essential field measurements or estimates of lava output and movement can be made easily, using simple, inexpensive, easily portable equipment. Swanson (1992) makes an eloquent and compelling case for the importance of geological field observations in volcano monitoring, including the making of repeated simple measurements to test interpretations and models suggested by the observations. Taking reasonable precautions, the observers can accomplish these tasks with relative safety. Moreover, the needed on-site observations can be made at low cost by intelligent, dedicated, appropriately instructed personnel without the need for much formal academic training – an important consideration in the monitoring of active lava at volcanoes located in developing countries, which do not have adequate scientific and economic resources.

Acknowledgements

The authors recognize the many contributions made by present and past observers of Hawaiian eruptive activity, too numerous to mention individually. In recent decades, these observers have included the staff members at the US Geological Survey's Hawaiian Volcano Observatory, as well as many colleagues from other research organizations in the USA and abroad. The authors thank all these individuals for their observations, enthusiasm and co-operation in contributing to the body of knowledge on how Hawaiian volcanoes work. In addition, the authors greatly appreciate the constructive criticism and suggestions received from our US Geological Survey colleagues Norman G. Banks, Wendell A. Duffield, James G. Moore, Donald A. Swanson and Thomas L. Wright – all experienced observers of Hawaiian eruptions – who reviewed an earlier version of this chapter. The comments of Harry Pinkerton (Lancaster University, UK) on the manuscript were also helpful.

References

Anderson, L. A. 1987. *Geoelectric character of Kilauea Iki lava lake crust.* US Geological Survey Professional Paper 1350 2, 1345–55.

Anderson, L. A., D. B. Jackson, F. C. Frischknecht, 1971. Kilauea volcano – detection of shallow magma bodies using the VLF and ELF induction methods. *Eos, Transactions of the American Geophysical Union* **52**, 383.

Bevens, D., T. J. Takahashi, T. L. Wright (eds) 1988. *The early serial publications of the Hawaiian Volcano Observatory.* Hawaii: Hawaii National Park, Hawaii Natural History Association.

Chouet, B. 1979. Sources of seismic events in the cooling lava lake of Kilauea Iki, Hawaii. *Journal of Geophysical Research* **84**, 2315–30.

Clague, D. A. & G. B. Dalrymple 1987. *The Hawaiian–Emperor volcanic chain: part I. Geologic evolution.* US Geological Survey Professional Paper 1350, 5–54.

Clague, D. A. & G. B. Dalrymple 1989. Tectonics, geochronology, and origin of the Hawaiian–Emperor volcanic chain. In *The geology of North America.* Vol. N, *The eastern Pacific Ocean and Hawaii*, E. L. Winterer, D. M. Hussong, R. W. Decker (eds), 188–217. Boulder: Geological Society of America.

Dana, J. D. 1849. Geology. *US Exploring Expedition, under the command of Charles Wilkes, USN, 1838–1842* **10**, 156–226.

Decker, R. W., T. L. Wright, P. H. Stauffer (eds) 1987. *Volcanism in Hawaii.* US Geological Survey Professional Paper 1350, 1–1667.

Duffield, W. A., R. L. Christiansen, R. Y. Koyanagi, D. W. Peterson 1982. Storage, migration, and eruption of magma at Kilauea volcano, Hawaii, 1971–72. *Journal of Volcanology and Geothermal Research* **13**, 273–307.

Dutton, C. E. 1884. Hawaiian volcanoes. *US Geological Survey, 4th Annual Report*, 75–219.

Ellis, W. 1825. *Narrative of a Tour through Hawaii, or, Owhyhee.* London: H. Fisher, Son, & P. Jackson. (Reprinted 1826, 1827; also reprinted by Hawaiian Gazette, Honolulu; 1827 London edition reprinted 1963 as *Journal of William Ellis* by Advertiser, Honolulu).

Finch, R. H. & G. A. Macdonald 1953. Hawaiian volcanoes during 1950. *US Geological Survey Bulletin* **996-B**, 1–89.

Greeley, R. 1987. *The role of lava tubes in Hawaiian volcanoes.* US Geological Survey Professional Paper 1350, 1589–1602.

Grigg, R. 1973. Fire under the sea. *Oceans* **6**, 6–11.

Heliker, C. 1988. Volcanic hazards and eruption forecasting in Hawaii. *Proceedings of the Kagoshima International Conference on Volcanoes, July 19–23, 1988, Kagoshima, Japan*, 598–601.

Heliker, C. & T. L. Wright 1991. The Pu'u 'O'o–Kupaianaha eruption of Kilauea. *Eos, Transactions of the American Geophysical Union* **72**, 521, 526, 530.

Helz, R. T. 1980. Crystallization history of Kilauea Iki lava lake as seen in drill core recovered in 1967–1979. *Bulletin Volcanologique* **43**, 675–701.

Helz, R. T. 1987. Differentiation behavior of Kilauea Iki lava lake, Kilauea volcano, Hawaii. An overview of past and current work. In *Magmatic processes: physicochemical principles*, B. O. Mysen (ed.). *Geochemical Society Special Publication* 1, 241–58.

Helz, R. T. & C. R. Thornber 1987. Geothermometry of Kilauea Iki lava lake. *Bulletin of Volcanology* **49**, 651–68.

Helz, R. T., H. Kirschenbaum, J. W. Marinenko 1989. Diapiric transfer of melt in Kilauea Iki lava lake, Hawaii: A quick, efficient process of igneous differentiation. *Geological Society of America, Bulletin* **101**, 578–94.

Hermance, J. F. & J. L. Colp 1982. Kilauea Iki lava lake: Geophysical constraints on present (1980) physical state. *Journal of Volcanology and Geothermal Research* **13**, 31–61.

Holcomb, R. T. 1980. *Kilauea volcano, Hawaii: chronology and morphology of the surficial lava flows.* Ph.D. Thesis, Stanford University (also US Geological Survey Open-File Report 81–354).

Holcomb, R. T., D. W. Peterson, R. I. Tilling 1974. Recent landforms at Kilauea volcano, a selected photographic compilation. In *Guidebook to the Hawaiian Planetology Conference, Washington, D.C.*, R. W. Greeley (ed.). NASA *TMX* 62362, 49–86.

Hon, K. & J. Kauahikaua 1991. The importance of inflation in formation of pahoehoe sheet flows. *Eos, Transactions of the American Geophysical Union* **72**, 557 (abstract).

Kauahikaua, J., T. Moulds, K. Hon 1990, Observations of lava tube formation in Kalapana, Hawai'i. *Eos, Transactions of the American Geophysical Union* **71**, 1711 (abstract).

Kelly, K. M., K. Hon, G. Tribble 1989. Bathymetric and submarine studies of an active lava delta near Kupapau Point, Hawaii. *Eos, Transactions of the American Geophysical Union* **70**, 1202 (abstract).

Kilburn, C. R. J. 1981. Pahoehoe and aa lavas: A discussion and continuation of the model of Peterson and Tilling. *Journal of Volcanology and Geothermal Research* **11**, 373–89.

Kilburn, C. R. J. 1990. Surfaces of aa flow-fields on Mount Etna, Sicily: morphology, rheology, crystallization and scaling phenomena. In *IAVCEI Proceedings in Volcanology*. Vol. 2, *Lava flows and domes: emplacement mechanisms and hazard implications*, J. H. Fink (ed.), 129–156. Berlin: Springer.

Klein, F. W. 1982. Earthquakes at Loihi submarine volcano and the Hawaiian hot spot. *Journal of Geophysical Research* **87**, 7719–26.

Lipman, P. W. & N. G. Banks 1987. *Aa flow dynamics, Mauna Loa 1984*. US Geological Survey Professional Paper 1350, 1527–67.

Lockwood, J. P., N. G. Banks, T. T. English, L. P. Greenland, D. B. Jackson, D. J. Johnson, R. Y. Koyanagi, K. A. McGee, A. T. Okamura, J. M. Rhodes 1985. The 1984 eruption of Mauna Loa volcano, Hawaii. *Eos, Transactions of the American Geophysical Union* **66**, 169–71.

Lockwood, J. P., J. J. Dvorak, T. T. English, R. Y. Koyanagi, A. T. Okamura, M. L. Summers, W. R. Tanigawa 1987. *Mauna Loa 1974–1984: a decade of intrusive and extrusive activity*. US Geological Survey Professional Paper 1350, 537–70.

Macdonald, G. A. 1947. Bibliography of the geology and water resources of the island of Hawaii. *Hawaii Division of Hydrography, Bulletin* **10**, 1–191.

Macdonald, G. A. 1953. Pahoehoe, aa, and block lava. *American Journal of Science* **251**, 169–91.

Malahoff, A. 1987. *Geology of the summit of Loihi submarine volcano*. US Geological Survey Professional Paper 1350, 133–44.

Moore, J. G. 1975. Mechanism of formation of pillow lava. *American Scientist* **63**, 269–77.

Moore, J. G., R. L. Phillips, R. W. Grigg, D. W. Peterson, D. A. Swanson 1973. Flow of lava into the sea, 1969–1971, Kilauea Volcano, Hawaii. *Geological Society of America, Bulletin* **84**, 537–46.

Moore, J. G., D. A. Clague, W. R. Normark 1982. Diverse basalt types from Loihi Seamount, Hawaii. *Geology* **10**, 88–92.

Neal, C. A., T. J. Duggan, E. W. Wolfe, E. L. Brandt 1988. Lava samples, temperatures, and compositions. In *The Puu Oo eruption of Kilauea volcano, Hawaii: episodes 1 through 20, January 3, 1983, through June 8, 1984*, E. W. Wolfe (ed.). US Geological Survey Professional Paper 1463, 99–126.

Peck, D. L. 1966. *Lava coils of some recent historic flows, Hawaii*. US Geological Survey Professional Paper 550-B, B148–51.

Peck, D. L. & T. Minakami 1968. The formation of columnar joints in the upper part of Kilauean lava lakes, Hawaii. *Geological Society of America, Bulletin* **79**, 1151–66.

Peck, D. L., T. L. Wright, R. W. Decker 1979. The lava lakes of Kilauea. *Scientific American* **241**, 114–28.

Peterson, D. W. 1976. Processes of volcanic island growth, Kilauea Volcano, Hawaii, 1969–1973. *Proceedings of the IAVCEI Symposium on Andean and Antarctic Volcanology Problems*, 172–89.

Peterson, D. W. & D. A. Swanson 1974. Observed formation of lava tubes during 1970–1971 at Kilauea volcano, Hawaii. *Studies in Speleology* **2**, 209–22.

Peterson, D. W. & R. I. Tilling 1980. Transition of basaltic lava from pahoehoe to aa, Kilauea volcano, Hawaii: field observations and key factors. *Journal of Volcanology and Geothermal Research* **7**, 271–93.

Peterson, D. W., R. L. Christiansen, W. A. Duffield, R. T. Holcomb, R. I. Tilling 1976. Recent activity of Kilauea Volcano, Hawaii. *Proceedings of the IAVCEI Symposium on Andean and Antarctic Volcanology*

Problems, 646–56.

Pinkerton, H. & R. S. J. Sparks 1976. The 1975 sub-terminal lavas, Mount Etna: a case history of the formation of a compound lava field. *Journal of Volcanology and Geothermal Research* **1**, 167–182.

Richter, D. H., J. P. Eaton, K. J. Murata, W. U. Ault, H. L. Krivoy 1970. *Chronological narrative of the 1959–60 eruption of Kilauea volcano, Hawaii.* US Geological Survey Professional Paper 537-E, 1–73.

Sato, M. & T. L. Wright 1966. Oxygen fugacities directly measured in magmatic gases. *Science* **153**, 1103–05.

SEAN 1988. Kilauea (Hawaii): Lava bench collapse at seacoast. *Bulletin of the Scientific Event Alert Network* [now called *Bulletin of the Global Volcanism Network*] **13**, 12.

Shaw, H. R., D. L. Peck, T. L. Wright, R. Okamura 1968. The viscosity of basaltic magma: an analysis of field measurements in Makaopuhi lava lake, Hawaii. *American Journal of Science* **266**, 225–64.

Swanson, D. A. 1973. Pahoehoe flows from the 1969–1971 Mauna Ulu eruption, Kilauea volcano, Hawaii. *Geological Society of America Bulletin* **84**, 615–26.

Swanson, D. A. 1992. The importance of field observations for monitoring volcanoes, and the approach of "keeping monitoring as simple as practical". In *Monitoring volcanoes: techniques and strategies used by the staff of the Cascades Volcano Observatory, 1989–90*, J. W. Ewert & D. A. Swanson (eds). *US Geological Survey Bulletin*, 1966, 219–23.

Swanson, D. A., W. A. Duffield, D. B. Jackson, D. W. Peterson 1972. The complex filling of Alae crater, Kilauea volcano, Hawaii. *Bulletin Volcanologique* **36**, 105–26.

Swanson, D. A., W. A. Duffield, D. B. Jackson, D. W. Peterson 1979. *Chronological narrative of the 1969–71 Mauna Ulu eruption of Kilauea volcano, Hawaii.* US Geological Survey Professional Paper 1056, 1–55.

Tepley, L. 1975. Fireworks erupt when hot lava pours into the sea, 1975. *Smithsonian* **6**, 70–5.

Tilling, R. I. 1987. Fluctuations in surface height of active lava lakes during 1972–1974 Mauna Ulu eruption, Kilauea volcano, Hawaii. *Journal of Geophysical Research* **92**, 13,721–30.

Tilling, R. I. & J. J. Dvorak 1993. Anatomy of a basaltic volcano: Kilauea, Hawaii. *Nature*, (in press).

Tilling, R. I., D. W. Peterson, R. L. Christiansen, R. T. Holcomb 1973. Development of new volcanic shields at Kilauea volcano, Hawaii 1969–1973. *International Union for Quaternary Research, Christchurch, New Zealand, December 2–10, Ninth Congress Program*, 366–7 (extended abstract).

Tilling, R. I., R. Y. Koyanagi, R. T. Holcomb 1975. Rockfall seismicity – correlation with field observations, Makaopuhi crater, Kilauea volcano, Hawaii. *US Geological Survey Journal of Research* **3**, 345–61.

Tilling, R. I., R. L. Christiansen, W. A. Duffield, R. T. Holcomb, D. W. Peterson 1986. *Determinations of the depth to the surfaces of active lava lakes at Mauna Ulu and Alae, Kilauea volcano, Hawaii.* US Geological Survey Open-File Report 86-367, 1–16.

Tilling, R. I., R. L. Christiansen, W. A. Duffield, E. T. Endo, R. T. Holcomb, R. Y. Koyanagi, D. W. Peterson, J. D. Unger 1987. *The 1972–1974 Mauna Ulu eruption, Kilauea volcano: an example of quasi-steady state magma transfer.* US Geological Survey Professional Paper 1350, 405–69.

Ulrich, G. E., E. W. Wolfe, C. C. Heliker, C. A. Neal 1987. Pu'u 'O'o IV: evolution of a plumbing system. *Abstract Volume, Hawaii Symposium on How Volcanoes Work, January 19–25, 1987, Hilo, Hawaii*, 259.

Wolfe, E. W. (ed.) 1988. *The Puu Oo eruption of Kilauea volcano, Hawaii: episodes 1 through 20, January 3, 1983, through June 8, 1984.* US Geological Survey Professional Paper 1463, 1–251.

Wolfe, E. W., C. A. Neal, N. G. Banks, T. J. Duggan 1988. Geologic observations and chronology of eruptive events. In *The Puu Oo eruption of Kilauea volcano, Hawaii: episodes 1 through 20, January 3, 1983, through June 8, 1984*, E. W. Wolfe (ed.). US Geological Survey Professional Paper 1463, 1–97.

Wright, T. L. & R. T. Okamura 1977. *Cooling and crystallization of tholeiitic basalt, 1965 Makaopuhi lava lake, Hawaii.* US Geological Survey Professional Paper 1004, 1–78.

Wright, T. L. & D. A. Swanson 1987. The significance of observations at active volcanoes: a review and annotated bibliography of studies at Kilauea and Mount St. Helens. In *Magmatic processes: physicochemical principles*, B. O. Mysen (ed.). Geochemical Society Special Publication 1, 231–40.

Field observation of active lava in Hawaii: some practical considerations

Wright, T. L. & T. J. Takahashi 1989. *Observations and interpretation of Hawaiian volcanism and seismicity: an annotated bibliography and subject index.* Honolulu: University of Hawaii Press.

Wright, T. L., W. T. Kinoshita, D. L. Peck 1968. March 1965 eruption of Kilauea volcano and the formation of Makaopuhi lava lake. *Journal of Geophysical Research* **73**, 3181–205.

Wright, T. L., D. L. Peck, H. R. Shaw 1976. *Kilauea lava lakes: natural laboratories for study of cooling, crystallization and differentiation of basaltic magma.* American Geophysical Union Geophysical Monograph 19, 375–92.

Zablocki, C. J. 1978. Application of the VLF induction method for studying some volcanic processes of Kilauea volcano, Hawaii. *Journal of Volcanology and Geothermal Research* **2**, 155–95.

CHAPTER SIX

Measuring the properties of flowing lavas

H. Pinkerton

Abstract

There have been few systematic measurements of the physical properties of active lava flows. Such measurements are important because they permit limits to be imposed on the properties of magmas prior to eruption and because realistic three-dimensional models of lava flows will require, as input data, accurate measurements of the rheological, thermal and related physical properties of the margins and isothermal interior at different stages in the development of a flow. Such models are essential in hazard assessment and mitigation studies during future eruptions of volcanoes whose lavas pose a threat to occupied areas. Measurements of temperatures, rheological properties, densities, flow dimensions and velocities using both new and existing techniques are reviewed.

Introduction

During the past 50 years there have been many lucid accounts of the behaviour of lava flows. The majority of geologists who visit active flows, however, make few, if any, measurements of the physical properties of the erupting lavas. While in many cases this is due to access problems or to a lack of time or personnel, in others it is due to a lack of information on the measurements which should be made and of the equipment required to perform these measurements. The recent eruptions of Puu Oo and Mauna Loa in Hawaii mark a turning point in the observation and measurement of lava flows (Fink & Zimbleman 1986, Lipman & Banks 1987, Moore 1987, Wolfe et al. 1987, 1988), and it is therefore an appropriate time to review the methods currently used in the collection of data from lava flows; to look at the potential of additional methods; and to summarize the most useful measurements that can be made.

175

Measurement methods

While some important properties of lava flows can readily be measured in the laboratory, e.g. thermal conductivity, coefficient of thermal expansion and electrical conductivity (see Williams & McBirney 1979, Norton et al. 1990, Norton & Pinkerton 1992), other measurements must be made in the field. Many workers have noted that there are significant differences between, for example, measurements of the rheological properties of lavas in the laboratory and similar measurements in the field (e.g. Shaw 1969, Macdonald 1972 pp 54–64). These differences are due, partly to differences in volatile contents (dissolved and in the form of vesicles), and partly to the changes which lavas undergo during reheating episodes. The only way in which we can obtain some of the most important measurements is to make them in the field.

Properties to be measured

During the last few years, considerable progress has been made in the development of theoretical models of lava flows (Hulme 1974, Danes 1972, Harrison & Rooth 1976, Huppert et al. 1982, Wilson & Head 1983, Park & Iverson 1984, Baloga & Pieri 1986, Head & Wilson 1986, Pieri & Baloga 1986, Pinkerton & Wilson 1988, Kilburn & Lopes 1988, Blake 1990, Crisp & Baloga 1990, Kilburn & Lopes 1991, Pinkerton &Wilson 1992). However, as discussed by Wilson et al. (1987), the definitive lava flow model has yet to be devised. The most important advances in lava flow modelling are likely to be made using appropriate finite difference models. These, and other mathematical models of volcanic processes, require accurate measurements of the properties considered in this section.

Rheological properties
The rates of most igneous processes are dependent on the rheological properties of the magma or lava involved (see reviews by Marsh 1987, Pinkerton 1993). Since apparent viscosities of lavas and magmas can change by a factor of 10^{16} as they cool through a temperature interval of only 200°C, the rates at which magmatic processes can proceed is more strongly influenced by this than by any other physicochemical property.

Density
The density of a lava flow is a function of composition and temperature. However, these factors are generally insignificant for any lava flow when compared with differences due to variations in gas content. The presence of bubbles and dissolved gas also affects the rheological properties of flows significantly.

Temperature distribution within lava flows
In addition to its effect on the density of a flow, temperature also alters the rheological properties of a flow by changing the degree of polymerization of the melt and by

influencing crystal and vesicle growth rates. Accurate temperature measurements will detect not only decreases in temperature due to radiative and conductive heat loss, but also possible increases in temperature due to viscous heating (Shaw 1969). The effects of degassing on crystal growth rates (Sparks & Pinkerton 1978, Lipman et al. 1985), and the resulting crystallization on temperatures within a flow, can also be determined by a comprehensive sampling and temperature measurement programme. These measurements can usefully be complemented by measurements of the convective heat flow within the lavas. An elegant method is described in Chapter 7.

Topographic surveys of flow development

During an ideal lava flow survey, accurate measurements should be made of the depth and planimetric development of flows at regular intervals. While it is relatively straight-forward to map the positions of a flow front and the outer edges of the levées, accurate depth measurements require detailed contour maps to be made of the ground before and after it is covered by lava. These maps will also permit the gradients of the underlying ground and the upper part of the flow to be made; these measurements are essential because of their importance in affecting advance rates and, consequently, flow length.

Thickness variations within a lava flow

The thicknesses of lava flows are controlled by the rheological properties of the lava (Hulme 1974), flow density and effusion rate, the slope of the underlying terrain and, particularly in the proximal regions, by pre-existing topographic depressions and, in some long-lived flows, by thermal erosion. The thickness of the resulting flows control the shear stress distribution, and hence the velocity profiles of the flow (i.e. there is positive feedback) and the temperature distribution, as well as influencing the potential for convection, turbulence and thermal erosion. During an eruption, the thickness of a lava flow can vary at any point because of surges, cooling and thermal erosion. Accurate depth measurements along active channels should therefore be made at regular intervals.

Effusion rates

Because the total effusion rate affects the morphology of lava flows (Walker 1973, Pinkerton & Sparks 1976, Wadge 1978, Lopes & Guest 1982), accurate measurements of this are essential during observations of flow field development. Ideally, this can be calculated from the shape of the proximal active channel, together with corresponding vertical and horizontal velocity profiles of lava of lava in the channel. Regular measurements of these properties will permit changes in effusion rates to be documented. In addition, because temporal and spatial variations in effusion rates may take place further downflow, for example by overflows upflow, similar measurements should be made downstream (see Baloga & Pieri 1986, Lipman & Banks 1987).

Equipment design

Given unlimited resources, all of the relevant physical properties of active lava flows could be measured accurately. Since such funding is not available for most volcano-logical research, designing most scientific equipment inevitably involves compromises. Equipment for measuring the physical properties of lava flows is also subject to weight restrictions, since there are few eruptions where it is possible to guarantee helicopter support. In addition, the equipment has to be sufficiently robust to survive transport over rough terrain and insertion through the crust on lava flows or lakes. Given the problems of access to lava channels or lakes, and the physical discomfort of the geologist who operates the equipment, the response time of the instrument should be as fast as can realistically be achieved. Finally, in common with all research equipment, the instrument should alter the properties being measured by the minimum amount possible.

Few instruments can satisfy all of the above design criteria completely. For example, if we consider the problems of making temperature measurements using a thermocouple, maximum response time is achieved with an unsheathed thermocouple. However, such a thermocouple cannot be inserted into the interior of many lava flows and hence could measure only surface or near surface temperatures. It is also liable to be damaged on all but the slowest moving of lava flows and it is subject to inaccuracies because the surface of a lava flow, in addition to losing large amounts of heat by radiation, is also liable to measure excessively high temperatures due to exothermic oxidation during crystallization of the upper part of the flow. This can increase the measured temperatures by several tens of degrees (Einarsson 1949). On the other hand, a thermocouple which is over-robust will take several minutes to attain equilibrium, during which time the operator may experience extreme discomfort. This commonly results in a measurement being terminated before equilibrium is achieved. Even when this is achieved, the thermocouple may not record the actual temperature of the flow interior, either because of the presence of a cooled skin which can adhere to a thermocouple sheath, or because of high conductive heat loss along the shaft. A major aspect of equipment design involves attempting to attain a satisfactory compromise between portability, robustness, durability, fast response time, high accuracy, and ease of use and maintenance under field conditions, all at minimum cost.

Methods which will be discussed in this chapter include a variety of geophysical and topographic surveys, as well as numerous remote sensing techniques and a description of some of the instruments which can be used, in the field to measure the physical properties of the flows directly. We will begin by reviewing the problems of measuring the temperature of lava flows. This will be followed by an outline of the problems associated with measuring the rheological properties of lavas. Finally, the problems of surveying flow fields and measuring flow velocities and channel dimensions will be discussed.

Temperatures of lava flows

There are various methods which can be used to measure temperatures of lava flows. The majority of measurements have been made using chromel/alumel thermocouples as the temperature sensors, in conjunction with a self-compensating meter. These are relatively inexpensive, easy to use, portable, and they can be protected to withstand the stresses imposed by being thrust into lava flows. Some of the problems of designing and using thermocouples to measure the temperatures of lava have already been mentioned, and others are discussed by Ault et al. (1962), Archambault & Tanguy (1976), Pinkerton (1978) and Lipman & Banks (1987). As mentioned earlier, a common problem is the development of a skin around the shaft and tip of the thermocouple. This can result in stable temperatures being recorded. However, these temperatures will be less than that of the interior of the lava. The skin can generally be removed by withdrawing the thermocouple from a flow using a twisting action, and quickly reinserting it into another part of the flow. Several insertions may be required before the temperature of the thermocouple prevents skin formation. During the 1984 Mauna Loa eruption, even when the thermocouples had been sufficiently preheated, a skin formed when measurements were being performed in rapidly moving flows (Lipman & Banks 1987). Lipman & Banks attribute this to the stretching of vesicles and a consequent increase in viscosity of the lava adjacent to the thermocouple tip. This problem was overcome by reducing the relative velocity of thermocouple and lava to below 0.5 m s^{-1}.

Due to the long times required to attain equilibrium (from 3 minutes for a thermo-couple grounded to a 1.5 mm diameter inconel sheath inserted into a slowly moving pahoehoe flow to 30 minutes when a 6 mm diameter sheathed thermocouple was inserted into aa flows (Lipman & Banks 1987)), adequate protection from radiation is required, either using a radiation shield or suitable protective clothing. When appropri-ate precautions are taken, a precision of 1°C is possible.

Archambault & Tanguy (1976) claim that the use of platinum/platinum rhodium "Temtip" elements will reduce the equilibrium times to 20–25 seconds. However, since they cannot be used on very viscous flows because of their fragility, they have restricted volcanological applications. Pinkerton & Sparks (1976) note that, using a thermocouple housed in a 20 mm diameter casing, insertion depths in excess of 20 cm were required if accurate internal flow temperatures were to be measured. Archambault & Tanguy (1976) required insertion depths of at least 30–50 cm in other Etna flows. Response times can obviously be reduced by preheating the thermocouple on the surface of the flow. Reduced insertion depths are possible either by using a thinner thermocouple, where conductive heat losses along the shaft are reduced, or by making measurements where the radiative heat losses from a flow are low, generally close to the source of the lava flow. Even at the source, however, if the upper surface of a flow has been in contact with the walls of the feeding system, conductive heat loss may have generated a marked temperature profile. Because of variations in the dimensions, conductivity and type of thermocouples used during eruptions, and because of variations in thermal profiles, vesicularity and velocities of the flows being measured, minimum insertion depths and times will vary, and they need to be determined individually for each part of the flow being measured. Accurate temperature/depth profiles are most readily made using multiple thermocouples of the type described and used by Archambault & Tanguy (1976).

Optical pyrometers are also used extensively, and they are very useful in permitting the temperatures of the crust or, if it is exposed, the incandescent interior, to be measured fairly accurately, without the need to approach the flow. Instruments of this type usefully complement measurements made using a thermocouple, though they can replace thermocouple measurements on active flows only when rapid convection or vigorous degassing are taking place. Active lava lakes and fire fountains are examples of situations where optical pyrometers can replace thermocouples, though inaccurate measurements may be recorded as a consequence of surface cooling, oxidation or absorption by magmatic gases. In addition, allowances have to be made for the fact that lavas are not perfect black bodies. The emissivity of lava has been determined experimentally by Gauthier (1971) to be 0.83 for a wavelength of 0.65 m. A mono-chromatic optical pyrometer which has been found to be particularly useful for volca-nological work is the OPTIX (Pyrowerk, Hanover) pyrometer described by Archambault & Tanguy (1976).

At temperatures above 1,100°C most of the energy radiated from a body is in the infrared region. Infrared pyrometers are therefore more useful than optical pyrometers for measuring maximum temperatures during most volcanic eruptions. A particularly successful series of temperature measurements of lava fountains were collected by Lipman & Banks (1987) using a hand-held two-colour infrared pyrometer. This allowed the emissivities and hence temperatures of the fountains to be calculated directly. Their Hotshot infrared pyrometer, when fitted with a telephoto lens, had a field of view of less than 1 m^2 at a distance of 30 m. Recent developments in portable infrared ther-mometers have increased the temperature range over which these instruments can be used. Minolta/Land have developed a number of Cyclops infrared radiometers which, between them, cover the temperature range –50°C to 3,000°C. These instruments have fast (~1 s) response times, a digital read-out system which can be connected to a data logger, and they have a measuring field of view of 1° and a claimed accuracy of 1°C if the correct emissivities are used. For qualitative temperature information, infrared film can be used in most cameras, though this method is suitable only when the film can be kept refrigerated until required. In addition, the film must be loaded and unloaded from the camera in a darkroom, and the film must be processed rapidly after use.

Temperature data can also be obtained either by satellite (Francis & Rothery 1987, Rothery et al. 1988, Kahle et al. 1988, Glaze et al. 1989) or, at a higher resolution, from an aeroplane. Airborne infrared pyrometers will undoubtedly provide us with extremely useful data on the surface temperatures of lava flows, from which heat loss models can be developed. The problems of spatial resolution, development of an appropriate algorithm for atmospheric absorption and other problems with satellite, and, to a lesser extent, airborne surveys, are currently being addressed by workers at the Jet Propulsion

Laboratory and the Open University. When these problems are overcome, this remote sensing method will remove the need to make surface temperature measurements of lava flows on the ground. In the meantime, models which determine heat losses from lava flows require measurements both of the internal temperatures of lava flows and their surfaces.

Rheological properties of lavas

At temperatures above the liquidus, most lavas are Newtonian, and it is possible to calculate their viscosities using a variety of geochemical models which have been developed partly from a knowledge of the structure of silicate melts and partly from existing measurements of the viscosities of these melts (see Ryan & Blevins 1987). During a significant part of their history, most lavas are at subliquidus temperatures; consequently, they contain suspended crystals and bubbles, both of which affect the rheological properties of these materials (Pinkerton & Stevenson 1992a,b,c). If there are more than a few per cent of bubbles or crystals present, lavas will behave as non-Newtonian fluids. The effects of suspended crystals can be investigated in the laboratory using rotating viscometers (Ryerson et al. 1988), although the combined effects of crystals, bubbles and structurally-bound volatiles can be investigated only in the field.

Since many lavas will be non-Newtonian on eruption, and all will become so as they cool and continue to vesiculate, this immediately places constraints on the design requirements of a viscometric system for lava flows. The instrument should be capable of making measurements over a range of strain rates. Thus, any method which makes measurements at single or ill-defined strain rates is not suitable for accurate measurements of the rheological measurements of lava flows. This restriction automatically casts doubt on the usefulness of the majority of viscosity measurements which are made on lava flows – those based on the well known Jeffreys equation (Williams & McBirney 1979), which is based on Newtonian rheology.

An additional method which is commonly used to obtain information on the rheological properties of lavas involves measuring the dimensions of stationary lava flows (e.g. Moore & Schaber 1975, Fink & Zimbleman 1986, Moore 1987). These are based on the assumption that lavas can be approximated by Bingham rheological models, and that their flow dimensions are controlled by this non-Newtonian behaviour. Some of the limitations of these methods are discussed by Sparks et al. (1976), Moore (1987), Pinkerton (1987) and Wilson et al. (1987). An additional field method can be used when flows turn sharp bends in channels (Heslop et al. 1989). Unfortunately, there are few situations where this method can be applied. Consequently, there remains a need to measure rheological properties using field-based instruments.

There are essentially two methods which have been used to measure the rheological properties of lavas in the field. One measures the resistance to penetration of an object which moves into the lava, and the other measures the torque required to rotate a shear vane at a variety of speeds.

Various types of penetrometer have been used during the past 20 years. Some are rapidly injected into a flow (Gauthier 1973), whereas others are injected slowly by hand (Pinkerton & Sparks 1978). The rapidly inserted penetrometers are termed, using soil mechanics terminology, dynamic penetrometers, and those which are emplaced slowly are known as static penetrometers. Static penetrometers are capable of determining the minimum force required to move the penetrometer into the material being tested, and hence measure the yield strength of a lava flow. If the rate of emplacement of dynamically emplaced penetrometers can be varied and measured in a controlled manner, they have the potential of permitting a range of the shear stresses and corresponding strain rates to be determined and hence they can permit complete rheological characterization of a lava.

The major disadvantage of many penetrometers is that they are inserted through the outer, cooled part of a flow (e.g. Einarsson 1949); thus the force required to penetrate the lava is the result of a summation of shearing stresses induced within the thickness penetrated, the major resistance to shearing being due to the more viscous outer regions. Such penetrometers give a semi-quantitative measurement of the shearing strength of the skin of a flow, and little indication of the rheological characteristics of the hot interior of a lava flow. This can be overcome partially using a preheated penetrometer which is inserted through the cooled outer regions before being activated, only the nose of the penetrometer moving into the lava. A penetrometer of this type (Fig. 6.1) has been used

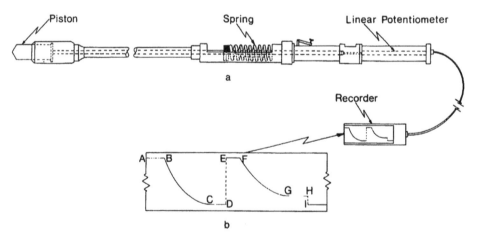

Figure 6.1 Penetrometer used by the author to make rheological measurements during the 1975 eruption of Mount Etna. (a) The viscometer and recorder; (b) an enlarged view of two typical traces. The viscometer is 2 m long, and it is constructed of high-temperature stainless steel. The stainless steel spring used on Etna had a strength of 0.155 N m^{-1}, and the traces were recorded on a hot-wire pen recorder. Trace A–D in part b is characteristic of a Newtonian fluid, and the horizontal distance A–D is a function of the Newtonian viscosity. Trace E–I is characteristic of a Bingham material, and G–H corresponds to the equilibrium position of the piston in the material, thus HI is related to the dynamic yield strength of the material.

to measure the rheological properties of the lavas erupted on Mount Etna in 1975 (Pinkerton & Sparks 1978). This instrument used a compressed spring as the energy source for penetration. The controlled reduction in axial force during penetration was recorded, together with the simultaneous piston advance rates. This permitted the shear stress/strain rate characteristics of the lava to be determined.

The first successful measurements of the rheological properties were obtained by H. R. Shaw and colleagues in 1968. They used a rotating shear vane to measure the rheological properties of the lava in Makaopuhi Lava Lake, and they confirmed that the lava was non-Newtonian. Their method was ideally suited to a stable lava lake; however, their equipment cannot be used on active lava flows. Pinkerton & Sparks (1978) used a hand-held shear vane to measure the yield strength of lavas on Mount Etna, and attempts were made, during the 1983 eruption of Mount Etna, to use a version of this equipment which could be rotated at different rates by hand. A motorized rotating shear vane has been constructed and used recently by the author (Fig. 6.2) to measure the rheological properties of natro-carbonatite lava flows on Oldoinyo Lengai (Dawson et al. 1990). This system, which utilizes a 24 volt d.c. motor with a speed controller coupled through a torque meter to a shear vane, measures the torque required to rotate the vane at different rates. Simultaneous rotation speeds are monitored using an optical ta- chometer.

There are various ways of analyzing the results from rotating viscometric systems in

Figure 6.2 Motor-driven shear vane being used to make measurements of the rheological properties of carbonatite lava during the eruption of Oldoinyo Lengai in November 1988.

non-Newtonian materials (e.g. see Van Wazer et al. 1963). Undoubtedly, the most successful methods are those which use, as a starting point, equations involving rotational speed and torque, and which make no prior assumptions regarding the rheological models to which the material belongs. In many cases it is valid to approximate the behaviour of lava to that of a Herschel–Bulkley model (essentially a Power Law material with a yield strength). This has an equation of the form

$$\tau = A \, (dv/dr)^n + \varphi,$$

where τ is the applied shear stress, dv/dr is the resulting shear strain rate and φ is the yield strength of the material. A least squares method allows the appropriate values of φ, A and n to be calculated (Pinkerton et al. 1985). Such a method allows a distinction to be made between simple power law pseudoplastic fluids ($n < 1$, $\varphi = 0$)), dilatant fluids ($n > 1$, $\varphi = 0$), Newtonian fluids ($n = 1$, $\varphi = 0$) and Herschel–Bulkley materials ($n \neq 1$, $\varphi \neq 1$).

Density

In order to measure accurate densities, samples need to be collected from the part of the flow being studied. While it is recognized that special undisturbed sampling tubes would be desirable, the majority of workers use either special scoops, or makeshift rods or, occasionally, geological hammers. On collection, the first precaution required is that the molten sample is quenched as rapidly as possible in order that the sample is not allowed to continue to vesiculate (Lipman & Banks 1987). Various methods of measuring the density of lava have been tried during the last few years, but only one appears to allow accurate densities to be measured: when cool, samples are cut into regular cuboids and their dimensions are noted, after which they are dried at 20°C and then weighed.

Survey methods

During eruptions, surveys of lava flows are generally restricted to the thickness and rates of movement of flow fronts. In addition, channel dimensions and maximum velocities are occasionally estimated or measured. These measurements can provide useful rheological information at different points in a flow and they permit *instantaneous* effusion rates to be calculated. In addition, *mean* effusion rates are commonly calculated using mean flow thicknesses multiplied by the measured planimetric area and divided by the total duration of the eruption. During some studies, accurate cross sections of the active channels have been made in order to permit the rheological properties of the flows to be calculated (Fink & Zimbleman 1986). The choice of equipment and survey methods is dependent on the accuracy required, the size of and access problems to the area to be surveyed, and the time and number of personnel available. Tacheometric surveys using automatic levels and invar rods inevitably take a considerable length of time if the survey area is large. Time can be reduced using automatic recording theodelites with computerized digital data recording facilities, stereophotographic facilities and reflectors. Alternatively, for large planimetric surveys, a rapid survey using a compass and an Abney level may provide data of sufficient accuracy. During compass and other rapid surveys, a baseline may be required, and other distance measurements may be needed. While stadia tacheometry can be used if a level or a theodelite is available, alternative methods are often necessary. Although tape measurements are traditionally used, there are some situations where such a method is unsuited; for example, where the width of an active channel is required, or where advance rates of rapidly advancing flow fronts are being measured. While these distances can be obtained using traditional surveying methods (resection from the ends of an accurately measured baseline), an alternative method utilizes a portable split image range-finder. The use of a range-finder in surveys of lava flow surveys can reduce the time required by an order of magnitude. The only problem with this method is that, in some circumstances, heat haze can reduce the accuracy with which distances can be measured, in which case an acoustic range-finder may be useful.

Measurements of the internal dimensions of channels

One of the most difficult measurements to be made on lava flows is the internal shape of the active proximal channel. During the majority of eruptions, the shape is assumed to be rectangular, and the depth is estimated using the height of the levees above the pre-flow topography. On many occasions, however, the pre-flow topography is poorly known in detail or has been covered by overflows from the main channel; and on others the flow may be advancing down a pre-existing depression. There is also the possibility that the proximal part of the channel is aligned along the active fissure. Thus estimates may be difficult to make, and they will often be unreliable. Alternative methods are therefore required.

One method relies on the assumption that the effusion rates at the vent are similar to those downflow, where depth and velocity measurements are more easily made. The disadvantages of this method are: (a) gas loss downflow may be significant and, consequently, although mass flow rates may be the same at both points, the volume flow rates may differ by up to a factor of 7, resulting in similar errors in depth measurements (Lipman & Banks 1987); (b) lava may overflow from the channel between the vent and the measuring point or, if the measurement is made in the distal regions, lava may be lost to the levées, or the lava flow may be thickening or actively spreading at a slow, but volumetrically significant amount; (c) lava in the channel may be undergoing a short- or long-term change in effusion rate, either in response to a blockage (Einarsson 1949, Frazzetta & Romano 1984, Guest et al. 1987), or as a consequence of changes in flux rate at the vent (e.g. Einarsson 1949, Wadge 1978). The slow response rate of this process makes it difficult to compare flux rates at different parts of the flow. In order to correctly monitor these processes, direct measurements of flow depths are invaluable.

The most reliable method of measuring the depths of lava in proximal channels involves using a depth probe of the type used by Pinkerton & Sparks (1976). However, this can be used only on relatively small (< 3 m wide), slowly moving (< 0.5 m s^{-1}) flows. During the 1984 eruption of Mauna Loa, attempts to drop 5 m long iron bars into the centre of large channels by helicopter failed, as did attempts to reach the bottom using 2–3 metre long ballistically inserted bars, and the depths of the channels were determined by assuming them to be half the diameter of large "lava boats" which were observed in the active channels (Lipman & Banks 1987). Measurements of the profiles of these channels was possible when they drained, and they showed that many of the lava boat estimates were 20–30% too low, and even these estimates may be inaccurate because channel dimensions can change significantly during flow, either by marginal cooling or thermal or mechanical erosion (Pinkerton et al. 1990) or because of incomplete drainage of the channels (Lipman & Banks 1987).

An acoustic method is a possible alternative to unreliable estimates or difficult direct methods. A technique which may be developed in the near future requires a gantry system suspended above the active channel on which transmitters and receivers are mounted. This method has the added potential of allowing the shapes of proximal channels to be determined during an eruption, thereby providing a check on the extent to which the shapes of long-lived active lava channels change with time due to thermal or mechanical erosion, or as a consequence of cooling of the flow margins. An acoustic method also has the potential of determining whether some of the marked fluctuations in depths of lava in channels can be attributed to variations in the vesicularity of different batches of lava (as suggested by Einarsson (1949)). A major limitation of this method is that the equipment will inevitably deteriorate during exposure to the high temperatures encountered above active lava channels. A pilot study above small channels will highlight the problems associated with this method. Finally, resistivity measurements may be used to measure the depths of lava channels in situations where it is possible to insert electrodes into the surface of the flow. This can readily be done on many large, slowly moving proximal flows, although in many cases, interpretation of the resulting data will be difficult due to the irregular shapes of many channels (Pinkerton 1978).

Velocity profiles and effusion rates

As mentioned earlier, accurate estimates of effusion rates require not only a knowledge of channel dimensions but also measurements of both vertical and horizontal velocity profiles. Unlike similar estimates in isothermal Newtonian fluids, calculations of the theoretical velocity profiles in lava flows are extremely difficult due to the thermal, and hence rheological, gradients referred to earlier. Indeed, if accurate measurements of the velocity profiles could be made, this could become a very useful indirect method of determining the rheological properties.

Measurements of the velocity profiles of small flows are readily made by placing a suitable tracer on the surface of a flow and recording its progress photographically, using cine, video or still cameras, taking appropriate precautions to ensure that all of the relevant dimensions, distances from the camera, camera focal length, etc., are recorded, and that precautions have been taken to reduce parallax and distortion. Distortion can be minimized by taking vertical photographs from a suitable vantage point, ideally a stationary helicopter positioned directly overhead. On larger flows, this method is clearly impractical, and measurements are made of the velocities of natural features (e.g. floating fragments of levée) or of blocks emplaced ballistically on to the central part of the flow. The main source of inaccuracy associated with this method is that the blocks being measured often migrate towards the margins while being timed. Several measurements are required if meaningful velocities are to be obtained.

While measurements of surface velocity profiles may, for some flows, be made with a reasonable degree of accuracy (e.g. see Lipman & Banks 1987, Moore 1987), similar measurements have yet to be made of vertical velocity profiles, and most workers assume, by comparison with isothermal Newtonian fluids, that the surface velocity is two-thirds that of the mean flow velocity (Einarsson 1949). Unfortunately, most of the equipment used to measure discharge in rivers appears to be unsuited for measurements in lava flows; it would be difficult to adapt horizontal-axis rotating-cup current meters for high-temperature work. Similar problems exist with the development of electromagnetic current meters, and an alternative method needs to be developed. When confronted with this problem on glaciers, glaciologists have successfully measured vertical velocity profiles by recording the displacement, from the vertical, of pipes inserted into the ice. Although there would be emplacement problems associated with such a method, it may be possible to project metal bars into some lava flows and to record changes in inclination as a function of time and of penetration depth. In addition, vertical velocity profiles of kaolin/water mixtures have been made in the author's laboratory at Lancaster University using a simple vertical-axis current meter, and such a system may have some potential on active lava flows. The measuring system could be mounted on a gantry which is bolted to the levées. A series of measurements can be made by lowering the measuring sensor to different depths within the flow. The gantry system required for these measurements could also be used for the acoustic depth measurement method mentioned earlier. In spite of its simplicity, the method suggested here is capable of generating more useful data than the currently prevalent method which involves estimating the velocities of blocks which are thrown ballistically on to the central part of a lava flow (see Einarsson (1949) for a discussion of some of the problems generally encountered using this method).

Finally, a novel method of measuring water discharge may be suitable for lava flows. This is based on the principle of the phase shift caused by the transportation of two cross-running sound waves from two transmitters. In common with electromagnetic current meters this has no moving parts, and it may be suitable for conversion for high-temperature work. A commercial acoustic current-meter is currently being marketed by ME Meerestechnik Elektronik GmbH.

During most eruptions, mass flux rates vary at different points along a flow. In order to understand the mechanisms responsible for these fluctuations it would be extremely useful to have survey teams located along the length of the flow. However, this can seldom be achieved, even for short periods. Consequently, there is a need for one or more automatic monitoring systems which can record short- and long-term changes along different parts of the flow. This can readily be achieved either using a cine camera with time lapse facilities or by using a video camera with telemetry facilities, both of which were used during the 1973 eruption of Heimaey, South Iceland, and at Mount St Helens in 1980.

Discussion

Given adequate funds, the shortage of data on lava flows could readily be resolved. There are a sufficient number of dedicated volcanologists who could monitor all of the important properties and phases of the development of a flow field, and the numerical data could be recorded on suitable data loggers (e.g. the latest Versatile Multi-channel Dataloggers marketed by Digitron). There is, of course, one potential problem which must be overcome if a large number of observers are making simultaneous measurements on flows – instruments must be recording accurate data. When pioneering measurements were being made on lava flows, the early observers took great care to ensure that their relatively crude instruments were correctly calibrated. The advent of advanced instruments with digital read-outs and associated data loggers could result in the generation of large amounts of useless information unless we follow in the footsteps of the founders of physical volcanology. Instrument calibration after a visit, by the author, to measure properties of carbonatite lavas on Oldoinyo Lengai confirmed that readings generated by a relatively new thermocouple were 14°C too low in the temperature range of interest; optical range-finder distances were similarly too high by 24%, and torque measurements also had to be altered after calibration. Calibration of all instruments, ideally before, during and after fieldwork is essential if we are to have any confidence in measured temperatures and other properties of erupting lavas.

   It is also essential that measurements are made as close as possible to the eruptive source if these measurements are to be used in modelling behaviour prior to eruption and if we are to compare the properties of lavas erupted from different volcanoes. Measurements should not, however, be restricted to the vent region. If we are to improve upon existing lava flow models, systematic measurements along the length of flows are required because of the effects of changes in dissolved and exsolved volatile content, temperature and crystallinity.

References

Archambault, C. & J.-C. Tanguy 1976. Comparative temperature measurements on Mount Etna: problems and techniques. *Journal of Volcanology and Geothermal Research* **1**, 113–25.

Ault, W. U., D. H. Richter, D. B. Stewart 1962. A temperature measurement probe into the melt of the Kilauea Iki lava lake in Hawaii. *Journal of Geophysical Research* **67**, 2809–12.

Baloga, S. M. & D. C. Pieri 1986. Time-dependent profiles of lava flows. *Journal of Geophysical Research* **91**, 9543–52.

Blake, S. 1990. Viscoplastic models of lava domes. In *IAVCEI Proceedings in Volcanology*. Vol. 2, *Lava flows and domes: emplacement mechanisms and hazard implications*, J. H. Fink (ed.), 88–126. Berlin: Springer.

Crisp, J. & S. M. Baloga 1990. A model for lava flows with two thermal components. *Journal of Geophysical Research* **95**, 1255–70.

Danes, Z. F. 1972. Dynamics of lava flows. *Journal of Geophysical Research* **77**, 1430–2.

Dawson, J. B., H. Pinkerton, G. E. Norton, D. M. Pyle 1990. Physico-chemical properties of alkali carbonate lavas from the 1988 eruption of Oldoinyo Lengai, Tanzania. *Geology* **10**, 260–3.

Einarsson, T. 1949. The flowing lava. Studies of its main physical and chemical properties. In *The eruption of Hekla 1947-1948*, T. Einarsson, G. Kjartansson, S. Thorarinsson (eds), IV (3), 1–70. Reykjavik: Visindafelag Islandinga & Museum of Natural History.

Fink, J. H. & J. R. Zimbelman 1986. Rheology of the 1983 Royal Gardens basalt flows, Kilauea volcano, Hawaii. *Bulletin of Volcanology* **48**, 87–96.

Francis, P. W. & D. A. Rothery 1987. Using the Landsat thematic mapper to detect and monitor active volcanoes: an example from Lascar volcano, northern Chile. *Geology* **15**, 614–7.

Frazzetta, G. & R. Romano 1984. The 1983 Etna eruption; event chronology and morphological development of the lava flow. *Bulletin Volcanologique* **47**, 731–62.

Gauthier, F. 1971. *Étude comparative des caracteristiques rheologiques de laves basaltiques en laboratoire et sur le terrain*. PhD thesis, University of Paris.

Gauthier, F. 1973. Field and laboratory studies of the rheology of Mount Etna lava. *Philosophical Transactions of the Royal Society, London* **A274**, 83–98.

Glaze, L., P. W. Francis, D. A. Rothery 1989. Measuring thermal budgets of active volcanoes by satellite remote sensing. *Nature* **338**, 144–6.

Guest, J. E., C. R. J. Kilburn, H. Pinkerton, A. M. Duncan 1987. The evolution of flow fields: observations of the 1981 and 1983 eruptions of Mount Etna, Sicily. *Bulletin of Volcanology* **49**, 527–40.

Harrison, C. G. A. & C. Rooth 1976. The dynamics of flowing lava. In *Volcanoes and tectonospheres*, H. Aoki & S. Iizuka (eds), 103–13. Tokyo: Tokai University Press.

Head, J. W. & L. Wilson 1986. Volcanic processes and landforms on Venus: theory, predictions and observations. *Journal of Geophysical Research* **91**, 9407–46.

Heslop, S. E., H. Pinkerton, L. Wilson, J. W. Head 1989. Dynamics of a confined lava flow on Kilauea volcano, Hawaii. *Bulletin of Volcanology* **51**, 415–32.

Hulme, G. 1974. The interpretation of lava flow morphology. *Geophysical Journal of the Royal Astronomical Society* **39**, 361–83.

Huppert, H. E., J. B. Shepherd, H. Sigurdsson, R. S. J. Sparks 1982. On lava dome growth, with reference to the 1979 extrusion of the Soufriere of St. Vincent. *Journal of Volcanology and Geothermal Research* **14**, 199–222.

Kahle, A. B., A. R. Gillespie, E. A. Abbott, M. J. Abrams, R. E. Walker, G. Hoover, J. P. Lockwood 1988. Relative dating of Hawaiian lava flows using multispectral thermal infrared images: a new tool for geological mapping of young terranes. *Journal of Geophysical Research* **93**, 15,239–51.

Kilburn, C. R. J. & R. M. C. Lopes 1988. The growth of aa lava fields on Mount Etna, Sicily. *Journal of Geophysical Research* **93**, 14759–72.

Kilburn, C. R. J. & R. M. C. Lopes 1991. General patterns of flow field growth: aa and blocky lavas. *Journal* of *Geophysical Research* **96**, 19,721–32.

Lipman, P. W. & N. G. Banks 1987. *Aa flow dynamics, Mauna Loa 1984.* US Geological Survey Professional Paper 1350, 1527–68.

Lipman, P. W., N. G. Banks, J. M. Rhodes 1985. Gas-release induced crystallisation of 1984 Mauna Loa magma, Hawaii, and effects on lava rheology. *Nature* **317**, 604–7.

Lopes, R. M. C. & J. E. Guest 1982. Lava flows on Etna, a morphometric study. In *The comparative study of the planets*, A. Coradini & M. Fulchignoni (eds), 441–58. Dordrecht: Reidel.

Macdonald, G. A. 1972. *Volcanoes.* Englewood Cliffs, New Jersey: Prentice-Hall.

Marsh, B. D. 1987. Magmatic processes. *Reviews in Geophysics* **25**, 1043–53.

Moore, H. & G. G. Schaber 1975. An estimate of the yield strength of Imbrium flows. *Proceedings of the 6th Lunar and Planetary Science Conference*, 101–18

Moore, H. G. 1987. *Preliminary estimates of the rheological properties of 1984 Mauna Loa lava.* US Geological Survey Professional Paper 1350, 1569–88

Norton, G. E. & H. Pinkerton 1992. The physical properties of carbonatite lavas: implications for planetary volcanism. *Lunar and Planetary Science* **XXIII**, 1001–2.

Norton, G. E., H. Pinkerton, J. B. Dawson 1990. New measurements of the physico-chemical properties of natrocarbonatite lavas. *Lunar and Planetary Science* **XXI**, 901–2.

Park, S. O. & J. D. Iverson 1984. Dynamics of flowing lava: thickness, growth characteristics of two dimensional flow. *Geophysics Research Letters* **11**, 641–4.

Pieri, D. C. & S. M. Baloga 1986. Eruption rate, area and length relationships for some Hawaiian flows. *Journal of Volcanology and Geothermal Research* **30**, 29–45.

Pinkerton, H. 1978. *Methods of measuring the rheological properties of lavas.* PhD Thesis, University of Lancaster.

Pinkerton, H. 1987. Factors affecting the morphology of lava flows. *Endeavour* **11**, 73–9.

Pinkerton, H. 1993. Rheological properties of geological materials. *Concise encyclopaedia of environmental systems.* Oxford: Pergamon Press (in press).

Pinkerton, H. & R. S. J. Sparks 1976. The 1975 sub-terminal lavas, Mount Etna: a case history of the formation of a compound lava field. *Journal of Volcanology and Geothermal Research* **1**, 167–182.

Pinkerton, H. & R. S. J. Sparks 1978. Field measurements of the rheology of flowing lava. *Nature* **276**, 383–5.

Pinkerton, H. & R. J. Stevenson 1992a. Rheological properties of basaltic lavas. *Lunar and Planetary Science* **XXIII**, 1079–80.

Pinkerton, H. & R. J. Stevenson 1992b. Rheological properties of silicic lavas. *Lunar and Planetary Science* **XXIII**, 1081–2.

Pinkerton, H. & R. J. Stevenson 1992c. Methods of determining the rheological properties of lavas from their physico-chemical properties. *Journal of Volcanology and Geothermal Research* **53**, 47–66.

Pinkerton, H. & L. Wilson 1988. The lengths of lava flows. *Lunar and Planetary Science* **XVIII**, 937–8.

Pinkerton, H. & L. Wilson 1992. The dynamics of channel-fed lava flows. *Lunar and Planetary Science* **XXIII**, 1083–4.

Pinkerston, H., R. H. Williams, L. Wilson 1985. The rheological properties of lavas erupted from Mount Etna. *IAVCEI 1985 Scientific Assembly Abstract Volume.*

Pinkerton, H., L. Wilson, G. E. Norton 1990. Thermal erosion – observations on terrestrial lava flows and implications for planetary volcanism. *Lunar and Planetary Science* **XXI**, 964–5.

Rothery, D. A., P. W. Francis, C. A. Wood 1988. Volcano monitoring using short wave infrared data from satellites. *Journal of Geophysical Research* **93**, 7993–8008.

Ryan, M. P. & J. Y. K. Blevins 1987. The viscosity of synthetic and natural silicate melts and glasses at high temperature and 1 bar (10^5 Pa) pressure and at higher pressure. *US Geological Survey Bulletin* **1764**, 1–563.

Ryerson, F. J., H. C. Weed A. J. Piwinski 1988. Rheology of subliquidus magmas 1. Picritic compositions. *Journal of Geophysical Research* **93**, 3421–36.

Shaw, H. R. 1969. Rheology of basalt in the melting range. *Journal of Petrology* **10**, 510–35.

Shaw, H. R., T. L. Wright TL, D. L. Peck, R. Okamura 1968. The viscosity of basaltic magma: an analysis of field measurements in Makaopuhi lava lake, Hawaii. *American Journal of Science* **226**, 225–64.

Sparks, R. S. J. & H. Pinkerton 1978. Effect of degassing on the rheology of lava. *Nature* **276**, 385–6.

Sparks, R. S. J., H. Pinkerton, G. Hulme 1976. Classification and formation of lava levées on Mount Etna, Sicily. *Geology* **4**, 269–71.

Van Wazer, J. R., J. W. Lyons, K. Y. Kim, R. E. Colwell 1963. *Viscosity and flow measurement*. New York: Interscience.

Wadge, G. 1978. Effusion rate and the shape of aa flow fields on Mount Etna. *Geology* **6**, 503–6.

Walker, G. P. L. 1973. Lengths of lava flows. *Philosophical Transactions of the Royal Society, London* **A274**, 107–18

Williams, H. & A. R. McBirney 1979. *Volcanology*. San Francisco: Freeman, Cooper.

Wilson, L. & J. W. Head 1983. A comparison of eruption processes on Earth, Moon, Mars, Io and Venus. *Nature* **302**, 663–9.

Wilson, L., H. Pinkerton, R. Macdonald 1987. Physical processes in volcanic eruptions. *Annual Reviews of Earth and Planetary Sciences* **15**, 73–95.

Wolfe, E. W., M. O. Garcia, D. B. Jackson, R. Y. Koyanagi, C. A. Neal, A. T. Okamura 1987. *The Puu Oo eruption of Kilauea volcano, episodes 1–20, January 3, 1983, to June 8, 1984*. US Geological Survey Professional Paper 1350, 471–508.

Wolfe, E. W., C. A. Neal, N. G. Banks, T. J. Duggan 1988. Geologic observations and chronology of eruptive events. In *The Puu Oo eruption of Kilauea volcano, Hawaii: episodes 1 through 20, January 3, 1983, through June 8, 1984*, E. Wolfe (ed.). US Geological Survey Professional Paper 1463, 1–98.

CHAPTER SEVEN

Convection heat transfer rates in molten lava

Harry C. Hardee

Abstract

Convective heat fluxes measure the rate at which lava carries thermal energy downstream. Flux measurements are thus important for monitoring the thermal evolution of a flow. They are also important for investigating the structural development of a flow, since lava rheology is strongly temperature-dependent. Fluxes measured in active basaltic flows are consistent with theoretical models assuming a power law pseudoplastic rheology. Such measurements can be made as easily as standard temperature measurements and may have immediate practical value for identifying potentially unstable lava margins.

Introduction

Convection heat transfer processes in molten lava are of interest in the study of the movement and solidification of lava. This information is vital in evaluating lava flow behavior and eruption hazards. For instance, changes in lava convection rates can be used to predict whether levées will become unstable and how rapidly descending lava flows will spread or proceed. Convection in lava and magma is also of interest for energy extraction (Hardee 1988).

The convection heat transfer rate or film heat transfer coefficient can be theoretically predicted if all the local thermal properties, their variation with temperature, and the geometry and nature of the flow process are accurately known. Because of non-Newtonian effects and accumulated uncertainties in the various thermal properties, the theoretical prediction of convection heat transfer in lavas is frequently inaccurate. A better approach sometimes is simply to measure the heat transfer rate or film coefficient directly. This result can often be scaled to other situations of interest. Geometrical scaling is particularly easy provided that the flow processes (e.g. natural convection, forced convection, or combined flow) remain the same. Field measurements of temperature are frequently made in lava flows but convection measurements are rarely made.

Non-Newtonian lava rheology

Non-Newtonian effects must be considered when making convection measurements or calculations for lava. The most common rheological models for lava are Bingham and power law fluid models. In the Bingham case, convection cannot occur until the critical yield stress is exceeded and then convection is similar to that of a Newtonian fluid. With a power law fluid the convection would begin initially and change in a continuous monotonic fashion. Experimental measurements of convection in lavas, both in the laboratory and in the field, indicate that the power law model is a better assumption and that Bingham effects of convection are not observed (Hardee 1981).

The general equation for the power law viscosity model is

$$\mu = K(du/dy)^n, \tag{7.1}$$

where μ is the viscosity, du/dy is the usual velocity gradient normal to the direction of flow, and K and n are empirical constants. Two viscosity measurements are needed to define the viscosity constants K and n, but a number of measurements may be needed to determine appropriate values of the constants that give a good fit over most of the subliquidus temperature range. Viscosity can be measured from remelted samples of previously erupted material, although these measurements may be in error unless efforts are made to reconstitute the original volatile contents. Viscosity can also be inferred from the dimensions of lava levées and flow channels (Sparks et al. 1976, Pinkerton & Sparks 1978). Viscosity measurements have been made in the field during volcanic eruptions. Gauthier (1973) used a ballistic penetrometer in 1969 to obtain field measurements of lava viscosity at Etna. Pinkerton & Sparks (1978) also used a penetrometer at Etna in 1975 to measure viscosity in active flows. Rotational viscometers are frequently used in the laboratory, but are difficult to use in the field. Shaw et al. (1968) made field measurements of lava viscosity in Makaopuhi lava lake (Hawaii) using a rotational viscometer. Pinkerton & Sparks (1978) used a simple rotational viscometer to measure the yield strength of lava in flows at Etna in 1975. The data of Shaw et al. (1968) and Pinkerton & Sparks (1978) show that both these lavas appear to be pseudoplastics which can be approximated as Bingham fluids. Hardee & Dunn (1981) showed that the same lava viscosity data can be approximated well by a power law relation (Table 7.1).

Table 7.1 Power law viscosity fits to rheological data
for basalts from Hawaii and Etna.

Volcano	Power law coefficient, K (N s^n m^{-2})	Power law exponent, n
Hawaii	634	0.531
Hawaii	845	0.867
Etna	3694	0.425

Hawaiian data: Shaw *et al.* (1968) and Hardee & Dunn (1981).
Etnean data: Pinkerton & Sparks (1978) and Hardee & Dunn (1981).

Rates of convection heat transfer

Of all the properties that affect convection heat transfer in lavas, the rheological properties are the most difficult to characterize and quantify. If the rheological and thermal properties of a flowing lava are accurately known, then the heat transfer rate can be calculated directly from equations such as those developed by Acrivos (1960). Acrivos (1960) gives the following result for natural convection to a vertical surface in contact with a power law viscosity fluid:

$$q_0 = 0.55[(3n + 1)/(2n + 1)]^{(2n + 1)/(3n + 1)}[k(T_\infty - T_0)/L]Ra*^{1/(3n + 1)}. \qquad (7.2)$$

Here q_0 is the convection heat transfer rate, n is the exponent from the power law viscosity model, k is the thermal conductivity of the lava, T_∞ is the ambient temperature of the lava, T_0 is the temperature of the vertical surface in contact with the convecting fluid and L is the height of the vertical surface. $Ra*$ is a modified Rayleigh number of the form

$$Ra* = \rho^2 cgb(T_\infty - T_0)L^3/[kK(\kappa/L^2)^{n-1}], \qquad (7.3)$$

where ρ is the lava density, c is the lava specific heat, g is the acceleration due to gravity, b is the volumetric expansivity of lava, κ is the thermal diffusivity of lava and the other parameters are as defined previously. For the special case where $n = 1$ (Newtonian fluid), K (from Eq. 7.1) becomes viscosity, and the Rayleigh number and the heat transfer rate reduce to the usual results for a high Prandtl number Newtonian fluid (Kuiken 1962, Lin & Chao 1974).

   In many cases the local thermal properties of lava, particularly viscosity, are not sufficiently well known to allow one to calculate the heat transfer rate directly. The solution then is to make field or laboratory measurements of the heat transfer rate or the film coefficient directly. Such laboratory and field measurements of heat transfer rate have been made using both steady-state heat flux probes and transient heat flux probes (Hardee 1979, Hardee & Dunn 1981, Hardee 1983). A typical steady-state probe consists of two concentric tubes separated by a thermal insulator (Fig. 7.1). The inner tube (at temperature T_1, Fig. 7.1) is cooled by a steady flow of gas or fluid and the outer

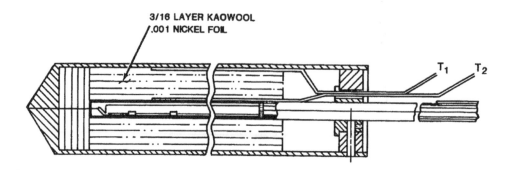

3/16 LAYER KAOWOOL
.001 NICKEL FOIL

T_1 T_2

Figure 7.1   Steady-state convection probe. See text for details.

tube (at temperature T_2, Fig. 7.1) is exposed to the lava. Temperature measurements at
the two tubes and a knowledge of the thermal conductivity of the insulator is sufficient
to determine the heat transfer rate at the lava interface (Hardee & Dunn 1981).
Steady-state probes of this type are difficult to use at eruption sites because of the coolant
requirement, although this difficulty can be overcome if the fluid coolant is replaced by
a large thermal heat sink (Hardee 1979).

Transient heat flux probes, properly designed, can be used for both laboratory and
field measurements. The transient probe consists of a high thermal conductivity metal
cylinder (or other geometrical shape) that is inserted into the molten lava. Usually, a
special metal alloy like tantalum/zirconium/molybdenum (TZM) must be used to resist
melting and corrosion while providing a high thermal conductivity. The transient
temperature rise of the metal cylinder is used to determine the heat transfer rate (Hardee
& Dunn 1981, Hardee 1983). Natural convection is induced adjacent to the surface of
the probe when it is inserted into the molten lava. Schneider (1957) gives the expected
temperature response for such a probe as

$$(T - T_\infty)/(T_0 - T_\infty) = \exp(-2ht/c\rho R), \qquad (7.4)$$

where T_∞ is the ambient lava temperature, T_0 is the nominal melt temperature of the
lava, usually taken as 1,050°C for Hawaiian basalt (Hardee & Dunn 1981), T is the
measured transient temperature at time t, c is the specific heat capacity of the probe, ρ
is the density of the probe, R is the radius of the probe and h is the film heat transfer
coefficient, where the heat transfer rate is

$$q_0 = h(T - T_\infty). \qquad (7.5)$$

In a typical test, the temperature T is recorded as a function of time t and the
dimensionless temperature is fitted to an exponential function of the form

$$(T - T_\infty)/(T_0 - T_\infty) = a \exp(-Bt). \qquad (7.6)$$

Normally only data above T_0 are used for the function fit and determination of the constants a and B. This simplifies data reduction and eliminates phase change effects due to the crust of solidified lava that initially forms on the probe but remelts as the probe is heated above T_0. The dimensionless temperature response of the probe is normalized to the condition that zero time corresponds to the point where the probe has reached the melt temperature of the lava. A typical set of normalized test data are shown in Table 7.2. An exponential curve fitting routine (Anon. 1976) can be used with the data in Table 7.2 to show that the best exponential fit leads to a B value of 0.01610 with a determination coefficient r^2 of 0.9981. In this case, the closeness of r^2 to unity indicates a high correlation of the data with the expected exponential fit. The film heat transfer coefficient can then be determined from the value of B as

$$h = Bc\rho R/2. \tag{7.7}$$

The heat flux at the melt/crust interface of a lava levée or heat exchanger surface is then

$$q_0 = h(T_\infty - T_0). \tag{7.8}$$

Using the value of B obtained above and appropriate values for $c\rho R$ for this probe results in a convective heat flux of $q_0 = 34$ kW m^{-2} (Hardee 1981). The transient probe gives good results when it is designed for and used in situations where the thermal response of the probe is rapid relative to the time required to establish convective flow

Table 7.2 Convection test of transient convection probes in remelted Hawaiian lava

Time (s)	T (probe) °C	t (s)	$(T - T_\infty)/(T_0 - T_\infty)$
60.1	1050	0	1
62.6	1056	2.5	0.9750
70.1	1075	10.0	0.8958
77.6	1096	17.5	0.8083
85.1	1117	25.0	0.7208
92.6	1135	32.5	0.6458
100.1	1154	40.0	0.5667
115.1	1184	55.0	0.4417
130.1	1208	70.0	0.3417
145.1	1225	85.0	0.2708
152.1	1231	92.0	0.2458
160.1	1236	100.0	0.2250
190.1	1258	130.0	0.1333
220.1	1272	160.0	0.0750
250.1	1280	190.0	0.0417
280.1	1283	220.0	0.0292
310.1	1285	250.0	0.0208
340.1	1287	280.0	0.0125

In this test the lava temperature T_∞ was 1,290°C and the equivalent melt temperature T_0 assumed to be 1,050°C. An exponential curve fitting routine (Anon. 1976) of the form $(T - T_\infty)/(T_0 - T_\infty) = a \exp (-Bt)$ was applied to these data, yielding the values $a = 1.04659$ and $B = 0.01610$, and a determination coefficient $r^2 = 0.99116$.

adjacent to the probe. This can be verified by calculation or the transient probe results can be checked by comparing them with measurements obtained using a steady-state probe.

Figure 7.2 shows heat flux data for molten basalt as determined by laboratory and field data. Included in this figure are laboratory data using the steady-state and transient heat flux probes, field data from the 1983 Kilauea eruption, and theoretical calculations using Equation 7.2 and thermal/fluid property data for Hawaiian and Etnean lavas. The data in Figure 7.2 show a definite change in slope at a lava temperature of 1,212°C, very close to the expected liquidus temperature of 1,220°C. This suggests that a change in convection occurs near the liquidus, probably because of different lava rheology above and below this temperature. Above the liquidus the trend of the heat flux is typical of a convecting Newtonian fluid, considering the expected property variations with temperature. Below the liquidus the heat flux rates are typical of a pseudoplastic power law fluid.

Applications to lava flows

As an example of how one convection test might be scaled to another situation, let us consider a case where a cylindrical transient probe is operated horizontally in a super-liquidus lava and the results are to be scaled to the case of convection to a vertical wall of a lava levée. At super-liquidus temperatures, we expect the lava to behave as a Newtonian fluid, for which convection to a horizontal cylinder in laminar natural convection yields a heat transfer coefficient given by (Chapman 1960)

$$h_c = 0.525[\rho gbk^4(T_c - T_0)/n\kappa]^{1/4}(1/D)^{1/4}, \qquad (7.9)$$

where the subscript c refers to a cylinder of diameter D. Chapman (1960) also shows that the heat transfer coefficient for laminar natural convection to a vertical wall is

$$h_w = 0.59[\rho gbk^4(T_w - T_0)/n\kappa]^{1/4}(1/L)^{1/4}, \qquad (7.10)$$

where the subscript w refers to a vertical wall of height L. The expected convective heat flux to the levée wall (q_w) in terms of the measured convective heat flux to the transient probe (q_c) is then

$$q_w = q_c(0.59/0.525)(L/D)^{1/4}. \qquad (7.11)$$

Similar results can be obtained for other more complicated situations. For instance, for the more common subliquidus lava temperature condition, convection correlations such as Equation 7.2 for a power law fluid must be used to arrive at a result similar to that of Equation 7.11.

As far as practical problems of field measurements are concerned, the measurement of heat flux or film coefficient is no more difficult than making field measurements of temperature. It is important to orient the probe in the flow in a selected manner in order to later interpret the geometry effect. For instance, if the probe is cylindrical, then it

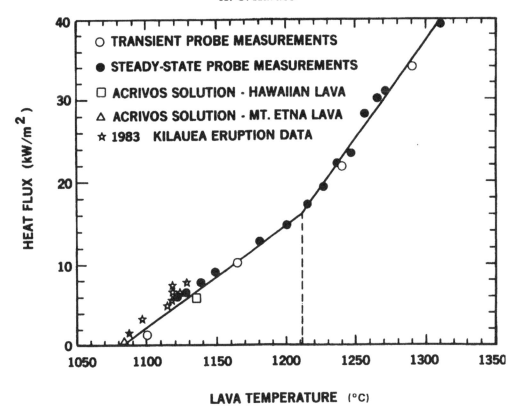

Figure 7.2 Convection heat flux data for molten Hawaiian basalt as determined by laboratory and field measurements. Analytical solutions for Hawaiian and Etnean lavas are included for comparison.

should be inserted either horizontally or vertically so that results can be compared easily with analytical solutions. If this is a problem, the probe can be constructed in a spherical shape that is independent of insertion angle. Another concern is the ability to separate the effects of forced and free convection and, occasionally, effects of thermal radiation heat transfer or simple conduction. If the probe is inserted into the main stream of flowing lava, forced convection effects will predominate and it will be important to know or to be able to measure the velocity of the lava flow. If it is desired to measure only free convection induced by insertion of the cold probe, it is best to insert the probe into a stagnant pool or eddy out of the main stream of the flow (Hardee 1981).

Free convection is important to estimate energy extraction rates (Hardee 1988). Forced convection is important in many situations such as the determination of lava levée stability. As an example, consider a thick levée subjected to a forced convection heat flux boundary condition such as provided by the flowing lava in contact with the levée. Ozisik (1968) gives an approximate solution for this situation. The solution shows that the factor which determines whether the solid reduces by melting or thickens by solidification depends on whether the dimensionless convection heat flux is greater or less than 4. This dimensionless heat flux (defined in Ozisik (1968)) is a function of the actual convection heat flux which is assumed to be known. Other analytical solutions for melting or solidification criteria can be used with other boundary or initial conditions, provided that the convection heat flux is known. These solutions can be used to predict the stability of different types of lava levée or the solidification rate of advancing flows.

Conclusions

Field and laboratory data on convective heat fluxes in basalt are consistent with models of heat transport in power law pseudoplastic fluids (Fig. 7.2). Flux measurements can be made as readily as those for lava temperature. The data obtained may be used to monitor the thermal and structural evolution of a flow and, in the future, may provide a basis for at least short-term forecasts of flow behavior.

References

Acrivos, A. 1960. A theoretical analysis of laminar natural convection heat transfer to non-Newtonian fluids. *American Institute of Chemical Engineering* **6**, 584–90.

Anon. 1976. *Hewlett-Packard HP-67 standard pac instruction manual.* Cornvallis: Hewlett-Packard.

Chapman, A. J. 1960. *Heat transfer.* New York: MacMillan.

Gauthier, F. 1973. Field and laboratory studies of the rheology of Mount Etna lava. *Philosophical Transactions of the Royal Society, London* **A274**, 83–98.

Hardee, H. C. 1979. Heat transfer in the 1979 Kilauea lava flow, Hawaii. *Journal of Geophysical Research* **84**, 7485–93.

Hardee, H. C. 1981. Convection heat transfer in magmas near the liquidus. *Journal of Volcanology and Geothermal Research* **10**, 195–207.

Hardee, H. C. 1983. Heat transfer mechanisms of the 1983 Kilauea lava flow. *Science* **222**, 47–8.

Hardee, H. C. 1988. Magma energy. *Geothermal Science and Technology* **1**, 165–224.

Hardee, H. C. & J. C. Dunn 1981. Convective heat transfer in magmas near the liquidus. *Journal of Volcanology and Geothermal Research* **10**, 195–207.

Kuiken, H. K. 1962. An asymptotic solution for large Prandtl number free surface convection. *Journal of Engineering Mathematics* **2**, 355–71.

Lin, F. N. & B. T. Chao 1974. Laminar free convection over two-dimensional and axisymmetric bodies of arbitrary contour. *Journal of Heat Transfer* **96**, 435–42.

Ozisik, M. N. 1968. *Boundary value problems of heat conduction.* Scranton: International textbook.

Pinkerton, H. & R. S. J. Sparks 1978. Field measurements of the rheology of lava. *Nature* **276**, 383–5.

Schneider, P. J. 1957. *Conduction heat transfer.* Reading, Mass.: Addison-Wesley.

Sparks, R. S. J., H. Pinkerton, G. Hulme 1976. Classification and formation of lava levées on Mount Etna, Sicily. *Geology* **4**, 269–71.

Shaw, H. R., T. L. Wright, D. L. Peck, R. Okamura 1968. The viscosity of basaltic magma: an analysis of field measuremets in Makaopuhi lava lake, Hawaii. *Journal of Geophysical Research* **266**, 225–64.

CHAPTER EIGHT

Remote sensing of active lava

David A. Rothery & David C. Pieri

Abstract

Remote sensing provides the only way of recording thermal data of a uniform quality instantaneously across an entire active lava flow, or other body of lava. Such data allow inferences to be made about the thermomechanical properties of the lava, and permit quantitative comparisons between different volcanoes. At a more basic level the images can detect vents and flows that might otherwise have been unobserved, and can be invaluable in post-eruption mapping of the extent of new lava and other deposits. In this chapter examples of remote sensing observations of active lava are summarized, a synopsis of the physics involved is provided, and what types of data are most suitable and where they can be obtained are suggested.

Introduction

Remote sensing is a "black art" as far as many volcanologists are concerned. Many have probably been impressed by spectacular satellite pictures showing eruptions, but have little appreciation of their scientific value. Probably the most severe obstacles to the more widespread adoption of remote sensing by volcanologists armoury are: lack of remote sensing training, ignorance of how to get hold of the data, and the sub-optimal nature of the data that are available. In this chapter we hope to contribute to the amelioration the first two of these. The extraction of thermal data from active lava flows and other lava bodies is concentrated on.

Why lava flows are difficult to handle

Active volcanic features are, in general, difficult to observe and measure for a variety of reasons, arising from their mobility, their high temperatures, and their propensity for forming or emplacing themselves on rough or poorly accessible terrain. Lava flows, in particular, offer their own set of observational difficulties. Active flows are usually in motion, and can extend for many kilometres across rough and steep terrain. They also emit large amounts of radiant energy, particularly at infrared wavelengths, most, but not all, of which is directed into the space above the flow.

In part because of practical problems in the field, and in part because of the nature of their physical and chemical properties, descriptions of the rheological and thermal behaviour of lavas have been poorly constrained, and lacking in reliable quantitative data. This is particularly notable as volcanoes and their lava flows have been extensively described in qualitative terms since Roman times. Such a paucity of hard facts is not due to a lack of effort by volcanologists; indeed, ground-based efforts to collect quantitative data such as temperatures and flow thicknesses have bordered on the heroic (e.g. Wolfe et al. 1987).

Measurements of the physical properties of lava flows can be elusive not only in their execution, but also in their meaning. For instance, what does one mean when one considers an apparently simple concept such as the temperature of an active lava flow? The upper surface of an active flow can be a complex tapestry of very rough, newly formed clasts (aa), or can be an icing-smooth swirl of pasty ropes (pahoehoe). Most of this material is radiating at some temperature well below its solidus, but some small percentage of the surface may be radiating at near- or above-solidus temperatures, at prodigious energy densities (Pieri et al. 1990, Flynn et al. 1991, 1993, Flynn & Mouginis-Mark 1992). Should one attempt to find an "average" temperature for the whole surface, and if so what does this mean physically? Can one decide on a sensible partition of temperatures based on a bimodal, trimodal or n-modal distribution of heat sources, and how closely does such a modelled distribution reflect physical reality? A further complication is that there is likely to be a systematic vertical and horizontal distribution of energy within a flow (e.g. Lipman & Banks, 1987, Crisp & Baloga, 1990; Ch. 7 this volume); how then does the thermal radiation from the top of the flow reflect and respond to such surface and internal flow temperature structure, and to the dynamics of regimes within the flow? Evidently even the conceptualization of "flow temperature" is fraught with extreme difficulty and requires careful qualification if it is to be used in a physically meaningful way.

After a decision has been made on what is a physically useful formulation of temperature to represent a particular situation, the problem of how to measure any such defined quantity then rears an often very ugly head. From the ground, the problem appears to be essentially insoluble. The classical approach whereby volcanologists attempt to insert thermocouples into the flow (what we may call "direct sensing", as opposed to "remote sensing") results in precise temperature measurements at a few well defined points in space and time. The positions at which temperature is measured are likely to be within a radiant crack at above-solidus temperature, or possibly within a layer of crystal mush if the probe can be forced into the flow. Some of the uses and limitations of such data have been discussed in Chapter 6. Recently, hand-held radiometers with small fields of view have become available (e.g. the Minolta/Land

Cyclops instruments described by Oppenheimer & Rothery (1991) and in Chapter 6). These make it possible to isolate the surface temperatures of each particular element of the lava flow, such as hot fractures, cool crusts, exposed flow fronts, and the like, provided that the whole of the region within the measuring circle (typically a degree of arc or less in diameter) is at the same temperature, and that absorption of radiation by atmospheric gases can be discounted or corrected for. How well these criteria can be satisfied depends to a large extent on how close it is possible to get to the target; at a range of a few metres concerns are slight, but they increase as the range rises beyond 10 m or so. Another shortcoming of these instruments is that their measurement of brightness temperature is based on the flux detected across a restricted range of wavelengths, and wavelength-dependent variations in the emissivity of surface conden- sates (which are poorly documented) could introduce significant errors if the brightness temperature of a surface calculated from radiance measured in one waveband is used as a basis for estimating radiant flux at other wavelengths. This problem can be partly circumvented by measuring temperatures using a spectroradiometer that measures radiated flux in many narrow channels across a wide spectral range (e.g. Flynn et al. 1991, Flynn & Mouginis-Mark 1992), but present models are heavy and are field-port- able only with care, and moreover they take of the order of a minute to scan through the spectrum, so rapidly changing surfaces cannot be properly documented in this way.

Measurements by radiometers and spectroradiometers can be thought of as "near sensing" or "ground-based remote sensing". As with direct sensing, these measurements made at a few scattered points can never fully address the dynamics of the truly three-dimensional nature of active lava flows, even in the context of an apparently simple quantity like temperature. Clearly, to fully characterize the problem for even a single flow by a ground-based approach would require a small army of researchers arrayed around an ever-enlarging flow perimeter, continually probing and measuring throughout the history of the emplacement of the flow. This has probably never been achieved, no matter how desirable. Personnel are often limited, terrain is difficult or totally inaccessible, there are risks involved in working near the flow front posed by fluid breakouts and (in densely vegetated areas) by methane explosions, and all of the separate instruments would need to be cross-calibrated. An elegant account of the potential hazards faced by those working on active lava even in the comparatively "safe" environment of Hawaii has been given in Chapter 5, and the rather different hazards posed by silicic lava flows have been discussed in Chapter 1. Thus, it is pertinent to ask whether adequate measurements can be made by remote sensing, while remembering that it remains desirable to have some ground-based data to allow validation and calibration of the remotely sensed measurements.

What is remote sensing?

The standard definition of remote sensing is "acquisition of information without being in physical contact". For our purposes, the information we gather is in the form of an image of part of the Earth's surface, and the medium through which this is obtained is electromagnetic radiation. Remote sensing can be done from an aircraft, or more usually from a satellite. An aircraft has the advantage that you may be able to specify exactly where and when it should fly, but this may be impracticable for volcanologists to arrange. Purchase of satellite data is usually cheaper than arranging for a special aircraft flight, because the satellite can be assumed to have been put into orbit to perform a wide variety of tasks and each user pays only a fraction towards this cost. Data are especially cheap from satellites operated on a non-commerical basis. However, the opportunities that an individual satellite has to image any particular target on the ground are limited by its orbital cycle. The practicalities of obtaining and using the various sorts of data will be discussed in later sections.

An airborne or spaceborne remote sensing device can record an image of the ground in any part of the electromagnetic spectrum to which the atmosphere is transparent. Radar can penetrate clouds, but sensors using shorter wavelengths require clear skies, and even then transmission is confined to "atmospheric windows", beyond which radiation is absorbed by molecular processes (Figure 8.1).

Usually an imaging system records several images of the same area simultaneously in different wavebands (or channels), producing what is known as a "multispectral image". When three of these channels are combined for display purposes, using red for one, green for another and blue for a third (red, green and blue being the three primary additive colours), the result is a "false-colour composite". It is "false" in that the colours do not correspond to true blue, green and red unless the channels fed into the display were recorded in these wavelengths in the first place.

Images are almost always recorded in digital form, and a complete picture in a single waveband is constructed from a series of picture elements, usually known as pixels, each of which represents (in most cases) an approximately square area of ground. The information recorded for each pixel is an integer, usually in the range 0–255 (which is 8 bits), and conventionally referred to as a digital number, or DN. This number is proportional to the amount of radiation received from the ground within the pixel in the waveband concerned. Strictly, the amount of radiation is the "spectral radiance", a quantity having the dimensions of power per area of surface per solid angle sensed per wavelength interval (usually expressed as watts per square centimetre per steradian per micrometre ($W\ cm^{-2}\ sr^{-1}\ \mu m^{-1}$)), and which, as will be explained later, can be used in the case of thermally emitted radiation to derive information about surface temperatures.

The geologist wishing to use such data usually buys them on magnetic tape (known as computer compatible tape, or CCT), but small extracts can sometimes be obtained on floppy disks. These digital data must be displayed on an image-processing system, which formerly required something like US$100,000 worth of computing hardware and software. There are now much cheaper options that can emulate these, especially when only small volumes of data need to be handled, such as a PC system costing around US$1,000, and taking advantage of cheap or even public-domain software. Images can also be bought already copied into photographic form; these are cheaper and are useful for simple mapping but cannot be used for quantitative studies of radiance. The basics

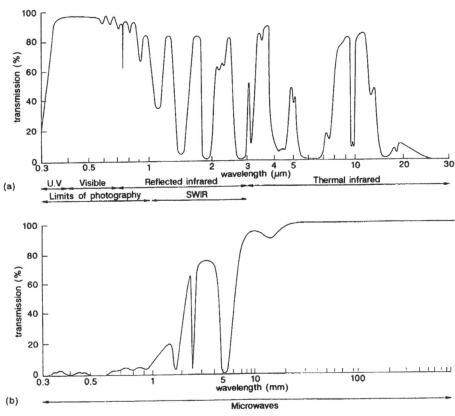

Figure 8.1 Transmission of electromagnetic radiation through the Earth's atmosphere. Remote sensing from high-altitude aircraft or a satellite can be done only in those regions of the spectrum where atmospheric transmission exceeds about 50%. The reflected infrared is sometimes divided into the very near infrared (wavelengths shorter than about 1 µm) and the short-wavelength infrared (wavelengths greater than about 1 µm).

of remote sensing and image processing are described in terms accessible to geologists in several textbooks, most usefully Drury (1987) and Sabins (1986).

The physics of thermal remote sensing

All matter emits radiation with a peak wavelength and spectral distribution that are characteristic of its surface temperature. The peak wavelength, λ_{max}, is given by Wien's law, which states

$$\lambda_{max} = k'/T, \tag{8.1}$$

where T is the surface temperature (expressed in kelvins), k' is a constant with the value 2,987 µm K, and λ_{max} is expressed in micrometres.

The spectral radiance, L_λ, detected from a hot body is given by Planck's formula, which states

$$L_\lambda = \varepsilon h c^2 \lambda^{-5}/[\exp(hc/k\lambda T) - 1], \qquad (8.2)$$

where h is Planck's constant (6.266×10^{-34} J s), c is the speed of light (3.0×10^8 m s^{-1}), k is Boltzmann's constant (1.38×10^{23} J K^{-1}) and ε is the emissivity of the surface at the appropriate wavelength (ε would be 1.0 for a black body, i.e. a perfect radiator, and is typically about 0.95 for basaltic lavas in the short-wavelength infrared).

Curves of radiance against wavelength, derived from Planck's formula, are shown in Figure 8.2 for a range of temperatures. These show that the greatest change in spectral radiance between 300 K (i.e. "normal" environmental temperature) and 1,400 K (the approximate temperature of molten lava) occurs in the region of the spectrum between 1 and 3 μm. This, therefore, is the most useful region of the electromagnetic spectrum for detecting and measuring magmatic temperatures by remote sensing.

The 3–5 μm atmospheric window is sometimes known as the "fire channel", because it is particularly useful for detecting fires (e.g. Matson et al. 1987). This channel is

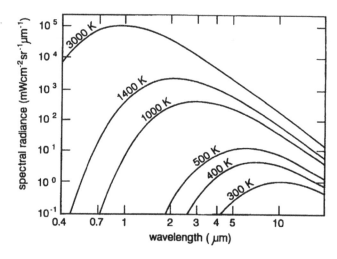

Figure 8.2 The spectral radiance emitted by surfaces at different temperatures, according to Planck's formula. These curves are drawn assuming the surface is a perfect radiator (i.e. a "black body"); for real surfaces the radiance has to be multiplied by a factor ε (<1), where ε is the emissivity of the surface. Emissivity is likely to vary with wavelength, but for most lavas it is about 0.95 for wavelengths between 1 and 3 μm. In practice the 1–3 μm region has proved most useful for measuring temperature distributions on lava bodies and hot fumarole fields. If the whole of a Landsat Thematic Mapper (TM) band 7 (2.08–2.35 μm) pixel were occupied by material at 1,400 K, the radiance received at the satellite would be more than a 1,000 times that necessary to saturate the pixel; however, the band 7 radiance from hot cracks at 1,400 K occupying less than a tenth of a per cent of the area of a crusted lava would be within the measurable range, allowing both the temperature and the total area of the cracks to be measured, provided (as discussed in the text) radiance could also be measured in a neighbouring region of the spectrum (such as TM band 5 at 1.55–1.75 μm).

recorded by a low-resolution (1 km pixel) sensor system known as AVHRR carried by a long-standing series of weather satellites, known as the NOAA polar orbiters (sometimes referred to as the TIROS-N series), that can provide daily data extremely cheaply. In August 1990 a team from the Planetary Geosciences Division at the University of Hawaii begain to monitor the development of the Kupianaha lava field several times a day using such data (Fig. 8.3; Mouginis-Mark et al. 1991b). However, from a volcano-logical perspective, the low spatial resolution and variable viewing geometry of AVHRR data make them unsuitable for anything other than simple detection of hot lava, and meaningful temperature measurements can rarely be made (e.g. Weisnet & D'Aguanno 1982, Rothery & Oppenheimer 1993). Instead, attention has focused on the use of data at shorter wavelengths, using the two short-wavelength infrared (or SWIR) channels (band 5, 1.55–1.75 µm, and band 7, 2.08–2.35 µm) recorded by the Thematic Mapper (TM) instrument of the Landsat satellite series, which operates at a pixel size of 30 m, except for band 6, which is a "thermal infrared" channel (10.4–12.5 µm) with 120 m pixels. Landsat satellites are in a regularly repeating pattern of orbits, allowing coverage of the same area by a single satellite once every 16 days (if there are no clouds in the way) near the Equator, and more frequently at higher latitudes where the degree of overlap between the 180 km wide image swaths increases.

A study of Láscar volcano, Chile, by Francis & Rothery (1987) demonstrated that an area within the summit crater was radiating strongly in both TM SWIR channels, and Rothery et al. (1988a) went on to show how by comparing the spectral radiance at these two wavelengths it could be proved that the radiant sources were actually only a tiny fraction of each anomalous pixel, and that this "dual-band method" could be used to estimate the size and temperature of the hot sources, by assuming that each pixel

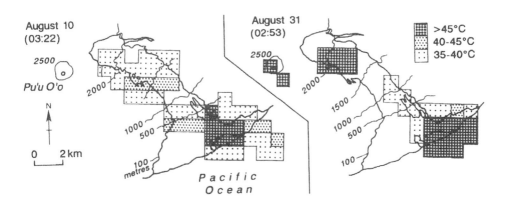

Figure 8.3 Outline maps of the the Kupianaha lava field, Hawaii, with anomalously hot pixels detected using AVHRR channel 3 (3–5 µm) data superimposed. The "temperatures" quoted are based on radiance averaged over entire pixels ("pixel-integrated" temperature, Rothery et al. 1988a) and are not simply related to the temperature of any hot lava. However, high "temperatures" do indicate the areas that are radiating most strongly. (Courtesy of Scott Rowland, Pierre Flament & Peter Mouginis Mark, University of Hawaii.)

contained just two thermal components: a hot (radiant) one and a cold (non-radiant) one. The methodology was refined by Oppenheimer (1991a,b) and Oppenheimer et al. (1993a), who examined the implications of the assumption that the cold component is non-radiant, and the extra information that could be obtained by relaxing this constraint, and by considering also the information provided by band 6. They also suggested that for undoubted lava bodies it is more realistic to assume the temperature of the hot component (molten lava approximately at the liquidus), and to use the dual-band method to determine the temperature of the crust and the proportion of the surface where the molten lava was exposed.

Future spaceborne remote sensing instruments capable of recording thermal radiance from hot lava in more than two wavebands will enable cross-checking of the dual-band solutions to two thermal components, and also more refined modelling where more than two thermal components need to be considered. The Japanese satellite JERS-1 (launched in February 1992, and renamed Fuyo-1) will record three SWIR channels in the 2.0–2.5 μm region and one at 1.6–1.7 μm and so will be a step in the right direction. The Earth Observing System, a series of satellites to be deployed by NASA and other space agencies in the late 1990s will provide even better spectral resolution (Mouginis-Mark et al. 1991a).

Deriving radiance from DN

Before temperatures can be calculated, by the dual-band method or otherwise, the DN recorded by the satellite in each band (DN_λ) has to be converted into spectral radiance (L_λ). The standard calibration for Landsat TM data is given by Markham & Barker (1986), in the following formula:

$$L_\lambda = LMIN_\lambda + [(LMAX_\lambda - LMIN_\lambda)/255]DN_\lambda, \tag{8.3}$$

where $LMIN_\lambda$ and $LMAX_\lambda$ are the spectral radiances corresponding to a DN of 0 and 255, respectively. Markham & Barker quote the following values for these (post 15 January 1984), in units of mW cm^{-2} sr^{-1} μm^{-1}: band 4 –0.15 and 20.62; band 5, –0.037 and 2.719; band 6, 0.1238 and 1.5600; band 7, –0.015 and 1.438. These values could change, and it is advisable to check the calibration data that are encoded in the header information on your tape (on which the data supplier should be able to advise), but in the authors' experience of Landsat 4 and 5 data no significant departures have been seen.

The raw value of DN_λ is not likely to represent the thermal radiance from the surface. There are two main reasons. First, the radiance leaving the ground may not be entirely thermal, except at long wavelengths. For example in a TM SWIR image recorded by day there will be a contribution to the upwelling radiance by sunlight reflected from the surface, unless the pixel happens to lie within deep shadow. The simplest way to estimate this correction is to compare a pixel that is evidently thermally radiant with one that is not, and subtract the DN of the latter from that of the former (e.g. Rothery et al. 1988a). More sophisticated corrections are described by Oppenheimer (1991a) and Oppenheimer et al. (1993a). The correction for reflected sunlight means reducing the value of DN_λ before applying Equation 8.3 (but see next paragraph). One way to avoid the sunlight correction is to use data recorded by night. Night-time recording is not done routinely by the Landsat TM, and data acquisition must be by special request (see later). The first volcanological examples of this are reported in Rothery & Oppenheimer (1991), Oppenheimer et al. (1993a) and Reddy et al. (1993).

The second reason is that the radiance leaving the ground will have been attenuated within the atmosphere before reaching the satellite, so it is advisable to correct for the atmosphere in some way. The standard method of atmospheric correction in remote sensing is to make use of public domain atmospheric models, such as Lowtran (Kneizys et al. 1988) that can be set according to altitude, latitude, season and certain other parameters. Unfortunately, the atmosphere over a volcano is unlikely to match a standard model, because of the enhanced concentrations of volcanogenic gases such as SO_2 but this is the best that can be done in the absence of simultaneous measurements of gas concentrations. Atmospheric correction to allow for attenuation of upwelling radiance typically means increasing the estimate of L_λ by up to 20% after applying Equation 8.3.

It is impossible to make a definitive statement about the magnitude of the uncertainties resulting from inadequate corrections. However, although the calculated spectral radiance is not often likely to be in error by more than about 10%, temperatures derived by the dual-band method could be wrong by as much as 200 K if the errors in spectral radiance have opposite signs in the two bands that were used (Rothery et al. 1988a). See Oppenheimer et al. (1993a) for an approach to limiting further uncertainties that are inherent in the dual-band assumptions themselves.

Examples of remote sensing of active lava

Lava flows
In the case of a moving lava flow, a description of its surface temperature distribution as just two thermal components, a chilled crust and a hot "core" temperature exposed in cracks, is a reasonable approximation. Crisp & Baloga (1990) developed a theoretical model for lava flows for which these two values were fundamental to the understanding of flow motion. Further discussion of the importance of temperature measurements in understanding and predicting flow behaviour may be found in Chapters 6, 7 and 9–11.

The first successful attempts to use Landsat TM data to measure dual-band temperatures on active flows were by Pieri et al. (1990) who studied an archived image which had recorded a 1984 lava flow near the summit of Mount Etna, and Oppenheimer (1991b) who studied a Landsat TM image recorded by special request of the 1989 lava flow at Lonquimay volcano, Chile (Fig. 8.4). These were both crusted flows, with only a small fraction of incandescent core material exposed. A rapidly flowing, channelized "river of fire" type flow whose surface was dominated by incandescent material would be a much stronger source of thermal radiance. Such a situation leads to saturation of the longer-wavelength channels, making them useless for temperature measurement, but, under the right conditions, there is measurable thermal radiance in the shorter-wave-

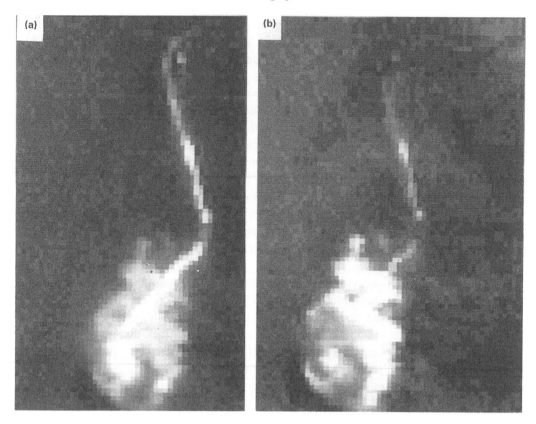

Figure 8.4 Extracts of Landsat TM images of the Lonquimay flow on 8 August 1989: (a) band 7 (2.08–2.35 μm), (b) band 5 (1.55–1.75 μm). The original data were obtained in Eosat A format, but for display purposes sets of 16 scan lines have been offset sideways (by 40–44 pixels) to remove the distortion caused by the alternating mirror sweeps. Each view is about 2.5 km from top to bottom.

length channels. Landsat TM images of a 1991 Kupianaha flow, Hawaii (P. Mouginis-Mark 1992, personal communication), and of a 1992 flow on Etna (Rothery et al. 1992) both recorded thermal radiance in TM band 4 in some pixels, but TM band 7 was saturated in these locations. In the Etna example, the band 4 radiance is confined to a 700 m length corresponding to a 10–15 m wide open channel at the source of the flow (Fig. 8.5d). Also, Gupta & Badarinath (1993) were able to detect the May 1991 basaltic eruption of Barren Island using TM bands 5 and 7, and we are not aware of thermal radiance having been detected by TM at wavelengths shorter than band 4. However, Rothery et al. (1988a) reported an example of a 10 km length of a basaltic flow from Sierra Negra, Galapagos, in 1979 radiating in the Landsat MSS 0.8–1.1 μm channel (Fig. 8.6), much of which was also radiant in the 0.7–0.8 μm channel and 1 km of which was also radiant at 0.6-0.7 μm. Unfortunately they were unable to derive a realistic temperatures from these data by means of the dual-band method.

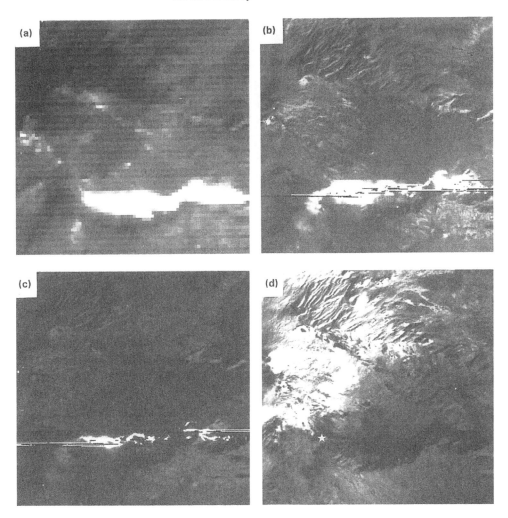

Figure 8.5 Extracts of a Landsat TM image of Mount Etna, Sicily, 2 January 1992, showing an area about 10 km across. (**a**) TM band 6 (10.4–12.5 μm); (**b**) TM band 7 (2.08–2.35 μm); (**c**) TM band 5 (1.55–1.75 μm); (**d**) TM band 4 (0.76–0.90 μm). In parts a–b the whole flow is bright, in part c only the more radiant portions are bright, and in part d the whole flow is seen by reflected sunlight (and looks black as a result of its very low albedo, whereas the snow is bright) apart from a 700 m length extending eastwards from the location marked by the star, where there is faint thermal radiance in this spectral band corresponding to an open channel. Note the bad scan lines in bands 7 and 5, which represent sensor overload as a result of the high radiances encountered. Thermal radiance from several summit craters can be made out as hot points at the extreme left of the image in bands 7 and 5 These data are Eurimage level 1.

We are aware of only two airborne data sets that show thermal radiance from an active lava flow. These are data from the National Aeronautics and Space Administration (NASA) Thematic Mapper Simulator or NS001 (NASA 1986) covering a 1987 eruption

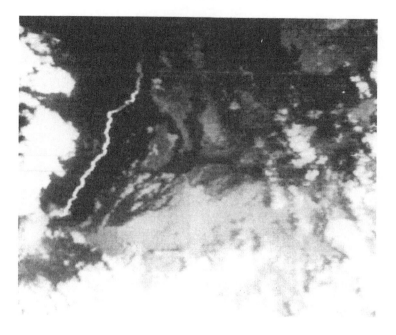

Figure 8.6 Extract of a Landsat MSS 0.8–1.1 μm image showing a radiant lava flow on Sierra Negra, Galapagos Islands 19 November 1979. The flow was evidently a dynamic open channel within which no crust had developed. The length glowing here at this wavelength is about 10 km.

of Kupianaha vent, Kilauea (Fig. 8.7), and a further eruption from the same vent recorded in 1990 (Fig. 8.8) by the same instrument and by the Daedalus Thermal Infrared Multispectral Scanner, or TIMS, which operates in the long wavelength thermal infrared (Pallucone & Meeks 1985). The characteristics of these two instruments are listed in Tables 8.1 & 2. Data recorded by the NS001 (D. A. Rothery & D. C. Pieri, unpublished data) and an imaging spectrometer (Oppenheimer et al. 1992) during the NASA Multispectral Airborne Campaign in Europe in July 1991 reveal SWIR radiance from active vents on both Etna and Stromboli, but there were no active lava flows at that time.

Figure 8.7 is revealing with regard to the thermal structure of this aa flow. At the flow front the temperatures estimated by the dual-band method are in the range 500–800°C, and major thermal losses are occurring there as a result of breakage of the overlying rubbly crust and exposure of the visibly incandescent (but not molten) interior as the flow advances. Hand-held field radiometer measurements of comparable flows from the Puu Oo vent indicate average surface temperatures of rubbly crusts over the actively moving parts of the flow to be in the range 250–350°C. At the flow fronts, where cool material is shed and hot material becomes exposed, temperatures as high as 750–800°C have been measured, using the same instruments. Such hot material, when exposed and freely radiating to space, typically cools by 400–500°C within a minute or less. All of these phenomena, observed on the ground over a period of time at a couple

Figure 8.7 This is an infrared image of an active 2 km long as flow on the Kupianaha shield, Kilauea volcano, Hawaii, in 1987. The image was obtained from an altitude of approximately 1,500 m using the Thematic Mapper Simulator (NS001) on the NASA Earth Survey Lockheed C-130B aircraft. Emission in three near-infrared channels between 1.13 and 2.38 μm is represented here, with the most intensely radiating areas (estimated temperature about 600–800°C) appearing as white. The source of the lava flow is in the upper left and the flow front is at lower right. Older dark (cold) flows appear black as do clouds, and the light blue grid pattern represents paved streets. Orange areas are cooler fresh flows (about 100–500°C). Vertical striping is due to sensor overload from the intense radiation. Images like this are useful for determining the thermal budgets of lava flows, but such data are rare.

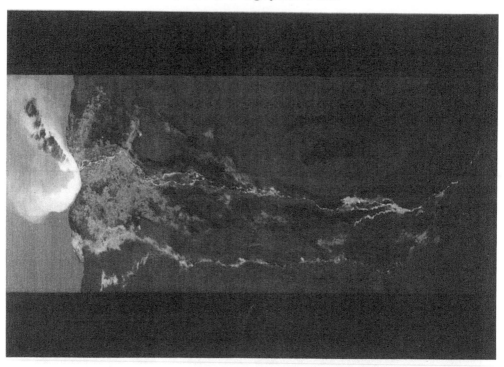

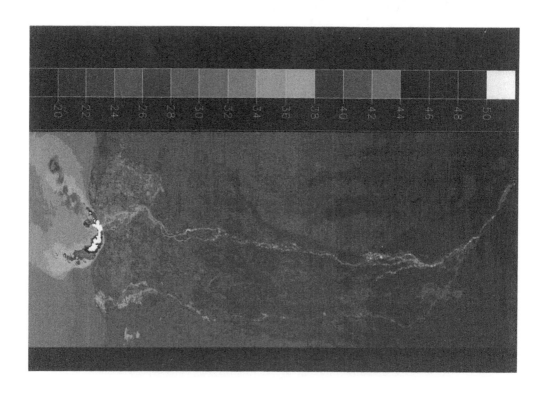

216

Figure 8.8a (Facing page) Temperature distribution over a 7 km long active lava tube system, Kupianaha flow field, Hawaii, October 1988. Data are from the TIMS aboard the NASA C-130B, taken at 3 am local time. A hot water plume resulting from lava ocean entry is at bottom. Colour scale for temperature in degrees centigrade is at left. Central trace is the active system, while a months-older inactive system is seen to the right.

Figure 8.8b Composite of combined emissivity/temperature distribution, as seen in three TIMS band-passes for the Kupianaha tube system (same as in (a)). Bright red areas are altered by downwind acidic gas emissions along the tube roof. Blue and red represent different surface states. Brighter areas are warmer (e.g. tube trace and ocean plume). Braided character of tube was unexpected from previous ground studies. (From Realmuto et al. 1992.)

---

of stations, are consistent with the overall airborne "snapshot" view shown in Figure 8.7.

In 1990, NS001 and TIMS data were recorded during night-time sorties over an active lava tube system fed from the Kupianaha vent, and stretching more than 7 km to the ocean (Fig. 8.8). TIMS images and the short-wavelength infrared bands of the NS001 data were particularly effective in delineating the surface trace of the active tubes, which could not, of course, be seen at visible wavelengths. Initial work on the TIMS data by Realmuto et al. (1992) has indicated up to 50 differential temperatures at the surface of the intact roof of the tube of normal thickness (estimated to be of the order of a few metres for most of its length). Where the tube roof had thinned or collapsed, TIMS data are saturated but NS001 data indicate dual-band temperatures in the range 500–800°C for the hot component (Glaze & Pieri 1989).

Lava lakes and lava domes

The longest-running series of observations of a hot lava phenomenon using TM data is of Láscar volcano, Chile (Fig. 8.9). This extends from December 1984 into 1992, having been begun by Peter Francis and continued by one of the authors (DAR). Recent work incorporating field observations (Oppenheimer 1991a, Oppenheimer et al. 1993a) suggests that the core of the radiant phenomenon since 1989, and probably prior to that, was a lava dome. The lava lake proposed by Francis & Rothery (1987) and Glaze et al. (1989a,b) as the radiant source prior to the September 1986 eruption now seems unlikely.

Genuine lava lakes have occasionally been observed by satellite. For example, the Mount Erebus lava lake can be detected using 1 km pixel AVHRR data in the 3–5 μm atmospheric window, although the data appear to be of no quantitative value (Weisnet & D'Aguanno, 1982; Rothery & Oppenheimer, 1993). The first report of the use of Landsat TM data on this target was by Rothery et al. (1988a), who also reported the presence of two active lava lakes on Erta 'Ale (Ethiopia) shown in bands 5 and 7 of a TM image recorded on 5 January 1986 (Fig. 8.10). It is noteworthy that this observation is the sole basis by which the continued presence of active lava lakes on Erta 'Ale is inferred during 1975–85 in the Smithsonian Institution decade summary of global

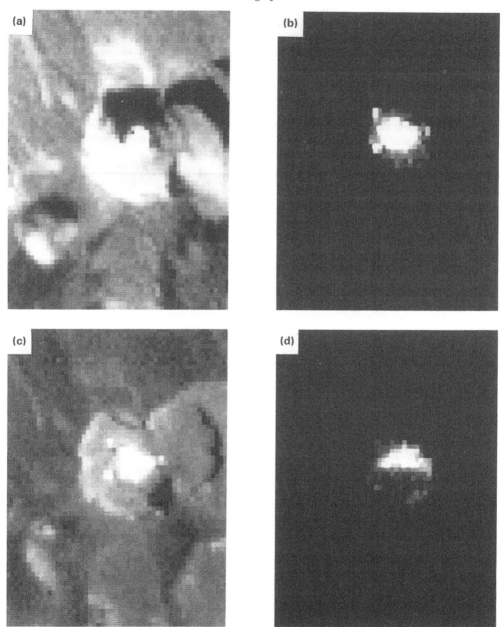

Figure 8.9 Extracts from Landsat ™ band 7 (2.08–2.35 μm) images showing thermal radiance from the lava dome within the active crater of Láscar volcano, Chile: (**a**) 21 July 1985; (**b**) 18 November 1989 (night-time); (**c**) 14 December 1989; (**d**) 26 March 1990 (night-time). The lava dome appears to have changed little between 18 November and 14 December 1989, but it can be more clearly seen on the night-time image, and thermal measurements derived from this night-time image are correspondingly more accurate. The night-time data have been rotated by 162° to allow them to be registered with the daytime data. The area shown is about 2 km from north to south.

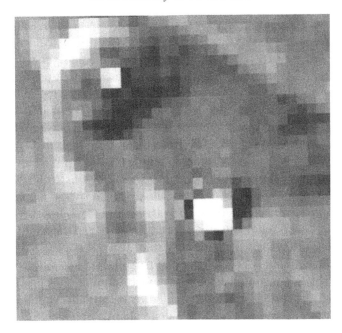

Figure 8.10   Extract of a Landsat ᴛᴍ band 5 image showing two lava lakes at Erta 'Ale, Ethiopia, 5 January 1986. Since the last known field observations in the mid-1970s, the northern of the two lakes appears to have shrunk and the southern one to have grown. The area shown is about 0.9 km from north to south.

volcanism (McClelland et al. 1989 pp. 16–28, 77). Rothery et al. (1988b) reported the same two lava lakes present on 9 Feburary 1987 on the basis of the examination of a photographic version of a ᴛᴍ image (which at the time was much cheaper to buy than digital data). This remained the most recent information on the continuity of this longest lived of all lava lakes (active since at least 1960) until a low-altitude overflight and subsequent field visit confirmed continuing activity in late 1992 (Smithsonian Institution 1992a,b).

Fumaroles
Data from ᴛᴍ bands 5 and 7 have been proved capable of detecting fumaroles as well as active lava, for example on Poàs in 1986 (Rothery et al. 1988b, Oppenheimer & Rothery 1991), around the walls of the active crater on Láscar 1987–91 (Oppenheimer et al. 1993a,b) and on the floor of the same crater in the aftermath of an explosive eruption in 1986 (Rothery et al. 1988a), at the summit lava dome of Colima in 1985 and 1986 (Abrams et al. 1991) and on Kinamura cone, Niyamuragira, in 1989 (Smithsonian Institution 1989). In the absence of supporting evidence from the field, it is by no means a simple matter to distinguish between a small lava-filled vent and a fumarole field on such data. Oppenheimer et al. (1992a,c) suggest some strategies to overcome this.

Ground-based thermal imaging

The dynamic aspects of lava flows are perhaps the most difficult to document and study. Recently, however, the easy accessibility of standard visible light camcorders and the advent of field-portable thermal infared video rigs has made it possible to record the dynamic nature of a variety of volcanic phenomena on video across a range of wavelengths. One thermal infrared video system, the I525 Thermal IR Scanner built by Inframetrics Inc. (Bedford, Massachusetts, USA) has been deployed by the Jet Propulsion Laboratory (JPL/NASA) volcanology group during eruptions in the USA and abroad.

The I525 is sensitive to a broad channel spanning the 8–12 μm atmospheric window, and this can be filtered selectively to choose narrow regions within this range using manually inserted filters. For lava flow work, the so-called "flame suppression" narrow-band filter (designed for firefighters) at 10.5–10.7 μm is ideal. Improved thermal infra-red video scanners with automatically selectable filters and capability in the 3–5 μm atmospheric window have now become available. The I525 is powered by rechargeable 12 V gel-cell batteries and is portable by one person. It stores video data at 256 grey levels (8 bits), which is recorded at standard video framing rates in NTSC format. The mercury–cadmium telluride detector is cooled by a refillable liquid nitrogen Dewar flask. This has a 10 cm^3 capacity, and lasts up to about 4 hours per filling. The need to have liquid nitrogen available in the field is probably the greatest constraint on the use of such equipment. For measurement of hot lava flows it is not feasible to provide a high-temperature (~1,000°C) calibration source within the instrument. Thus, absolute accuracies, particularly at high temperatures, are only as good as whatever reference source the experimenter can provide in the field. In practice, this is not too difficult: such things as sky temperature, ground temperature or the actual vent temperature can be measured independently using portable Cyclops-type radiometers (see earlier). For I525 records of flowing lava calibrated in this way, absolute accuracies of 25–50 K are not too difficult to achieve. Relative precision is high, ranging from 0.1 K at ambient temperature to less than 5 K at elevated temperatures.

Using this instrument, video sequences were recorded from low-altitude (30 m above ground level) helicopter flights during the 1984 eruption of Mauna Loa in Hawaii over the main 27 km long active aa flow (Pieri et al. 1984). Additional (unpublished) I525 observations have been made in Hawaii at the Kupianaha flow field of Kilauea during eruptions. The I525 has also been deployed outside the USA in collaboration with various international groups, namely: in Italy in 1984–5 (with the co-operation of the Consiglio Nazionale delle Richerche, Osservatorio Vesuviano, and Ministero della Protezione Civile) to observe hot fractures associated with the 1984 volcanic crisis at Campi Flegrei, again in Italy, in 1989 in various collaborative efforts (involving the Gruppo Nazionale per la Vulcanologia and the Open University) to record activity during the October 1990 eruption of Etna and fumarolic activity at Vulcano, and in 1991 on board an Aeroflot Mi-8 helicopter and An-2 biplane for thermal infrared measurements of active summit craters at Tolbachik, Bezymianny, and Avachinski volcanoes on the Kamchatka peninsula (as a result of collaborative agreements between the Institutes of Volcanology/Volcanic Geology and Geochemistry (Russian Academy of Sciences), JPL/NASA, and the US Geological Survey).

Remote sensing strategies for observing lava flows

Using aircraft

Aircraft offer the means to cover wide areas quickly. Remote sensing performed by fixed-wing aircraft can provide spatially detailed data (with pixels around 1–5 m across) at will. Generally speaking, a fixed-wing aircraft operates at a constant altitude during a given data run, the success of which may be compromised by clouds at a lower altitude than the aircraft. However, such an aircraft provides a stable observation platform resulting in image data of predictable geometry. The geometric distortions in images from all but the highest-altitude of aircraft are significantly greater than in Landsat-type satellite images, because of the wide scan angle needed to image a broad swath (e.g. Tables 8.1 & 2). First-order scale changes across the scan width can be removed automatically, but this does not correct for relief displacement such as the shifting of the summits of hills away from the centre of the flight line.

A helicopter offers more flexibility than a fixed-wing aircraft in terms of varying operational altitudes, the ability to hover over a given spot, and the capability of flying below all but the lowest clouds, but these options introduce strong geometric variability into image data. Helicopters, however, are very useful for ultra-low (<30m above ground level) observations, and for crossing otherwise impassable terrain to reach flows for ground-based study.

Both fixed- and rotary-wing aircraft, while useful, are moderately to very expensive, and sometimes introduce formidable logistic problems. The largest aircraft are the most logistically complicated and expensive. For example, at the time of writing a small single piston-engined aircraft (e.g. Cessna-172 class) usually costs less than about US$100 per

Table 8.1 The NS001 multispectral scanner, or Thematic Mapper Simulator, as used on the NASA C-130B aircraft records in the seven channels used by the Landsat 4 and 5 TM, plus an extra channel (1.13–1.35 µm)

Band number	Band-pass (µm)
1	0.458 – 0.519
2	0.529 – 0.603
3	0.633 – 0.697
4	0.767 – 0.910
5	1.13 – 1.35
6	1.57 – 1.71
7	2.10 – 2.35
8	10.9 – 12.3
Instantaneous field of view	2.5 mrad
Total scan angle	100°
Pixels/scan line	699

Table 8.2 The TIMS uses a dispersive grating and a six-element mercury–cadmium telluride detector array to record six discrete channels in the 8.2–12.2 μm region

Band number	Band-pass (μm)
1	8.2 – 8.6
2	8.6 – 9.0
3	9.0 – 9.4
4	9.4 – 10.2
5	10.2 – 11.2
6	11.2 – 12.2
Instantaneous field of view	2.5 mrad
Total scan angle	76.6°
Pixels/scan line	638
Noise equivalent ΔT	< 0.3°C (all channels)

hour (including pilot and fuel) to operate. A large four-engine transport turboprop aircraft (e.g. NASA Lockheed C-130B) can cost of the order of US$3,000 per hour, while the largest class of turbine engined jet aircraft (e.g. NASA McDonnell Douglas DC-8) can cost of the order of US$10,000 per hour (including crew and fuel). Lear jet class aircraft costs are somewhere between those of the C-130 and DC-8, while typical commercial helicopter (e.g. Hughes 500D) costs are in the region of US$500 per hour (including pilot and fuel).

Fuel logistics are often an important consideration for aircraft operations. Light piston-engined aircraft require high-octane aviation gasoline, which is widely available (worldwide) at fixed-base airstrips. However, most turbo-engined helicopters and all large turbine or turbo-prop aircraft require jet-fuel grade kerosene (e.g. JP-4), and this is available only at large airports, or by special arrangement in the field in the case of helicopters. Would-be researchers working on government grants or contracts, using large government-owned aircraft, should be prepared for substantial negotiation on logistical considerations before any deployment. For example, the range of deployment for sorties with a given aircraft from country A operating in country B may have to be calculated from the nearest country A military base where jet fuel is available at discounted rates, rather than from a nearer country B commerical airport where fuel is available only at higher, commerical, rates. On the other hand, it is possible that one of the conditions under which an aircraft from country A will be allowed to operate in country B may be that the aircraft must buy its fuel from commerical sources within country B. With fuel capacities in the thousands of litres (for the largest class of aircraft) such financial considerations are not insubstantial. Additionally, when operating in a country not of your origin, it is important to be sympathetic to security considerations involved in aircraft overflights. The researcher must be prepared to have data scrutinized or embargoed even after official permission to collect data has been granted. Areas including, or near to, military installations are especially sensitive.

Airborne data on active lava flows are still rare, but, as discussed in the previous section, some have been acquired in Hawaii by JPL/NASA, in co-operation with the US Geological Survey's HVO, over the last 5 or 6 years. The NS001 and TIMS instruments were carried by the NASA Earth Survey Lockheed C-130B aircraft, which is operated by the NASA Ames Research Center. It has been a goal of a number of researchers in the remote sensing/volcanology area to capture a high spatial resolution (5–10 m pixels) multispectral infrared image in the wavelength range between 2 μm (the peak of emission at 1,150°C) and 10 μm (25°C peak emission) over the entire course of development of a lava flow. Such data would be invaluable as boundary conditions for a variety of models of lava flow thermal–rheological behaviour. However, in spite of continuing eruptions at Kilauea it has proved difficult, in a practical sense, to deploy a very expensive and complicated instrument platform such as the C-130B to Hawaii any more often than about once every 3 years for a 2 week period. Often, when such a deployment is finally carried out, active lava flows have not been present during the observational period, or full instrumentation was not available as a result of maintenance problems. Thus, it is truly fortuitous when a coincidence of natural and technical events occurs, to allow collection of appropriate data sets, such as those illustrated in Figures 8.7 & 8.

Instrumentation like the NS001 and TIMS systems has, until recently, been available only on board large expensive government-owned aircraft, and has thus been inaccessible, in the prospective sense, for many would-be users. However, data are usually made available retrospectively very cheaply, although data users must make do with the coverage requested by the original principal investigators. Now and in the future, general reductions in cost and size due to instrumental improvements are leading to capable instruments that can be carried by smaller, light twin- and single-engined aircraft, that are much more economical and flexible (e.g. a 0.4–5.0 μm airborne Fourier spectrometer currently being developed at the University of Hawaii (P. Mouginis-Mark 1992, personal communication)). Such improvements in flexibility and cost allieviate some of the main problems in attempting to carry out lava flow monitoring, and ought to greatly increase the chances of carrying out systematic thermal measurements throughout the lifetime of many different types of active lava flow.

Despite the often considerable difficulties, if an aircraft with appropriate instrumentation can be deployed at an active eruption in time to take data in good weather (or, for imaging radar, in virtually any weather), this will almost always be the observational platform of choice. This is because of its flexibility to record data continuously or at will during the eruption, the high spatial resolution (i.e. small pixel size) that can be achieved by low-altitude sorties, and the potential to co-ordinate aircraft observations with simultaneous ground operations. However, it is necessary to ensure that the aircraft can avoid flying into any airborne ash, which can cause engine failure and abrade the forward windshield to the point of invisibility.

Although the foregoing discussion may seem out of the range of normal volcanological considerations, anyone planning aircraft observations of an active eruption needs to take these factors into account. Such planning generally has to be done at the last minute, and we hope our practical advice will at least serve to focus thoughts usefully.

Using spacecraft

Spacecraft are in many ways more convenient than aircraft for the remote sensing of volcanoes. Their most important advantage is that they can provide repeated coverage of volcanoes worldwide. While their initial cost is very high (US$50 million – 1,000 million, exclusive of launchers), when pro-rated over the life of a typical Earth orbital mission (5–10 years) their cost per hour is equal to or less than that for most aircraft. The actual cost paid for the data by the user depends on whether the satellite is operated on a commerical basis (e.g. Landsat since 1985) or non-commercially (e.g. AVHRR data).

Earth orbital spacecraft generally fall into several classes: (a) imaging reconaissance satellites, (b) communications satellites, (c) non-imaging experimental satellites, and (d) manned vehicles. Any of these can be deployed into either a low Earth orbit or a geosynchronous orbit, and may be dedicated to either civilian or military applications. Of primary concern here is class (a), though astronauts have made visual observations that constituted the first reports of eruptions at remote sites.

For many global geological investigations the most useful type of low Earth orbit spacecraft is one with its orbit inclined at a large angle to the Equator, so that it passes close to the poles. A polar orbiting satellite with downward-looking image instrumentation can repeatedly build up image mosaics covering the globe, the recurrence frequency of which depends on the angular field of view of the instruments and the period of the orbit, as determined by orbital altitude. For remote sensing satellites, the orbit is almost invariably chosen to be circular with an altitude in the range 500–1000 km. Usually, the orbital period and inclination are so arranged that the satellite completes a whole number of orbits during each day, and always makes its daytime crossings of the Equator at the same local (solar) time. This is known as a Sun-synchronous orbit, and Equator-crossing times are often chosen to be in the mid–late morning (0900–1100 hours). Using an imaging system with a wide field of view, the whole globe can be imaged during a single day. However, the field of view is usually traded against resolution, in order to capture and store images with a small pixel size (10–100 m for geological purposes), which means that the field of view of the imaging system becomes narrower than that which would permit daily global coverage. In the case of Landsats 4 and 5, it takes 16 days to complete global coverage, upon which the orbital pattern begins to repeat. In the following section, information and advice on obtaining the most suitable data for lava monitoring from high resolution polar orbiting satellites, such as Landsat, is provided.

A geostationary satellite is placed directly over the Equator at such an altitude (40,000 km) that a single orbit takes a whole day, thereby maintaining the satellite permanently overhead the same point on the equator. Such a platform offers a number of advantages and disadvantages in the context of monitoring effusive eruptions. The main advantage is that observations can be virtually continuous, so that an eruption can, in principle, be monitored throughout its entire duration. The main disadvantages are the intrinsically lower resolution engendered by the long viewing distance, which requires very large optics (about 1 m aperture for 30–40 m pixels in the visible range). Although high performance geostationary military satellites have been used to track even the thermal infrared signatures of objects as small as individual vehicles, no thermal imaging system of sufficient mass has been, or is, contemplated in the civilian sector. Another disadvantage of geostationary platforms is that (contrary to popular perception) they are incapable of imaging a complete hemisphere, because they are too close to image areas beyond about 55° from the sub-spacecraft point, so for complete global monitoring (polar regions excepted) a girdle of 3–5 satellites is required.

Current civilian imaging systems in geostationary Earth orbit include the two American GOES satellites, the European Meteosat, the Japan Meteorology Satellite, and a similar Indian weather satellite. Data from these are inadequate to detect or monitor lava, although they have proved invaluable in detecting and monitoring eruption plumes (Sawada 1989, Glaze et al. 1989b). Soviet civilian capabilites in this sphere were not well known in the West, and the present capability of the Russian and other successors to the Soviet space programme remain uncertain. Contemplated future capabilites as part of national and multinational civilian ventures include multispectral imaging systems in geostationary orbits with pixel sizes in the 50–100 m range at visible wavelengths.

Satellite data: where to get them, what to ask for and what they cost

When buying remotely sensed data, it is prudent, if possible, to make a visual inspection first. Although catalogues usually provide an estimate of the total amount of cloud cover within an image, only by checking visually can you be sure that cloud or fumes are not obscuring the portion of the volcano that is of interest. It is cheaper to buy an image in photographic form than in digital form. Photographs can be interpreted visually, providing a simple way of recording changes between one eruption and the next, but only digital data can be used to derive quantitative radiance values necessary for thermal modelling. If you intend to measure radiances to derive temperatures by means of the dual-band method then you should try to avoid using data that have been resampled in any way, as this destroys the integrity of the original pixels. This means never using any data that are sold as "geocoded", because such an image has been resampled to fit a map projection, and although this is useful for mapping purposes it introduces irreversible radiometric degradation.

Some information on how data can be ordered is given below. Many countries have national remote sensing centres (too numerous to list here) that will give further advice. If your research is funded by a national research council then this body may already have established procedures for enquiring after and buying satellite data. This is the case, for example, with the Natural Environmental Research Council in the UK.

Data from the Landsat TM (and its less useful, but cheaper, MSS instrument) are held at many centres worldwide. Landsat's orbital pattern repeats every 16 days, allowing any volcano to be imaged once in each period (clouds permitting). Sideways overlap between image swaths increases towards the poles, so there is potential for imaging more frequently than once in each 16 day period for targets at high latitudes. Opportunities to record data are doubled if the satellite is operated at night.

Some Landsat data are received directly at ground stations over a range of about 3,000 km, and are available from local distribution centres. Directly transmitted data covering North America, and data covering other parts of the globe transmitted via relay satellites, are held centrally in the USA. You can expect to pay over US$3,000 for a quarter of a full TM image, covering an area of approximately 90 × 90 km. A black and white photographic transparency covering a full scene at 1:1 million scale costs about US$2,700, but photographic versions of simultanously recorded images from Landsat's Multispectral Scanner instrument (80 m pixels) are a much better buy at US$120, and should be adequate for cloud and fume assessment. In addition, it is worth checking with your supplier to see if any kind of "quick-look" product is available for this purpose. Some old data are available as 15 x 15 km extracts on PC-readable floppy disks, but only in resampled format, for US$600. Landsats 4 and 5 were operated on a commercial basis by the Eosat company. Landsat 6 is due to be launched in 1993, and sees the beginning of de-commercialization of the Landsat programme. It will be operated jointly by the US Defense Department and Eosat, and this may offer the opportunity of significantly reduced prices for scientific research, at least after July 1994 when Eosat relinquish exclusive rights to TM data.

In place of the TM, Landsat 6 will carry an Enhanced Thematic Mapper (ETM), having identical spectral bands to the TM with the addition of a panchromatic (0.50–0.90 μm) band with a pixel size of 15 m. More importantly, it will be possible for users to request imaging at a low-gain setting that will reduce the extent of saturation experienced in bands 7 and 5 over strongly radiant sources. Landsat 7 will probably be similar, and is due to be launched in the late 1990s and may be run on a non-commercial basis, thereby reducing the costs of the data to the scientist.

A general rule, whenever buying data from which you intend to derive temperatures, especially if you intend to use the dual-band technique, and whichever data supplier you deal with, is to state clearly what you intend to do with the data and to seek advice on the most suitable format for your purpose. Bear in mind that the salespersons who handle most requests for data are unlikely to be familiar with the problems facing a satellite volcanologist, so you should ask to consult somebody able to give sound technical advice. Data meeting this requirement was formerly supplied by Eosat (see below) under the description of "A-format" data, as opposed to the more commonly used "P-format" or geocoded varieties. A-format data did not have forward and reverse sweeps of the scan mirror lined up, so the untreated image has alternate sets of 16 lines displaced, but this was the only format in which the radiometric quality of the data are fully to be trusted (Rothery et al. 1988a, Glaze et al. 1989a). A-format was discontinued in 1991, but data continued to be available (though only as full scenes) in raw form under the designation of "level 0" (zero) data. The analogous format supplied by European stations is called "level 1".

An alternative strategy is to buy geometrically corrected image data in which the original pixels have been resampled by nearest-neighbour resampling only (avoiding images that have been sampled by, say, cubic convolution resampling in which the DN of each output pixel is a linear or non-linear function of several neighbouring original pixels). A small fraction of pixels may have been duplicated or omitted in the nearest-neighbour resampling procedure, but, in principle, the band-to-band radiometry of each pixel should be identical to that on the original image. Unfortunately, Glaze et al. (1989) found that pixels had been replicated in different positions in TM bands 5 and 7 in nearest-neighbour resampled TM data, so that these two bands were locally displaced from one another by one pixel, thereby undermining the basis of the dual-band technique at unpredictable locations. Eosat now claim that TM data supplied with a 30 m resampling interval and produced by nearest-neighbour resampling will not exhibit replicated pixels (D. Fischel (Eosat Chief Scientist) December 1991, written communication). The only drawback to using such data is that a small proportion (less than 1%) of the original pixels will not be represented in the resampled data set, and in consequence there is a slight chance that a thermal anomaly confined to a single pixel could be missed. However, for some users the improved geometric quality of the image compared to level 0 data may outweigh this deficiency.

Initial enquiries, requests for a free computer printout of all centrally held Landsat data over any area you care to specify, and special requests for the satellite to record an image (e.g. of an on-going eruption, or by night) should be directed to:

EOSAT, c/o EROS Data Center, Sioux Falls, SD 57198, USA
Telephone (1) 605-594-6511

or

EOSAT Customer Services, 4300 Forbes Boulevard, Lanham, MD 20706, USA
Telephone (1) 800-344-9933 or (1) 301-552-0537, fax (1) 301-522-0507.

A recently established means of interrogating a global data base of Landsat and other satellite data (including AVHRR) is provided by the Global Land Information System (GLIS), which can be reached using an alphanumeric terminal or terminal emulator package on a PC as follows:

from NSI/DECNET: $SET HOST GLIS
                 USERNAME: GLIS
from INTERNET: $TELNET glis.cr.usgs.gov
                 or
                 $TELNET 152.61.192.54

Direct dial: set modem to 8 bits, no parity, 1 top bit. Dial (605)-594-688 or FTS 753-7888.

Direct dialing is not recommended from outside the USA, but the Internet connection works well. Assistance and information about GLIS can be obtained by telephoning (1) 800-252-4547 or (1) 605-594-6099.

Enquiries for Landsat images covering Europe and north Africa, may also go to:

ESA-ESRIN, Earthnet User Services, Via Galileo Galilei, 00044 Frascati, Italy
Telephone (39) 69401360, fax (39) 694180361.

From within the UK, enquiries about all varieties of satellite data can be directed to the following address (but also seek research council advice if you are so funded):

National Remote Sensing Centre Ltd (NRSCL), Delta House, Southwood Crescent,
Southwood, Farnborough, Hants GU14 0NL, UK
Telephone 0252 541464, fax 0252 375016.

A Japanese satellite JERS-1 (also known as Fuyo-1) launched in 1992 records images by means of its OPS instrument in four wavebands in the SWIR region, and has considerable potential (untested at the time of writing) for thermal measurements on lava bodies. Prices will probably be in line with those for Landsat TM. Enquiries may be directed to:

Remote Sensing Technology Center of Japan, Uni-Roppongi Building 7-15-17,
Roppongi, Minato-Ku, Tokyo 106, Japan
Telephone (81) (0)3-3403-1761, fax (81) (0)3-3403-1766, telex 02426780
RESTEC J.

Data from the SPOT satellite are expensive, subject to stringent copyright rules, and do not extend into the thermally useful part of the SWIR spectrum. However, by combining two SPOT images of the same area that were recorded when the satellite was looking to either side of its ground track it is possible to derive a stereoscopic view of the terrain. By using this sideways-looking facility, a single area can be imaged several times during SPOT's 26 day orbital repeat period. There are franchises selling SPOT data in most countries. Initial enquiries should be directed to:

SPOT Image, 16 bis, av. Edouard Belin, 31030 Toulouse Cedex, France
Telephone (33) 61 53 99 76, fax (33) 61 28 18 59

or, within the Americas, to:

SPOT Image Corporation, 1897 Preston White Drive, Reston, VA 22091, USA
Telephone (1) 703-620-2200, fax (1) 703-648-1813.

Photographs taken by astronauts on the Space Shuttle, which may have captured an eruption in progress, or provide the first post-eruption views of an area, are very cheap; currently US$3 for a 35 mm slide and US$16 for a 9×9 inch (22×22 cm) colour print. Unlike automated satellites, oblique as well as near-vertical views are taken and, because the astronauts respond to requests to photograph eruptions, and will photograph anything they think might be interesting on their own initiative, there is a better than random chance of an eruption being recorded. However, Space Shuttles are in orbit for only a few weeks during every year. Catalogues are available by mail that give the latitude and longitude co-ordinates of all Space Shuttle photographs, and there are several viewing centres in the USA (but no comprehensive collections elsewhere) where users can check whether or not listed pictures are free of cloud and fumes over critical areas. Initial queries about Space Shuttle photography should be directed to:

Media Services Branch, Still Photography Library, NASA-JSC, PO Box 58425, Mail
   Code AP3, Houston, TX 77258, USA
Telephone (1) 713-483-4231.

The Space Shuttle Earth Observations photo database containing references to over 120,000 photographs of the Earth made from space over the past three decades is accessible to the public on VAX node SSEOP. The username and password are both PHOTOS. This can be reached through the SPAN network (SET HOST SSEOP or SET HOST 9299), through the Internet network (TELNET sseop.jsc.nasa.gov or FTP sseop.jsc.nasa.gov or FTP 146.154.11.34) or via a telephone line (713-483-2500) that connects to a Xyplex terminal server. Local computer centres should be able to advise on how to access SPAN and Internet. Interactive sessions are likely to be slow for users based outside north America.

A small number of Space Shuttle photographs have been digitized into a band sequential format with a 512 byte header followed by 512×512 bytes of red, then 512×512 bytes of green then 512×512 bytes of blue. These can be downloaded and viewed through FTP or SPAN (see above) using the account name ANONYMOUS and the password GUEST.

Conclusions

The price of the simplest image-processing facilities is now less than the commerical price of a digital image from most satellites. Data costs to scientists may become less in the future as a result of de-commercialization of the Landsat programme, and can sometimes be avoided through participation in airborne and spaceborne projects as a principal investigator. Remotely sensed data offer a means of determining temperatures of both "crust" and "core" of an active lava flow, and of gathering this information at 30 m intervals along the whole length of a flow instantaneously. To achieve this depends on the good luck of there being cloud-free conditions and (for short-lived or rapidly changing events) of the satellite overpassing the target area on the appropriate date. It is important to beware of the pitfalls posed by geometrically corrected image data, if comparisons between spectral channels are to be used as the basis of temperature determinations. Despite the many difficulties, remote sensing looks set to play an increasingly valuable role in studies of lava flows and other eruptive activity in the future.

Acknowledgements

This work was carried out in part under contract to the NASA Geology Program at the Jet Propulsion Laboratory, California Insitute of Technology, Pasadena, California. Other parts were facilitated by support from the UK Overseas Development Adminis- tration and the Natural Environmental Research Council (GR9/14 and GR3/8006). The authors are grateful to Peter Mouginis-Mark, Clive Oppenheimer and Geoff Wadge for their careful reviews of an early version of this paper, and subsequent discussions, which encouraged the authors to make many much-needed improvements.

References

Abrams, M., L. Glaze, M. Sheridan 1991. Monitoring Colima volcano, Mexico, using satellite data. *Bulletin of Volcanology* **53**, 571–4.

Crisp, J. & S. M. Baloga 1990. A model for lava flows with two thermal components. *Journal of Geophysical Research* **91**, 9543–52.

Drury, S. A. 1987. *Image interpretation in geology*. London: Allen & Unwin.

Flynn, L. P. & P. J. Mouginis-Mark 1992. Cooling rate of an active Hawaiian lava flow from nighttime spectroradiometer measurements. *Geophysical Research Letters* **19**, 1783–6.

Flynn, L. P., P. J. Mouginis-Mark, J. C. Gradie, P. G. Lucey 1991. Radiative temperature measurements taken at Kupianaha lava lake: final results and implications for satellite remote sensing. *Eos, Transactions of the American Geophysical Union: AGU 1991 Fall Meeting Program & Abstracts*, 562.

Flynn, L. P., P. J. Mouginis-Mark, J. C. Gradie, P. G. Lucey 1993. Radiative temperature measurements at Kupianaha lava lake, Kilauea volcano, Hawaii. *Journal of Geophysical Research* (in press).

Francis, P. W. & D. A. Rothery 1987. Using the Landsat Thematic Mapper to detect and monitor active volcanoes. *Geology* **15**, 614–7.

Glaze, L. S. & D. C. Pieri 1989. Thermal energy losses from skylights of the New vent lava tube. *Eos, Transactions of the American Geophysical Union* **70**, 1410.

Glaze, L. S., P. W. Francis, D. A. Rothery 1989a. Measuring thermal budgets of active volcanoes by satellite remote sensing. *Nature* **338**, 144–6.

Glaze, L. S., P. W. Francis, S. Self, D. A. Rothery 1989b. The 16 September 1986 eruption of Láscar volcano, north Chile: satellite investigation. *Bulletin of Volcanology* **51**, 149–60.

Gupta, R. K. & K. V. S. Badarinath 1993. Volcano monitoring using remote sensing data. *International Journal of Remote Sensing* (in press.)

Kneizys, F. X., E. P. Shettle, L. W. Abreu, J. H. Chetwynd, G. P. Anderson, W. O. Gallery, J. E. A. Selby, S. A. Clough 1988. *Users' guide to LOWTRAN 7*. Air Force Geophysics Laboratory Environmental Research Paper 1010. Mass.: Hanscom AFB.

Lipman, P. W. & N. G. Banks 1987. *Aa flow dynamics, Mauna Loa, 1984*. US Geological Survey Professional Paper 1350, 1527–68.

McClelland, L., T. Simkin, M. Summers, E. Nielsen, T. C. Stein (eds) 1989. *Global volcanism 1975–1985*. Englewood Cliffs, New Jersey: Prentice-Hall.

Markham, B. L., J. L. Barker 1986. Landsat MSS and TM post-calibration dynamic ranges, exoatmospheric reflectances and at-satellite temperatures. *EOSAT Landsat Technical Notes* 1, 3–8. Lanham, MD: Earth Observation Satellite Company.

Matson, M., G. Stephens, J. Robinson 1987. Fire detection using data from NOAA-N satellites. *International Journal of Remote Sensing* **8**, 961–70.

Mouginis-Mark, P., S. Rowland, P. Francis, T. Freidman, H. Garbeil, J. Gradie, S. Self, L. Wilson, J. Crisp, L. Glaze, K. Jones, A. Kahle, D. Pieri, H. Zebker, A. Kreuger, L. Walter, C. Wood, W. Rose, J. Adams, R. Wolf 1991a. Analysis of active volcanoes from the Earth Observing System. *Remote Sensing of the Environment* **36**, 1–12.

Mouginis-Mark, P., S. Rowland, H. Garbeil, P. Flament 1991b. AVHRR observations of the Kupianaha eruption, Hawaii. *Eos, Transactions of the American Geophysical Union: AGU 1991 Fall Meeting Program & Abstracts*, 562.

NASA 1986. *C-130B Aircraft Experimenter's Handbook*. Washington, DC: NASA.

Oppenheimer, C. 1991a. *Volcanology from space: applications of infrared remote sensing*. PhD thesis, Open University.

Oppenheimer, C. 1991b. Lava flow cooling estimated from Landsat Thematic Mapper infrared data: the Lonquimay eruption, Chile, 1989. *Journal of Geophysical Research* **96**, 21,856–78.

Oppenheimer, C. M. M. & D. A. Rothery 1991. Infrared monitoring of volcanoes by satellite. *Geological Society of London Journal* **148**, 563–9.

Oppenheimer, C., D. Pieri, V. Carrere, M. Abrams, D. Rothery, P. Francis 1992. Volcanic thermal features observed by AVIRIS. In *JPL Publication 92–14*. Vol. 1, *Summaries of the Third Annual JPL Airborne Geoscience Workshop June 1–5, 1992*, R. O. Green (ed.), 41–3. Pasadena: Jet Propulsion Laboratory.

Oppenheimer, C., P. W. Francis, D.A. Rothery, R. W. Carlton, L. S. Glaze 1993a. Infrared image analysis of volcanic thermal features: Láscar volcano, Chile 1984-1991. *Journal of Geophysical Research* **98**, 4269–86.

Oppenheimer, C, D. A. Rothery, P. W. Francis 1993b. Thermal distributions at fumarole fields: implications for infrared remote sensing of active volcanoes. *Journal of Volcanology and Geothermal Research* **55**, 97–115.

Palluconi, F. D. & G. R. Meeks 1985. *Thermal Infrared Multispectral Scanner (TIMS): An investigator's guide to TIMS data. Jet Propulsion Laboratory Publication 85–32.*

Pieri, D. C., A. R. Gillespie, A. B. Kahle, S. M. Baloga 1984. Thermal infrared observations of lava flows during the 1984 Mauna Loa eruption, reviewed abstract. *Geological Society of America, Abstracts with Program* **16**, 623.

Pieri, D. C., L. S. Glaze, M. J. Abrams 1990. Thermal radiance observations of an active lava flow during the June 1984 eruption of Mount Etna. *Geology* **18**, 1018–22.

Realmuto, V., A. Kahle, D. Pieri, K. Hon 1992. Thermal observations of the Kupianaha flow field, Kilauea volcano, Hawaii. *Bulletin of Volcanology* (in press).

Reddy, C. S. S., A. Battacharya, S. K. Srivastav 1993. *Night time TM short wavelength infrared data analysis of Barren Island volcano, south Andaman, India.* (In preparation.)

Rothery, D. A. & C. M. M. Oppenheimer, 1991. Monitoring volcanoes using short wavelength infrared images. *Proceedings of the 5th International Colloquium on Spectral Signatures of Objects in Remote Sensing, Courchevel, France 14–18 January 1991*, ESA SP-319, 513–6.

Rothery, D. A., & C. M. M. Oppenheimer 1993. Monitoring Mount Erebus by remote sensing. In *Volcanic studies of Mount Erebus, Antarctica. Antarctic Research Series*, P. Kyle (ed.). Washington, DC: American Geophysical Union (in press).

Rothery, D. A., P. W. Francis, C. A. Wood 1988a. Volcano monitoring using short wavelength infrared data from satellites. *Journal of Geophysical Research* **93**, 7993–8008.

Rothery, D. A., P. W. Francis, C. A. Wood 1988b. Volcano monitoring by short wavelength infrared satellite remote sensing. In *Proceedings of the 6th Thematic Conference on Remote Sensing for Exploration Geology, Houston, Texas, May 16–19, 1988*, 283–92. Ann Arbor: Environmental Research Institute of Michigan.

Rothery, D. A., A. Borgia, R. W. Carlton, C. Oppenheimer 1992. The 1992 Etna lava flow imaged by Landsat TM. *International Journal of Remote Sensing* **13**, 2759–63.

Sabins, F. F. 1986. *Remote Sensing:- Principles and Interpretation*, 2nd ed. New York: Freeman.

Sawada, Y. 1989. The detection capability of explosive eruptions using GMS imagery, and the behaviour of dispersing eruption clouds. In *IAVCEI Proceedings in Volcanology*. Vol. 1, *Volcanic hazards: assessment and monitoring*, J. H. Latter (ed.), 233–45. Berlin: Springer.

Smithsonian Institution 1989. Niyamuragira. *Scientific Event Alert Nework (SEAN) Bulletin* **14(9)**, 21.

Smithsonian Institution 1992a. Erta 'Ale. *Global Volcanism Network Bulletin* **17(8)**, 5.

Smithsonian Institution 1992b. Erta 'Ale. *Global Volcanism Network Bulletin* **17(11)**, 2.

Wiesnet, D. R. & J. D'Aguanno 1982. Thermal imagery of Mount Erebus from the NOAA-6 satellite. *Antarctic Journal of the US* **17**, 32–4.

Wolfe, E. W., M. O. Garcia, D. B. Jackson, R. Y. Koyangi, C. A. Neal, A. T. Okamura 1987. *The Puu Oo eruption of Kilauea volcano, episodes 1–20, January 3, 1983, to June 8, 1984.* US Geological Survey Professional Paper 1350, 471–508.

Part III

MODELLING

Preface

A key aim in flow modelling is to describe changes in flow dimension during eruption, that is, to know how the three velocity components of a flow vary with time and position. Ideally, since there are three unknown variables, only three independent equations are needed involving these variables alone. This is not possible, however, because the derived equations must also satisfy the recognized laws of nature and these, in turn, require knowledge of additional variables to describe the physical properties of the lava and its environment.

In practice, the task of calculating lava velocities becomes that of resolving *at least* the conservation equations (for mass, momentum and energy), the equation of state (to relate density, temperature and pressure), and the constitutive equations (for describing lava rheology).

Fortunately, many of these equations can be reasonably simplified for first-order modelling (e.g. assuming constant lava density and negligible inertial forces). Unfortunately, a major complication remains. Because of cooling and crystallization, a lava may change from a viscous fluid to a brittle solid during emplacement. As a result, it is extremely difficult to characterize the resistance of a lava to motion.

Such difficulty is reflected by the first three chapters in this part, all of which consider flow resistance: Chapter 9 reviews the factors controlling lava rheology and illustrates the use of increasingly complex flow models; Chapter 10 emphasizes the role of lava crusts; and Chapter 11 describes how the strong temperature dependence of lava rheology in the crystallization range may lead to extreme thermal instability. Chapter 12 takes a radically different view, avoiding explicit use of the governing equations and modelling flow field growth with cellular automata techniques. Finally, Chapter 13 describes the fundamental assumptions in fluid mechanics and, though not oriented specifically to lava flows, it provides a background for setting the previous chapters in context.

CHAPTER NINE

Modelling the rheology and cooling of lava flows

Michele Dragoni

Abstract

The rheological properties of lavas are of major importance in determining the dynamics of lava flows. Such properties reflect the inherent structures of molten silicates and the fact that lava is a multiphase system. Below its liquidus temperature, lava behaves as a non-Newtonian fluid and is commonly described as a pseudoplastic fluid. A simpler model which is often used is the Bingham fluid, characterized by a yield stress and a plastic viscosity. These rheological parameters depend on the composition of lava (basic or acidic) and are strongly temperature-dependent. Many other factors, including the presence of water, gas loss, crystal content and degree of polymerization, control the rheology of lava. Surface effects, such as surface tension, the strength of solid crust and ground erosion by melting, also influence flow morphology and dynamics. The roles of these factors are examined on the basis of existing studies. To better understand lava flow dynamics such factors must be investigated systematically, both theoretically and in the laboratory, supported by more accurate measurements in the field.

Notation

A	constant in Equation 9.4	r_c	critical radius of curvature
a	parameter in Equation 9.5	T	absolute temperature
b	parameter in Equation 9.6	T_l	liquidus temperature
c_p	specific heat	T_s	solidus temperature
E	surface emissivity	t	time
e	Neper's number	v	flow velocity
f	a function	x, y, z	Cartesian co-ordinates
g	acceleration of gravity	a	slope angle
h	flow thickness	ε	strain
h_c	crust thickness	$\dot{\varepsilon}$	strain rate
h_p	plug thickness	η	viscosity
I	rheological index	η_0	viscosity of the liquid fraction
i, j, k, l	tensor indices	η_1	viscosity at $T = T_l$
K_{ijkl}	viscosity tensor	η_p	plastic viscosity
n	power law exponent	κ	thermal conductivity
p	pressure	ρ	mass density
p_0	atmospheric pressure	Σ	Stefan constant
p_c	pressure of curvature	σ	stress
Q	activation energy	σ_y	yield stress
q	mass flow rate per unit width	σ_y'	yield stress at $T = T_l$
R	gas constant	τ	surface tension
R_v	volumetric ratio	φ	crystal concentration
r	radius of curvature	χ	thermal diffusivity

Introduction

Rheology is the study of the deformation and flow properties of bodies under an applied stress. The rheological properties of lava are of major importance in determining the dynamics of lava flows. Lava flows show great variations in size, shape and surface features, but in all cases they have a characteristic behaviour. It is observed that lava flows construct their levées by themselves and come to rest on a slope when the supply of fresh lava ceases. Flow fronts are often high and steep, although unconfined by topographic features. In spite of the fluid-like aspect of flowing lava, objects thrown on to the flow surface do not always sink nor remain buoyant, even if their density is greater than the density of lava, but can be supported as if they were on a rigid surface. It is even possible to walk across many flows, even when a solid crust has not yet formed at their surface (Ch. 3). Hulme (1974) has argued that solidification of lava due to cooling can limit the motion of a flow front to a certain distance from the effusion vent, thus limiting the length of the flow, but cannot prevent either lateral or downhill movement at any other point along the flow. Therefore the observed behaviour of lava flows must be a consequence of the rheological properties of lava at the high temperatures at which effusion takes place.

The rheological properties of lavas reflect the inherent structures of molten silicates (Hess 1980, Spera 1980). From a thermodynamic point of view, lava is a multiphase system, made of solid, liquid and gaseous components. The transition from liquid to solid lava occurs within a temperature interval (the melting range) delimited by the liquidus and the solidus temperatures. The liquidus is the temperature T_1 at which fusion is completed during heating. The solidus is the temperature T_s at which solidification is completed during cooling. At a constant load pressure, the liquidus and solidus depend on the chemical composition of a lava. Since lavas erupt at temperatures close to the liquidus and cool during emplacement, the temperature T of an active flow usually lies in the melting range, i.e. $T_s < T < T_1$. For most magmas, the difference between the solidus and liquidus is 150–200°C.

The rheology of lava

Newtonian and non-Newtonian fluids

The rheological behaviour of a viscous fluid is generally expressed by a "constitutive equation", that is, a relation between viscous stress and strain rate $\dot{\varepsilon}$ (see the Appendix). If the components of σ are linear functions of the components of $\dot{\varepsilon}$, the fluid is called Newtonian. If the fluid is isotropic and incompressible and a one-dimensional flow is considered, the constitutive equation can be written as

$$\sigma = \eta\dot{\varepsilon}, \tag{9.1}$$

where η is the viscosity.

In a graph of σ versus $\dot{\varepsilon}$, the flow curve of a Newtonian fluid is a straight line through the origin (Fig. 9.1). All the fluids for which the flow curve is not linear through the origin, at a given temperature and pressure, are said to be non-Newtonian (Skelland 1967, Böhme 1987). We have particular non-Newtonian fluids if the constitutive equation can be written as a power law:

$$\sigma = \sigma_y + \eta_p\dot{\varepsilon}^n, \tag{9.2}$$

where σ_y, η_p and n are constants.

If $\sigma_y = 0$ and $n < 1$, the fluid is called pseudoplastic. The majority of non-Newtonian fluids are pseudoplastic. A typical flow curve for these fluids is shown in Figure 9.1. When the flow curve is non-linear, as in this case, a unique value of viscosity, as defined for Newtonian fluids, does not exist. If viscosity estimates are made on a non-Newtonian fluid, as if it were Newtonian, different viscosity values are obtained, according to the strain rates involved: what is measured is the apparent viscosity. In a pseudoplastic fluid, the apparent viscosity decreases with increasing strain rate.

The opposite behaviour is found in dilatant fluids (Fig. 9.1), where the apparent viscosity increases with increasing shear rate. These fluids can be again described by

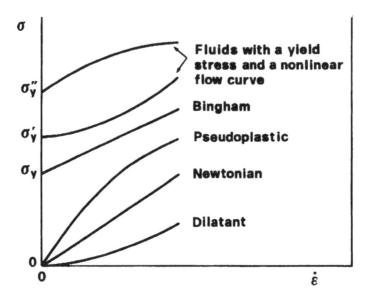

Figure 9.1 Flow curves for various types of fluids, showing variation in strain rate $\dot{\varepsilon}$ with shear stress σ.

Equation 9.2 with $\sigma_y = 0$ and $n > 1$. The microscopic mechanisms which may account for the rheological behaviour of pseudoplastic and dilatant fluids are discussed by Skelland (1967).

A further possibility is that of fluids having a yield stress, i.e. $\sigma_y \neq 0$ in Equation 9.2: σ_y is the minimum shear stress required to initiate permanent deformation. The existence of a yield stress is usually ascribed to an internal structure which is capable of preventing movement for values of shear stress less than the yield value. When $\sigma > \sigma_y$, the internal structure collapses, allowing shearing movement to occur. The internal structure is considered to be reformed instantaneously when σ becomes less than σ_y.

The simplest fluid with a yield stress is the Bingham fluid, which is characterized by two parameters, yield stress σ_y and plastic viscosity η_p. Its flow curve is linear, as for a Newtonian fluid, but intercepts the stress axis at a value σ_y. Its constitutive equation is given by Equation 9.2 with $n = 1$. The Bingham fluid can be considered as an approximation to a pseudoplastic fluid characterized by a high viscosity at infinitesimally small shear rate.

Laboratory experiments show that, at super-liquidus temperatures, common igneous melts behave as Newtonian fluids. Below their liquidus, lavas are instead non-Newtonian (Shaw et al. 1968, Sparks et al. 1977, Pinkerton & Sparks 1978). Many complex fluids, such as suspensions and emulsions, are non-Newtonian. The main reasons for this change in lava behaviour below the liquidus are the presence of dispersed crystals and gas bubbles, as well as some polymerization in the silicate melt. Non-Newtonian behaviour has many implications, both in understanding flow dynamics and in interpreting flow morphology.

The Bingham fluid

Robson (1967) first proposed that lava has an approximately Bingham (or plastic) rheological behaviour, in order to explain the relation found by Walker (1967) between flow thickness and ground slope among Etnean lavas. The existence of a yield stress in basaltic lava was assumed by Shaw et al. (1968) after measuring the rheological properties of a lava lake. The actual rheological behaviour of lava includes time- and shear rate-dependent effects which are not considered by the two-parameter Bingham model, but the assumption of Bingham rheology has proven to be useful for the interpretation of field observations (Hulme & Fielder 1977, Borgia et al. 1983, Wilson & Head 1983, Fink & Zimbelman 1989, Borgia & Linneman 1990) and has been extensively used in flow modelling (Hulme 1974, 1982, Park & Iversen 1984, Dragoni et al. 1986, Dragoni 1989). Examples of fluids with a yield stress may be found in many materials, like oil well drilling muds, sand in water, coal, cement, margarine, greases, toothpaste, soap slurries and other. Like other rheological models, the Bingham fluid can be represented by a mechanical model. In addition to the dashpot, representing viscous behaviour, a frictional contact, called a Saint-Venant element, is introduced to represent the yield stress (Fig. 9.2). The contact is made between a heavy block sliding on a rough surface, the block moving only when a frictional stress threshold is overcome. A more elaborate version of the Bingham substance has a spring in front of the block, taking into account an initial elastic response (Jaeger & Cook 1976).

Once a rheological model (e.g. the Bingham fluid) has been adopted as an acceptable approximation for lava flows, it is necessary to establish which values the rheological

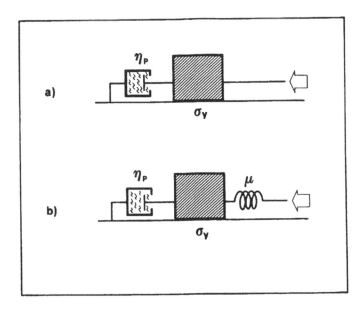

Figure 9.2 Mechanical models of the Bingham fluid, characterized by plastic visosity η_p and yield stress σ_y. Model (b) has an initial elastic response with rigidity μ (represented by the spring).

parameters (plastic viscosity and yield stress) may assume in different situations. From a theoretical point of view, one should determine the main factors affecting the rheological parameters and then derive equations for calculating such parameters from the chemical composition of lava, temperature, pressure and so on. At the same time, laboratory and field measurements should provide estimates of such parameters and of other quantities appearing in the model equations.

Unfortunately, only a few measurements have been made concerning the plastic viscosity and yield stress of lava (Shaw et al. 1968, Shaw 1969, Pinkerton & Sparks 1978, McBirney & Murase 1984). Viscosities of silicate melts can be estimated from their chemical composition, temperature and pressure using empirical formulae (Bottinga & Weill 1972, Shaw 1972, Murase & McBirney 1973, Scarfe 1973, Kushiro 1976), but the validity of these estimates has been established only at super-liquidus temperatures, where Newtonian behaviour is found. There is also an extensive literature concerning the rheology of hot subsolidus silicate materials (e.g. Goetze & Brace 1972, Ashby & Verrall 1978), but much less is known about silicates in the melting range, where non-Newtonian behaviour is observed. The most striking feature of available data is the wide variation of the rheological parameters, which are strongly dependent on several factors, including silica and water content, temperature, crystallinity, vesiculation and polymerization. The effect of pressure will not be discussed, since we are considering subaerial lava flows under constant, atmospheric pressure.

Factors controlling rheology

Silica content

The rheological parameters of a lava are strongly dependent on its silica content. Some measurements of the apparent Newtonian viscosities of lavas and melts, ranging in chemical composition from rhyolite to basalt, are summarized in Figure 9.3. At a given temperature and strain rate, acidic lavas have a much higher viscosity than basic lavas: the difference may be as large as five orders of magnitude. This has a significant effect on the dynamics of lava flows of different compositions. For instance, at $T = 1,200°C$, a tholeiitic basalt may have $\eta = 10^2$ Pa s, while a rhyolite has $\eta = 10^6$ Pa s (1 Pa s = 10 P). For comparison, the viscosity of water at room temperature is about 10^{-3} Pa s. Although basaltic melts at superliquidus temperatures behave essentially as Newtonian fluids, this is not true for single-phase (no crystals or vapour bubbles) melts more than about 65 wt% silica-rich (Spera et al. 1982). Viscometric data obtained on a rhyolitic composition show a definite deviation from Newtonian behaviour and a pseudoplastic flow curve with $n = 1.25$.

A graph of yield stress versus silica content was presented by Hulme (1974) and is shown in Figure 9.4. A trend of increasing yield stress with silica content may be seen, similar to the variation of viscosity. However these estimates of yield stress, determined from the morphology of halted lava flows, do not provide a way of separating the effects of silica content from those of temperature and crystallinity. The silica content also

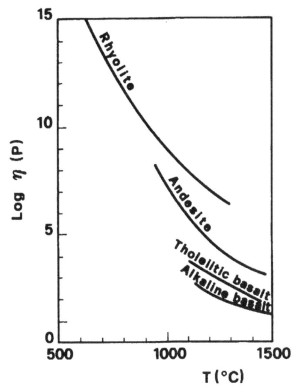

Figure 9.3 Apparent viscosity η as a function of temperature T for some common igneous rocks. (After McBirney & Murase 1984.)

affects the surface tension of lava (McBirney & Murase 1970), lavas with lower silica content having higher surface tensions. The surface tension of basaltic lava at its liquidus temperature is about 0.35 N m^{-1} (350 dyn cm^{-1}). For comparison, the surface tension of water at 20°C is about 0.07 N m^{-1}.

Because of its influence on magma rheology, the silica content can be used to classify the rheological behaviour of lavas (Bottinga & Weill 1972). For example, Scarfe (1973) used as a rheological index (I) the ratio of the molecular percentage of non-bridging oxygens to network-forming cations:

$$I = O/(Si + Al + P). \tag{9.3}$$

Basalts and ultrabasic compositions have the highest I-values, ranging from 2.2 to 3.5, and the lowest viscosities at a given temperature.

Water content
Dissolved water significantly reduces the viscosity of lava. This effect is shown in Figure 9.5 for basaltic compositions at various temperatures and in Figure 9.6 for melts of obsidian glasses. The effect of water is stronger for silica-rich liquids where, at constant temperature, 1 wt% H_2O lowers the viscosity by nearly an order of magnitude.

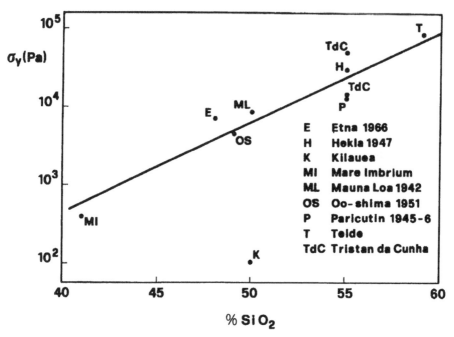

Figure 9.4 Yield stress σ_y of lava versus silica content. (After Hulme 1974.)

In liquids of lower silica content, the effect is less pronounced. The reason is probably that water reduces the polymerization of Si–O bonds and so is more effective in silica-rich liquids (McBirney & Murase 1984). Moreover, it can be seen from the graphs that the effect of water in reducing viscosity is greater at lower temperatures. Data on the effect of water were collected by Scarfe (1973) and Khitarov & Lebedev (1978) for basaltic lavas and by Shaw (1963), Burnham (1963) and Carron (1969) for rhyolitic lavas.

Temperature
As shown in Figures 9.3, 5 & 6, both yield stress and plastic viscosity are strongly temperature-dependent (Johnson & Pollard 1973, Pinkerton & Sparks 1978, Spera et al. 1982). Above the liquidus the dependence of viscosity on temperature is given by the Arrhenius equation:

$$\eta(T) = A \, e^{Q/RT}, \tag{9.4}$$

where A is a constant, Q is the activation energy of viscous flow and R is the gas constant. Both A and Q depend on composition. Activation energies for viscous flow at liquidus temperatures ($1{,}200°C$) decrease with increasing I value (Scarfe 1973), suggesting that ultrabasic melts are the least polymerized of natural melts.

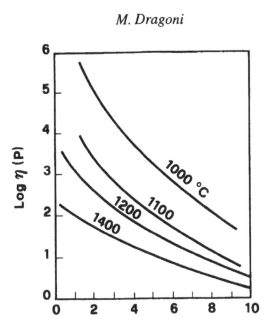

Figure 9.5 The effect of water on the apparent viscosity η of basaltic liquids at various temperatures. (After Khitarov & Lebedev 1978.)

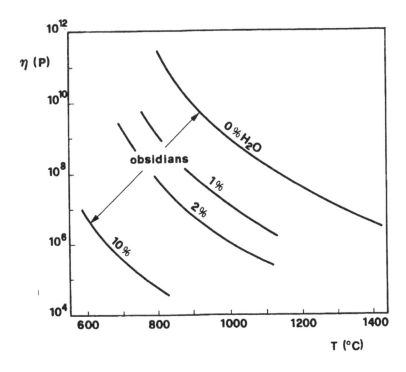

Figure 9.6 Apparent viscosity η as a function of temperature T for melts of obsidians: curves for different water contents are shown. (After Johnson & Pollard 1973.)

243

At subliquidus temperatures, viscosities of crystal-free liquids can be calculated from empirical formulae such as (Spera et al. 1982)

$$\eta(T) = \eta_1 \, e^{a(T_1 - T)} \quad \text{(for } T < T_1), \tag{9.5}$$

where η_1 is the viscosity at the liquidus and a is a parameter measuring the sensitivity of η on temperature. The parameter a is a complicated function of lava composition, crystal content, cooling rate and strain rate. Experimental data indicate that a generally lies between 0.02 and 0.10 K^{-1} (Shaw et al. 1968, Shaw 1969, Murrell & Chakravarty 1973, Pinkerton & Sparks 1978, Arzi 1978). As a consequence of composition and temperature variations, the viscosity of lava varies by at least 10 orders of magnitude in the crystallization interval (Sakuma 1953, Shaw et al. 1968, McBirney & Murase 1984). It is evident that rheology is dominated by these extremely large variations of the apparent viscosity in the melting range.

There are less experimental data for yield stress. Available measurements of σ_y (Shaw et al. 1968, Pinkerton & Sparks 1978, McBirney & Noyes 1979) suggest that it is an exponentially decreasing function of temperature in the melting range. Accordingly, an expression for σ_y can be written in analogy to Equation 9.5 (Chester et al. 1985, Dragoni 1989):

$$\sigma_y(T) = \sigma_y{}_1 \, e^{b(T_1 - T)} \quad \text{(for } T < T_1), \tag{9.6}$$

where $\sigma_y{}_1$ is the yield stress at the liquidus and b is a parameter measuring the sensitivity of σ_y on temperature. No unambiguous evidence exists that crystal-free liquids above T_1 have measurable yield stresses (McBirney & Murase 1984). Data from Murase & McBirney (1973) and Williams & McBirney (1979) seem to show, however, that yield stress tends to a constant after the lava has cooled more than about 100°C below the liquidus. As for surface tension, laboratory experiments have shown that it increases with increasing temperature for all lavas (McBirney & Murase 1970).

Crystal content

Most lavas have temperatures below their liquidus and contain different proportions of crystals in suspension. The crystallization of a magma body has been studied by Kirkpatrick (1976), Brandeis et al. (1984) and others. The viscosity η of a suspension is greater than that of the liquid alone. Considering a Newtonian liquid, η can be estimated for small crystal concentrations using the Einstein–Roscoe equation (Shaw 1969, Landau & Lifshitz 1971)

$$\eta = \eta_0(1 - R_v\varphi)^{-5/2} \quad \text{(for } \varphi \ll 1) \tag{9.7}$$

where η_0 is the viscosity of the liquid alone, φ is the volume fraction of the suspended solids and R_v is the volumetric ratio of solids at maximum packing ($R_v = 1.35$ for spheres of uniform size). Other formulae (Sherman 1968) consider the mean diameter of suspended solids. See Chester et al. (1985) for a detailed bibliography. The observed effect of phenocrysts on the apparent viscosity for Hawaiian lavas is shown in Figure 9.7, compared with Equation 9.7.

There is a great difference between measured viscosities of crystal-rich lavas and those predicted by Equation 9.7. The discrepancy is partly due to the sizes and high concentrations of crystals, both of which may exceed the range of values for which Equation 9.7 was derived (McBirney & Murase 1984). Moreover, the composition of the liquid phase is not independent of φ, but changes with progressive crystallization. Accordingly, η_0 changes and the magnitude of this effect may be greater than that of the suspended crystals alone. Secondly, in the experiments the temperature of the melt decreases with increasing crystallization and this again affects η_0 (Shaw 1969).

The concentration of crystals will also affect the yield stress σ_y. In the melting range, where the liquid is saturated with one or more crystalline phases, the bulk crystal/liquid assemblages have a yield stress which increases with increasing crystallinity (McBirney & Murase 1984), even if no simple relation has been found between σ_y and φ.

In summary, we may expect that the rheological properties will not be greatly affected by the crystal content if this is less than about 25% of the total volume: in this case, the rheology will be primarily controlled by composition and temperature of the melt. Crystals may markedly influence the rheology only when the concentration exceeds 25% of the volume (Johnson & Pollard 1973, Ryerson et al. 1988). Since crystallization changes the rheological properties of lava, it also contributes to the morphology of the flow surface and is one factor determining whether it has a pahoehoe or aa crust (Kilburn 1990).

Vesiculation and degassing

Degassing has a significant effect on the rheology of lava (Sparks 1977, Sparks & Pinkerton 1978, Lipman et al. 1985, Westrich et al. 1988). This effect is most important

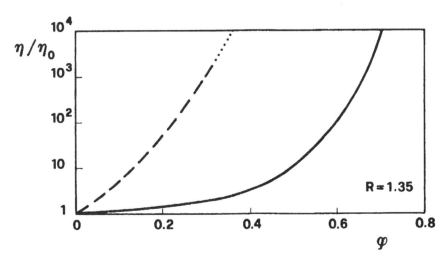

Figure 9.7 Apparent viscosity η of a liquid/crystal suspension relative to viscosity η_0 of the liquid alone, as a function of the volume fraction φ of crystals, according to the theoretical relation 9.10 (solid curve), compared with data from Hawaiian lava (dashed curve). (After Shaw 1969 and Johnson & Pollard 1973.)

during the early stages of flow and may cause drastic rheological changes as magma approaches the Earth's surface.

Before eruption, most magmas lose volatiles, such as water, carbon dioxide, chlorine, fluorine and sulphur. With the important exception of carbon dioxide, the loss of most common volatile components increases the viscosity of silicate melts (Shaw 1963, Scarfe 1973). As a result, the rheology of lava flows does not reflect the rheology of magma before eruption. The viscosity change is dramatic (several orders of magnitude) for rhyolitic magma and much less for basaltic magmas. There are two main processes by which magma can lose volatiles (Westrich et al. 1988): isothermal decompression, wherein degassing occurs because of decreasing solubility of volatiles in magma with decreasing pressure, and isobaric crystallization (or second boiling), wherein volatiles are forced out as the magma crystallizes. The two processes can be considered to occur in sequence.

A consequence of gas loss is a sudden undercooling, triggering a pervasive crystal growth (Sparks & Pinkerton 1978, Lipman et al. 1985): this leads to a rapid increase in viscosity and the development of a high yield stress. Gas loss may thus be the main contributor to the non-Newtonian behaviour of lava flows, as well as a major cause of downflow changes in rheology. Sparks & Pinkerton (1978) have suggested that this mechanism is far more effective in changing lava rheology than is atmospheric cooling, which affects lavas only slowly because of their low thermal conductivity. In addition, crystallization changes the composition of the residual melt which, in most cases, becomes more siliceous and hence increases in both viscosity and yield stress.

Polymerization

In the most common silicate minerals, the Si^{4+} ion occurs in tetrahedral co-ordination with oxygen over a wide range of temperatures and pressures. Each oxygen surrounding a central Si^{4+} ion has the potential of bonding to another Si^{4+} ion, thus forming silicate structures. Polymerization is the linking of silicate tetrahedra by bridging oxygens.

In a silicate melt, there is no fixed lattice and the positions of the ionic groups change continuously due to thermal motions. However the arrangement of atoms is not completely random. X-rays and neutron diffraction studies (e.g. Mozzi & Warren 1969, Konnert & Karle 1972) demonstrate that the local structure around a central Si^{4+} ion is similar to that existing in the crystalline state: silicate melts are solutions containing a wide distribution of silicate ionic groups of various sizes and shapes (Hess 1980). The average structure of a melt is a unique function of composition, temperature and pressure. As the silica content of a metal-rich liquid increases, the SiO_4 units polymerize and progressively increase in size and complexity.

Although polymerization affects lava rheology, the nature and importance of its influence is not fully understood. A basic question is whether part of the non-Newtonian behaviour of lava below the liquidus is inherent in the polymerized liquid or is only a bulk property of the crystal/liquid suspension. For example, the large yield stresses measured in complex alumina-silicates containing only a few per cent of small crystals suggest an extensive polymerization (McBirney & Murase 1984). At larger crystal concentrations, however, it may be crystal–crystal interactions which dominantly influence the rheology of a lava.

Surface effects

In a lava flow, surface effects add to rheology in controlling flow morphology and dynamics. Surface tension enters the equation for the shape of the flow surface only at a very small scale (e.g. Huppert 1982), while cooling produces a thermal boundary layer at the surface of the flow, both the free surface and the interface with the ground.

Surface tension

Normally, the surface of a liquid is kept smooth by the action of surface tension: if a plane liquid surface is distorted, surface tension acts to restore the surface to a plane. An interesting effect results for a Bingham fluid: since lava possesses a yield stress σ_y, the pressure of curvature p_c must overcome σ_y before any smoothing of the surface takes place. For distortions with radius of curvature r,

$$p_c = 2\tau/r, \tag{9.8}$$

where τ is the surface tension. Hulme (1974) suggested that for any lava there is a certain radius of curvature for distortions, above which they are not removed by surface tension. Hence surface tension removes small-scale perturbations of the surface, but not large-scale ones. The critical radius of curvature is

$$r_c = 2\tau/\sigma_y, \tag{9.9}$$

Lavas of low silica content at high temperature have the lowest yield stresses and highest values of surface tension. Both of these factors lead to a high value of r_c and so the surfaces of these lavas remain smooth up to large scales. Cooling such lavas increases their yield stress and lowers their surface tension: both of these changes cause r_c to become smaller and so r_c is very sensitive to temperature. When r_c falls to small values, a lava surface may become spiny. Therefore, the inability of the pressure of curvature to overcome the yield stress may have a role in the transition from pahoehoe to aa surfaces.

Tensile strength

As lava leaves the eruption vent and flows downhill, a thin solid crust starts developing at the surface of the flow. Its thickness h_c increases with increasing time t approximately as (e.g. Ozisik 1968)

$$h_c = (8\chi t)^{1/2}, \tag{9.10}$$

where χ is the thermal diffusivity of lava. Hulme (1974) argued that the tensile strength of chilled lava cannot be responsible for the formation of flow levées. However, the mechanical properties of the crust may become important in the evolution of a lava flow.

In the proximity of the eruption vent, the crust is continually being distorted and broken but, in the later stages of flow, cooling produces marked rheological gradients in the flow margins, the tensile strength of which may become sufficient to stop the flow front (Wadge 1978). On the other hand, the hot lava in the interior of a mature flow may find a weakness in the chilled front, draining out to form a new bocca and a new flow (e.g. Pinkerton & Sparks 1976, Kilburn & Lopes 1988).

Erosion by melting

As lava advances, it loses heat to the ground and a thermal boundary layer develops close to the bed and thickens downstream. Since the ground material is commonly made of solidified lava, its solidus temperature is below the temperature of flowing lava. If a flow remains active for a long enough period, it may be able to raise the temperature of the ground to the solidus and melt its own bed. It follows that the thickness of a flow may be greater than is obtained by measuring the height of levées from outside.

Hulme (1982) studied an approximate model for melting, finding that a lava flow may begin to melt its bed at a given point after a time which is about two orders of magnitude greater than the time originally taken to travel from the vent to that point. The model also implied typical erosion rates of the order of 1 m per month.

Flow modelling

Theoretical modelling of lava flows is still at an early stage. Some analytical models have been proposed combining dynamic, rheological and thermal equations. The typical behaviour of a lava flow can be grossly reproduced by two-dimensional models of a Bingham fluid flowing down a constant slope. Such models are, of course, simplifications of actual conditions. They consider steady, laminar flows, so that transient phenomena, caused by rapid changes in model parameters, cannot be described. These models only represent the flow well behind the front, where levées can be taken as fixed, since they are cooler and have a higher yield stress than fresh lava flowing between them. Processes at the front, including levée formation, the change in shape of the flow and the choice of flow path, are not taken into account.

Even if analytical models cannot reproduce the details of any specific flow, they are important because they allow us to understand the complex relationships between the numerous factors influencing the rheology and dynamics of lava flows. Hence the results of analytical models can serve as a guide for more elaborate, numerical flow models (e.g. Crisci et al. 1986, Ishihara et al. 1990, Ch. 12 this volume).

Some relatively simple flow models are illustrated in this section. They are two-dimensional models, describing an infinitely wide fluid layer, flowing on an inclined plane. The steady flow of an incompressible Bingham liquid is considered in the following cases: (a) isothermal; (b) non-isothermal, with constant temperature in any given flow section and temperature decrease in the downstream direction; and (c)

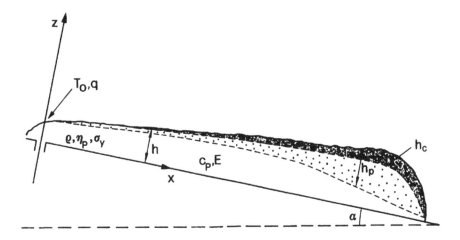

Figure 9.8 Sketch of a two-dimensional flow model: a downslope section of lava flow is shown with the co-ordinate system and the relevant parameters. See the notation list for an explanation of the symbols.

non-isothermal, with temperature decreases both downstream and towards the top of the flow.

Dynamics of a Bingham fluid

A typical two-dimensional model considers an infinitely wide fluid layer, flowing on an inclined plane. Assume that the flow occurs in the x direction and that z is perpendicular to the inclined plane (Fig. 9.8). No changes in the flow are considered in the y direction. For an isotropic and incompressible Newtonian fluid, the equation of motion is (e.g. Landau & Lifshitz 1971)

$$\rho\left(\frac{\partial v_i}{\partial t} + v_j\frac{\partial v_i}{\partial x_i}\right) = -\frac{\partial p}{\partial x_i} + \frac{\partial}{\partial x_j}\left[\eta\left(\frac{\partial v_i}{\partial x_j} + \frac{\partial v_j}{\partial x_i}\right)\right] + \rho g_i \qquad (9.11)$$

where ρ is the density, v_i is the velocity, p is the pressure and g_i is the acceleration due to gravity (the subscripts i and j correspond to values in the x and z directions). Summation over repeated indices i and j is assumed.

On the left-hand side of Equation 9.11, the time derivative of v_i vanishes if we consider a steady flow, and the non-linear term $v_j\,\partial v_i/\partial x_j$ can be neglected since we are dealing with a low-velocity flow. We are mainly interested in the x component of Equation 9.11, which becomes (neglecting pressure gradients)

$$2\frac{\partial}{\partial x}\left(\eta\frac{\partial v_x}{\partial x}\right) + \frac{\partial}{\partial z}\left[\eta\left(\frac{\partial v_x}{\partial z} + \frac{\partial v_z}{\partial x}\right)\right] + \rho g\sin a = 0, \qquad (9.12)$$

where a is the slope angle. This equation can be further simplified if one assumes that: (a) the change of velocity along the flow is much slower than the change of velocity

Modelling the rheology and cooling of lava flows

with depth, so that $\partial v_x/\partial x$ can be ignored with respect to $\partial v_x/\partial z$; and (b) viscosity is constant with depth. Equation 9.12 then reduces to the lubrication theory approximation (Batchelor 1967, Huppert 1982, Emerman & Turcotte 1983):

$$\eta\frac{\partial^2 v_x}{\partial z^2} + \rho g \sin a = 0. \tag{9.13}$$

Boundary conditions for Equations 9.11–13 are (a) $v_x = 0$ at $z = 0$ and (b) vanishing shear traction and $p = p_0$ (where p_0 is the atmospheric pressure) at the free surface of the flow, $z = h$.

The Newtonian equations 9.11–13 can be employed for a Bingham fluid owing to the fact that the Bingham flow curve is still linear once the yield stress is overcome. To be used for a Bingham fluid, the plastic viscosity η_p replaces η in Equations 9.11–13 and an extra condition is applied that there is no shear deformation if $\sigma < \sigma_y$. On these grounds, analytical expressions can be derived for the flow velocity v_x of a Bingham fluid with constant viscosity and yield stress (Skelland 1967, Johnson 1970, Johnson & Pollard 1973, Dragoni et al. 1986):

$$v_x(x,z) = (1/2\eta)\rho g \sin a\, z(2h - z) - 2\sigma_y z \quad (0 \leqslant z \leqslant h - h_p), \tag{9.14}$$

$$v_x(x,z) = (1/2\eta)\rho g \sin a\, h^2[1 - (\sigma_y/\sigma_b)]^2 \quad (h - h_p \leqslant z \leqslant h),$$

where

$$\sigma_b = \rho g h \sin a \tag{9.15}$$

is the shear stress at the base of the flow.

Role of density
Density is a fundamental parameter of the equation of motion and is dependent on both lava composition and temperature. There have been many determinations of the densities of silicate melts and it is possible to predict the density of a melt of given composition and temperature with an accuracy better than 1% (Nelson & Carmichael 1979, Bottinga et al. 1982). These studies show that small changes in composition may have relatively large effects on density. In general, the effects on density of fractionating minerals are much greater than are those of temperature (Huppert & Sparks 1984). However, density values lie almost always between 2.5×10^3 and 3×10^3 kg m^{-3}. Density measurements on molten lava samples were made by Swanson (1973). Flow models (e.g. Dragoni 1989) show that the effect of density changes on the dynamics is much smaller than are those of rheological parameters.

Flow cross section and levées
In a Bingham fluid, there may be regions where the shear stress is lower than the yield value. In such regions lava will not "flow", i.e. it will not undergo shear deformation.

If the unsheared lava is in contact with the ground, it will not move at all: this occurs along the sides of a flow, leading to the formation of stationary levées.

Hulme (1974) showed that a flow ceases to spread laterally when it achieves a certain width and this width is maintained as long as conditions do not change. He also gave an approximate expression for the width of the levées. From observations on Mount Etna, Sparks et al. (1976) described various types of levées and proposed several mechanisms by which the initial cross-sectional shape of a flow may be modified.

The aspect ratio is defined as the ratio of flow height to flow width. It has been observed in the field that lavas with low values of yield stress, such as basalts, produce flows of low aspect ratio (<1), while more acidic lavas produce flows with higher aspect ratios, but usually always less than 1 (Walker 1973).

The plug
A region where shear stress is lower than the yield value also forms at the top of the flow. This region, which is called the plug, is surrounded by flowing lava, since shear stress increases towards the interior of the flow. The plug is thus carried passively by the underlying lava, as if it were solid. The plug, however, is subject to compressive stresses and these may cause the formation of fold structures which are commonly observed on the surface of cooled flows (Dragoni et al. 1992). In a two-dimensional model, considering a fluid layer with vertically uniform viscosity and yield stress, the thickness of the plug is given by (Hulme 1974, Dragoni et al. 1986)

$$h_p = \sigma_y/(\rho g \sin a). \tag{9.16}$$

The plug, which is a characteristic feature of Bingham fluids, is very thin in the proximity of the vent, where lava has a temperature close to the liquidus and behaves like a Newtonian fluid, but may become a significant fraction of the flow thickness in distal parts of the flow. As a rheological boundary layer, the plug is quite distinct from the crust, which is a thermal boundary layer. The plug exists even in a vertically isothermal flow, while a crust develops only if there is a vertical temperature variation and the top of the flow is below the solidus temperature. The thickness of the crust is usually much smaller than that of the plug (Hulme 1982, Dragoni & Pondrelli 1991). A description of the plug, as observed in the field, is given by Borgia et al. (1983).

Isothermal flow

The temperature gradient along a lava flow is usually small: once the crust has formed, the heat loss is minimal due to the insulating effect of the crust itself. Further heat is also generated by viscous dissipation in the flow, but is concentrated along the flow base. Most heat loss occurs in the frontal zone, where lava becomes directly exposed to the air and flows over a cold surface. Cooling is the main factor that limits the downslope flow of lava (Wadge 1978), but solidification of lava is a slow process. Consequently, isothermal models of lava flows (model 1) have been considered as a first approximation

(Hulme 1974, 1982, Dragoni et al. 1986). The value of such models is that they illustrate how a Bingham rheology may affect flow dynamics, independently of other factors.

In particular, such models have shown that, for specific ranges of model parameters, such as yield stress and flow rate, a Bingham fluid behaves much as a Newtonian fluid, at least as far as flow thickness and velocity are concerned. At low values of yield stress ($\sigma_y < 10^2$ Pa), the flow is essentially in a Newtonian regime. As σ_y increases, the flow enters the Bingham regime: h increases rapidly, while v_x decreases sharply (Dragoni et al. 1986). For the same flow rate, therefore, the Bingham regime is generally characterized by greater thicknesses and lower velocities than the Newtonian regime.

Non-isothermal flow

Cooling of a lava flow must be taken into account in any realistic model. Among the various processes of heat loss, conduction to the atmosphere is negligible. Convection in the atmosphere is responsible for some heat loss, but numerical estimates indicate that its contribution is much smaller than that due to black body radiation (Murase & McBirney 1970, Daneš 1972). As to the effect of conduction to the ground, it has been shown (Hulme 1982) that flow lengths are generally much less than the distances at which this contribution to cooling can significantly affect the flow. Moreover, most of the heat generated by dissipation will be transferred to the ground. Heat loss by radiation is therefore the dominant process of cooling. The heat flow equation for a homogeneous viscous fluid can be written as (Landau & Lifshitz 1971)

$$\rho c_p \left(\frac{\partial T}{\partial t} + v_i \frac{\partial T}{\partial x_i} \right) = \frac{\partial}{\partial x_i} \left(\kappa \frac{\partial T}{\partial x_i} \right) + \sigma_{ij} \frac{\partial v_i}{\partial x_j}, \tag{9.17}$$

where c_p is the specific heat at constant pressure and κ is the thermal conductivity. The left-hand side of Equation 9.17 is the total derivative of temperature multiplied by ρc_p, while the right hand side represents the heat exchanged by the unit volume of the fluid: the heat received by conduction plus the heat generated by viscous dissipation.

If the temperature is not uniform, the fluid cannot be considered as incompressible, since density may vary as a consequence of temperature variations. However, variations in ρ, as well as in κ and c_p, are neglected to a first approximation. On the contrary, the rheological parameters of a lava (σ_y and η) are strongly temperature-dependent and so their variations must be accounted for in any realistic model.

Downflow temperature variation
Analytical models have been proposed for describing the downslope flow of a Bingham fluid cooling by radiation. The simplest condition is to consider only temperature variations along the flow direction (model 2). If we consider the steady flow of a viscous fluid layer with constant flow rate, neglecting viscous dissipation, the heat equation becomes (Shaw & Swanson 1970, Daneš 1972, Harrison & Rooth 1976)

$$c_p q \frac{dT}{dx} + E\Sigma T^4 = 0, \qquad (9.18)$$

where q is the mass flow rate per unit width, E is the surface emissivity and Σ is the Stefan constant. Equation 9.18 yields the absolute temperature T as a function of the co-ordinate x along the flow. The T^4 term takes into account the radiation boundary condition at the free surface of the flow. Since the temperature changes along the flow, formulae such as Equations 9.5 & 6, which relate T to lava rheology, must be introduced. The steady-state solution of the equation of motion (Eq. 9.13) can then be obtained analytically, assuming a slow downslope change in flow parameters.

For a given flow rate and eruption temperature, Equation (9.18) can be used to calculate the temperature decrease due to heat radiation and the consequent changes in rheological parameters along a flow. For example, Park & Iversen (1984) have investigated numerically the role played by yield stress in flow thickening, while Dragoni (1989) has studied the downslope evolution of the rheological and dynamic parameters of a lava. Such models allow us to estimate the sensitivity of flow dynamics to changes in initial conditions (at the vent), ground slope and rheology.

The change in lava temperature with distance from the vent according to Equation 9.18 is shown in Figure 9.9a. The corresponding changes in rheological parameters (using Eqs 9.5 & 6 are shown in Figure 9.9b: both viscosity and yield stress increase by orders of magnitude along the flow. Figure 9.9c & d show the dependence of flow thicknesses and velocity on the mass flow rate per unit width q. At high temperatures, close to the liquidus, the flow is still in the Newtonian regime. As the temperature decreases, the flow thickness increases exponentially, while the velocity decreases exponentially and the plug thickness h_p approaches the total flow thickness h (Fig. 9.9c). It is evident from such models that higher effusion rates lead to slower cooling, which results in slower rates of increase of viscosity and yield stress, thus producing greater flow lengths. A remarkable increase in the thickening rate of the flow occurs at some distance from the vent: it corresponds to a strong deceleration in the lava motion (Fig. 9.9d) and indicates that the flow has entered the Bingham regime. These results are consistent with the observation that lava flows often show an initial rapid advance, followed by a marked deceleration (e.g. Borgia et al. 1983, Lockwood et al. 1985).

Vertical temperature variation
A more realistic description of lava flows requires that temperature variations with depth within the flow are also considered (model 3). At some distance from the eruption vent, when a thick thermal boundary layer has developed at the surface of the flow, significant differences between surface and interior temperatures may arise. Discrepancies of as much as 50–100°C have been found in very slow lava flows (Archambault & Tanguy 1976). At such distances, the temperature in the interior of the flow will be higher than that predicted by vertically isothermal models. Accordingly, the mean velocity of the flow will be higher, so that the flow deceleration resulting from vertically isothermal models may be exaggerated. In fact some observations (e.g. Pinkerton & Sparks 1976, Wolfe et al. 1988) suggest that, by the time many individual lava flows come to rest,

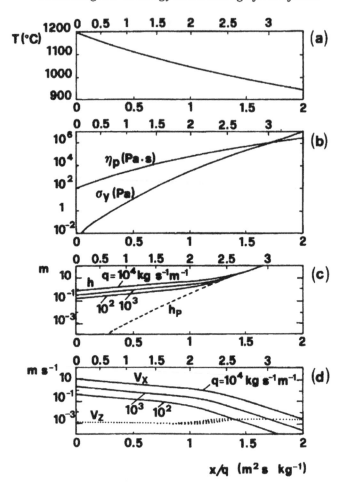

Figure 9.9 Lava flow evolution according to the model of Dragoni (1989). Effusion temperature T_0 = 1,200°C. The lower horizontal scale is for vertical thermal mixing; the upper scale is for a thermally unmixed model with surface temperature T_s = 900°C. (a) Temperature T along the flow, as a function of the parameter x/q; (b) yield stress σ_y and plastic viscosity η_p; (c) flow thickness h, for three different values of flow rate per unit width q (dashed curve is plug thickness h_p); (d) flow velocity components v_x (solid curves) and v_z (dashed curves), for three different values of flow rate per unit width q.

the temperature drop in the interior is less than 100°C with respect to the effusion temperature.

Pieri & Baloga (1986) have suggested that the presence of a cooler crust might be taken into account by assuming that radiative heat loss occurs at an effective temperature which is smaller than the temperature in the interior of the flow. This possibility was also considered by Dragoni (1989). However it is difficult to estimate beforehand what the effective temperature should be; for example, longer-lived flows may have lower effective temperatures. A vertical temperature variation has been introduced by Crisp & Baloga (1990), considering an upper crust cooling by radiation and an inner core

which is vertically isothermal. Dragoni & Pondrelli (1991) describe a vertical temperature variation that is dictated by the presence of the plug, which is not subject to shear
deformation during the flow. Accordingly, heat is transferred by conduction across the
plug, which has a thermal insulating effect, in addition to that of the thinner solid crust.
Below the plug, the temperature is taken as vertically constant, while the upper surface
of the plug loses heat by radiation to the atmosphere. All these models show a rapid
decrease in surface temperature after the formation of a significant crust or plug, while
the inner temperature remains high (Fig. 9.10). Due to the strong temperature dependence of viscosity and yield stress, there is a close interplay between thermal and
rheological parameters. For instance, the temperature distribution at a certain cross
section of the flow determines the plug thickness at that point. However, temperature
is controlled by the thickness and the thermal conductivity κ of the plug, as well as by
the specific heat c_p of lava.

Final remarks

In spite of their approximations, theoretical models can already yield insights into
several aspects of lava flow behaviour. The data which have been collected so far on
active lavas appear to be broadly consistent with model results, but it is normally

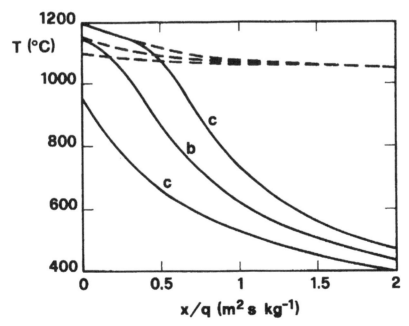

Figure 9.10 Surface (solid curves) and inner (dashed curves) temperatures of a thermally layered lava
flow model (model 3 in the text), as functions of distance from the eruption vent, for three different
values of the effusion temperature: **(a)** $T_0 = 1,200°C$, **(b)** $1,150°C$, **(c)** $1,100°C$. (After Dragoni &
Pondrelli 1991.)

impossible to compare with reasonable accuracy the numerical results of models with *in situ* observations. Measurements of lava flow parameters are in fact too sparse and insufficient to verify theoretical predictions. For instance, temperature measurements are often limited to a single point in the flow, while rheological parameters are seldom estimated. Simultaneous measurements of several quantities, such as flow dimensions and velocity, temperature, viscosity and yield stress, taken at different points in the flow, would be necessary for a comparison with theoretical models. Measuring strategies are further discussed in Pinkerton (1992).

The measurement of the physical quantities of a lava, such as temperature and viscosity, requires the immersion of instruments into a lava body. Such measurements have been performed only rarely (e.g. Archambault & Tanguy 1976, Pinkerton & Sparks 1978, Wolfe et al. 1988), due to the severe difficulties which are met in operating over a moving flow. The depth to which an instrument can penetrate a lava is controlled by the rheological properties of the lava at high temperatures, and these should be carefully considered during instrument design and operation. The resistance of lava to penetration depends on several factors. If lava is completely fluid, surface tension is the first barrier opposing penetration. However, if a solid crust has already formed, the resistance of the crust would predominate as soon as its thickness is more than a fraction of millimetre, even if its tensile strength is smaller than that of cold rock (Hulme, 1974). Furthermore, since penetration implies shearing, it is necessary that the yield stress is overcome. After penetration has started, viscosity controls the rate at which the motion takes place. If the average density of the object is less than the density of lava, buoyancy forces will also be present, hindering penetration. The horizontal drag force due to the flow velocity will be superimposed on these effects.

Ideally, a model should be able, on the basis of initial and boundary conditions, to yield the evolution of a lava flow as a function of time. Initial conditions include effusion rate, temperature and chemical composition at the vent. Moreover, since lava is a multiphase system, the concentrations of different phases and their chemical composi-tions should be known. Of course, initial conditions may vary with time. Boundary conditions include the physical properties and state of the environment (ground and atmosphere).

The evolution of a flow is given by a set of coupled equations, including continuity, dynamic, constitutive, thermal and chemical equations. Density, rheological and ther-mal parameters must be known in order to use the governing equations. Further equations must be added, relating such quantities to temperature and chemical compo-sition, which vary as functions of time and position.

Provided with an accurate physical model, deterministic predictions of lava flow behaviour may be feasible. Even though at present one cannot predict how long an eruption will last, it would be extremely useful to know how far a given flow will go assuming that the effusion rate remains constant. Even this would be a formidable task, since it requires a better knowledge of a set of physical quantities (e.g. lava density, viscosity, crystallinity and vesicularity) than is currently available. It is important, however, to establish how much the uncertainties on the values of such quantities affect the prediction of the path of a flow.

Understanding the rheological behaviour of lava is also relevant to attempting flow diversion for civil defence purposes. For instance, if lava were a Newtonian fluid, any break through a levée would permit a part of the flow to follow a new direction. This is not the case, however, for non-Newtonian fluids, for which specific conditions must be fulfilled in order that a diversion may occur. As Walker (1967) observed, there is a minimum size for a lava stream to flow and reach a significant distance from its origin. In fact, no downhill movement occurs if the flow thickness is less than the thickness of the plug. This means that the diversion of lava through a breached levée becomes more difficult at greater distances from the vent. As a result, once a lava flow has constructed its own channel, any Bingham behaviour will make it reluctant to change direction. Whatever method is considered to halt or divert a flow (levée breaching, artificial cooling of the flow, or construction of embankments), the rheological behaviour of lava cannot be neglected for such actions to be successful.

Conclusions

The rheology of lava is of paramount importance in controlling lava flow dynamics. Indeed, the main characteristics of lava at subliquidus temperatures are its non-Newtonian behaviour and the strong dependence of rheological parameters on temperature. Some aspects of lava flows have been reproduced by simple analytical models, while others must await more sophisticated analyses. In addition to temperature, many other factors have an effect on rheology, including chemical composition, crystal content, polymerization, vesiculation and degassing. Typically, we still lack accurate physical laws relating these factors to the rheological parameters. Since it is difficult to obtain such laws on theoretical grounds, empirical relationships must be sought from laboratory experiments and field observations.

Acknowledgement.

The author wishes to thank Christopher Kilburn for useful comments on the manuscript of this paper.

Appendix. Constitutive equation

The rheological behaviour of a viscous fluid is generally expressed by a "constitutive equation", that is a relation between viscous stress σ and strain rate $\dot{\varepsilon}$:

$$\sigma_{ij} = f_{ij}(\dot{\varepsilon}), \tag{A9.1}$$

where f_{ij} indicates a generic tensor function.

If the components of σ are linear functions of the components of $\dot{\varepsilon}$, i.e.

$$\sigma_{ij} = K_{ijkl}\dot{\varepsilon}_{kl}, \tag{A9.2}$$

the fluid is called Newtonian and K_{ijkl} is the viscosity tensor. If the fluid is isotropic and incompressible, K_{ijkl} reduces to a single coefficient, the viscosity η, and Equation (A9.2) reads

$$\sigma_{ij} = 2\eta\dot{\varepsilon}_{ij}. \tag{A9.3}$$

Furthermore, if the flow occurs in a fixed direction, neglecting indices, Equation (A9.3) can be written as

$$\sigma_{ij} = \eta\dot{\varepsilon} \tag{A9.4}$$

and viscosity can be defined as the ratio of shear stress to strain rate. If f is a generic function, the fluid is called non-Newtonian and a unique viscosity cannot be defined.

References

Archambault, C. & J. C. Tanguy 1976. Comparative temperature measurements on Mount Etna lavas: problems and techniques. *Journal of Geophysical Research* **1**, 113–25

Arzi, A. A. 1978. Critical phenomena in the rheology of partially melted rocks. *Tectonophysics* **44**, 173–84.

Ashby, M. F. & R. A. Verrall 1978. Micromechanisms of flow and fracture and their relevance to the rheology of the upper mantle. *Philosophical Transactions of the Royal Society, London* **A288**, 59–95.

Batchelor, G. K. 1967. *An introduction to fluid dynamics.* Cambridge: Cambridge University Press.

Böhme, G. 1987. *Non-Newtonian fluid mechanics.* Amsterdam: North-Holland.

Borgia, A. & S. Linneman 1990. On the mechanics of lava flow emplacement and volcano growth: Arenal, Costa Rica. In *IAVCEI Proceedings in Volcanology.* Vol. 2, *Lava flows and domes: emplacement mechanisms and hazard implications*, J. H. Fink (ed.), 208–43. Berlin: Springer.

Borgia, A., S. Linneman, D. Spencer, L. D. Morales, J. B. Andre 1983. Dynamics of lava flow fronts, Arenal Volcano, Costa Rica. *Journal of Volcanology and Geothermal Research* **19**, 303–29.

Bottinga, Y. & D. F. Weill 1972. The viscosity of magmatic silicate liquids: a model for calculation. *American Journal of Science* **272**, 438–75.

Bottinga, Y., D. F. Weill, P. Richet 1982. Density calculations for silicate liquids. I. Revised method for alumino-silicate compositions. *Geochimica et Cosmochimica Acta* **46**, 909–20.

Brandeis, G., C. Jaupart, C. J. Allègre 1984. Nucleation, crystal growth and the thermal regime of cooling magmas. *Journal of Geophysical Research* **89**, 10,161–77.

Burnham, C. W. 1963. Viscosity of a water rich pegmatite melt at high pressures. *Geological Society of America Special Paper* **76**, 1–26.

Carron, J. 1969. Recherches sur la viscosite et les phenomenes de transport des ions alcalins dans les obsidiennes granitiques. *Travaux du Laboratoire de Géologie, École Normale Supérieure, Paris* 1–112.

Chester, D. K., A. M. Duncan, J. E. Guest, C. R. J. Kilburn 1985. *Mount Etna. The anatomy of a volcano.* London: Chapman & Hall.

Crisci, G. M., S. Di Gregorio, O. Pindaro, G. A. Ranieri 1986. Lava flow simulation by a discrete cellular model: first implementation. *International Journal of Modelling and Simulation* **6**, 137–40.

Crisp, J. & S. Baloga 1990. A model for lava flows with two thermal components. *Journal of Geophysical Research* **95**, 1255–70.

Daneš, Z. F. 1972. Dynamics of lava flows. *Journal of Geophysical Research* **77**, 1430–32.

Dragoni, M. 1989. A dynamical model of lava flows cooling by radiation. *Bulletin of Volcanology* **51**, 88–95.

Dragoni, M. & S. Pondrelli 1991. Lava flow dynamics with vertical temperature variation. *Acta Vulcanologica* **1**, 1–5.

Dragoni, M., M. Bonafede, E. Boschi 1986. Downslope flow models of a Bingham liquid: implications for lava flows. *Journal of Volcanology and Geothermal Research* **30**, 305–25.

Dragoni, M., S. Pondrelli, A. Tallarico 1992. Longitudinal deformation of a lava flow: the influence of Bingham rheology. *Journal of Volcanology and Geothermal Research* **52**, 247–54.

Emerman, S. H. & D. L. Turcotte 1983. A fluid model for the shape of accretionary wedges. *Earth and Planetary Science Letters* **63**, 379–84.

Fink, J. H. & J. Zimbelman 1989. Longitudinal variations in rheological properties of lavas: Puu Oo basalt flows, Kilauea volcano, Hawaii. In *IAVCEI Proceedings in Volcanology.* Vol. 2, *Lava flows and domes: emplacement mechanisms and hazard implications*. J. H. Fink (ed.), 157–73. Berlin: Springer.

Goetze, C. & W. F. Brace 1972. Laboratory observations of high temperature rheology of rocks. *Tectonophysics* **13**, 583–600.

Harrison, C. G. A. & C. Rooth 1976. The dynamics of flowing lavas. In *Volcanoes and tectonospheres*, H. Aoki & S. Iizuka (eds), 103–13. Tokyo: Tokai University Press.

Hess P. C. 1980. Polymerization model for silicate melts. In *Physics of magmatic processes*, R. B. Hargraves (ed.), 3–48. Princeton: Princeton University Press.

Hulme, G. 1974. The interpretation of lava flow morphology. *Geophysical Journal of the Royal Astronomical Society* **39**, 361–83.

Hulme, G. 1982. A review of lava flow processes related to the formation of lunar sinuous rilles. *Geophysical Surveys* **5**, 245–79.

Hulme, G. & G. Fielder 1977. Effusion rates and rheology of lunar lavas. *Philosophical Transactions of the Royal Society, London* **A285**, 227–34.

Huppert, H. E. 1982. Flow and instability of a viscous current down a slope. *Nature* **300**, 427–9.

Huppert, H. E. & R. S. J. Sparks 1984. Double-diffusive convection due to crystallization in magmas. *Annual Reviews in Earth and Planetary Science* **12**, 11–37.

Ishihara, K., M. Iguchi, K. Kamo 1990. Numerical simulation of lava flows on some volcanoes in Japan. In *IAVCEI Proceedings in Volcanology*. Vol. 2, *Lava flows and domes: emplacement mechanisms and hazard implications*, J. H. Fink (ed.), 174–207. Berlin: Springer.

Jaeger, J. C. & N. G. W. Cook 1976. *Fundamentals of rock mechanics*. London: Chapman & Hall.

Johnson, A. M. 1970. *Physical processes in geology*. San Francisco: Freeman, Cooper.

Johnson, A. M. & D. D. Pollard 1973. Mechanics of growth of some laccolithic intrusions in the Henry Mountains Utah, I. *Tectonophysics* **18**, 261–309.

Khitarov, N. I. & E. B. Lebedev 1978. The peculiarities of magma rise in presence of water. *Bulletin Volcanologique* **41**, 354–9.

Kilburn, C. R. J. 1990. Surfaces of aa flow-fields on Mount Etna, Sicily: morphology, rheology, crystallization and scaling phenomena. In *IAVCEI Proceedings in Volcanology*. Vol. 2, *Lava flows and domes: emplacement mechanisms and hazard implications*, J. H. Fink (ed.), 129–56. Berlin: Springer.

Kilburn, C. R. J. & R. M. C. Lopes 1988. The growth of aa lava flow fields on Mount Etna Sicily. *Journal of Geophysical Research* **93** 14,759–72.

Kirkpatrick, R. J. 1976. Towards a kinetic model for the crystallization of magma bodies. *Journal of Geophysical Research* **81**, 2565–71.

Konnert J. H. & J. Karle 1972. Tridymite-like structure in silica glass. *Nature* **236**, 92–4.

Kushiro, I. 1976. Changes in viscosity and structure of melt of $NaAlSi_2O_6$ composition at high pressures. *Journal of Geophysical Research* **81**, 6347–50.

Landau, L. & E. Lifshitz 1971. *Mecanique des Fluides*. Moscow: MIR.

Lipman, P. W., N. G. Banks, J. M. Rhodes 1985. Gas-release induced crystallization of 1984 Mauna Loa magma, Hawaii, and effects on lava rheology. *Nature* **317**, 604–7.

Lockwood, J. P., N. G. Banks, T. T. English, L. P. Greenland, D. B. Jackson, D. J. Johnson, R. Y. Koyanagi, K. A. McGee, A. T. Okamura, J. M. Rhodes 1985. The 1984 eruption of Mauna Loa Volcano, Hawaii, *Eos, Transactions of the American Geophysical Union* **66**, 169–71.

McBirney, A. R. & T. Murase 1970. Factors governing the formation of pyroclastic rocks. *Bulletin Volcanologique* **34**, 372–84.

McBirney, A. R. & T. Murase 1984. Rheological properties of magmas. *Annual Reviews in Earth and Planetary Science* **12**, 337–57.

McBirney, A. R. & R. M. Noyes 1979. Crystallization and layering of the Skaergaard intrusion. *Journal of Petrology* **20**, 487–554.

Mozzi, R. L. & B. E. Warren 1969. The structure of vitreous silica. *Journal of Applied Crystallography* **2**, 164–72.

Murase, T. & A. R. McBirney 1970. Viscosity of lunar lavas. *Science* **167**, 1491–3.

Murase, T. & A. R. McBirney 1973. Properties of some common igneous rocks and their melts at high temperatures. *Geological Society of America, Bulletin* **84**, 3563–92.

Murrell, S. A. F. & S. Chakravarty 1973. Some new rheological experiments on igneous rocks at temperatures up to 1120°C. *Geophysical Journal of the Royal Astronomical Society* **34**, 211–50.

Nelson, S. A. & I. S. E. Carmichael 1979. Partial molar volume of oxide components in silicate liquids. *Contributions to Mineralogy and Petrology* **71**, 117–24.

Ozisik, M. N. 1968. *Boundary value problems of heat conduction*. Scranton: International Textbook Co.

Park, S. & J. D. Iversen 1984. Dynamics of lava flow: thickness growth characteristics of steady two-dimensional flow. *Geophysical Research Letters* **11**, 641–4.

Pieri, D. & S. Baloga 1986. Eruption rate area and length relationships for some Hawaiian lava flows. *Journal of Volcanology and Geothermal Research* **30**, 29–45.

Pinkerton, H. & R. S. J. Sparks 1976. The 1975 sub-terminal lavas, Mount Etna: a case history of the formation of a compound lava field. *Journal of Volcanology and Geothermal Research* **1**, 167–82.

Pinkerton, H. & R. S. J. Sparks 1978. Field measurements of the rheology of lava. *Nature* **276**, 383–5.

Robson, G. R. 1967. Thickness of Etnean lavas. *Nature* **216**, 251–2.

Ryerson, F. J., H. C. Weed, A. J. Piwinskii 1988. Rheology of subliquidus magmas 1. Picritic compositions. *Journal of Geophysical Research* **93**, 3421–36.

Sakuma, S. 1953. Elastic and viscous properties of volcanic rocks at high temperatures. *Bulletin of the Earthquake Research Institute, Tokyo University* **31**, 291–303.

Scarfe, C. M. 1973. Viscosity of basic magmas at varying pressure. *Nature* **241**, 101–2.

Shaw, H. R. 1963. Obsidian–H_2O viscosities at 1000 and 2000 bars in the temperature range 700 to 900°C. *Journal of Geophysical Research* **68**, 6337–42.

Shaw, H. R. 1969. Rheology of basalt in the melting range. *Journal of Petrology* **10**, 510–34.

Shaw H. R. 1972. Viscosities of magmatic silicate liquids: an empirical method of prediction. *American Journal of Science* **272**, 870–93.

Shaw, H. R. & D. A. Swanson 1970. Eruption and flow rate of flood basalts. *Proceedings of the 2nd Columbia River Basalt Symposium* 271–99. Cheney: Eastern Washington State College.

Shaw, H. R., T. L. Wright, D. L. Peck, R. Okamura 1968. The viscosity of basaltic magma: an analysis of field measurements in Makaopuhi Lava Lake, Hawaii. *American Journal of Science* **266**, 225–4.

Sherman P. 1968. *Emulsion Science*. New York: Academic Press.

Skelland, A. H. P. 1967. *Non-Newtonian flow and heat transfer*. New York: John Wiley.

Sparks, R. S. J. 1977. The dynamics of bubble formation and growth in magmas: a review and analysis. *Journal of Volcanology and Geothermal Research* **3**, 1–37.

Sparks, R. S. J. & H. Pinkerton 1978. Effect of degassing on rheology of basaltic lava. *Nature* **276**, 385–6.

Sparks, R. S. J., H. Pinkerton, G. Hulme 1976. Classification and formation of lava levées on Mount Etna, Sicily. *Geology* **4**, 269–71.

Sparks, R. S. J., H. Pinkerton, R. MacDonald 1977. The transport of xenoliths in magmas. *Earth and Planetary Science Letters* **35**, 234–8.

Spera, F. J. 1980. Aspects of magma transport. In *Physics of magmatic processes*, R. B. Hargraves (ed.), 265–323. Princeton: Princeton University Press.

Spera, F. J., D. A. Yuen, S. J. Kirschvink 1982. Thermal boundary layer convection in silicic magma chambers: effects of temperature-dependent rheology and implications for thermogravitational chemical fractionation. *Journal of Geophysical Research* **87**, 8755–67.

Swanson, D. A. 1973. Pahoehoe flows from the 1969-1971 Mauna Ulu eruption Kilauea volcano Hawaii. *Geological Society of America, Bulletin* **84**, 615–26.

Wadge, G. 1978. Effusion rate and the shape of aa lava flow-fields on Mount Etna. *Geology* **6**, 503–6.

Walker, G. P. L. 1967. Thickness and viscosity of Etnean lavas. *Nature* **213**, 484–5.

Walker, G. P. L. 1973. Lengths of lava flows. *Philosophical Transactions of the Royal Society, London* **A274**, 107–18.

Westrich, H. R., H. W. Stockman, J. C. Eichelberger 1988. Degassing of rhyolitic magma during ascent and emplacement. *Journal of Geophysical Research* **93**, 6503–11.

Williams, H. & A. R. McBirney 1979. *Volcanology*. San Francisco: Freeman, Cooper.

Wilson, L. & J. W. Head III 1983. A comparison of volcanic eruption processes on Earth, Moon, Mars, Io and Venus. *Nature* **302**, 663–9.

Wolfe, E. W., C. A. Neal, N. G. Banks, T. J. Duggan 1988. Geologic observations and chronology of eruptive events. In *The Puu Oo eruption of Kilauea volcano, Hawaii, episodes 1 through 20, January 3, 1983, through June 8, 1984*, E. W. Wolfe (ed.). US Geological Survey Professional Paper 1463 1–97.

261

CHAPTER TEN

Lava crusts, aa flow lengthening and the pahoehoe–aa transition

Christopher R. J. Kilburn

Abstract

Lava flows can be considered as hot viscous cores within much thinner and colder solidified crusts. Interaction between crust and core determines the morphological and dynamic evolution of a flow. When the lava core dominates, flow advance approaches a steady state. When crusts are the limiting factor, advance is more irregular. These two conditions can be distinguished by a timescale ratio comparing rates of flow deformation and crustal formation. Aa and pahoehoe lavas are used as examples of core- and crustal-dominated flows, respectively. With this assumption, a simple model previously developed for aa and blocky lavas is used to examine the influence of mean discharge rate on both flow morphology (aa or pahoehoe) and aa flow length. The quantitative expressions derived agree well with established empirical relations and offer the hope of forecasting flow evolution from initial eruption conditions.

Introduction

Surface crusts are essential features of active lavas. They tear and heal during emplacement to produce inward-growing cool boundary layers (Tanguy & Biquand 1967, Pinkerton & Sparks 1976) and act as thermal insulators for the lava interior (Williams & McBirney 1979). A flow can thus be treated as a hot lava core beneath a colder, thin crust (Kilburn & Lopes 1988, Crisp & Baloga 1990, Fink & Griffiths 1990, Kilburn & Lopes 1991).

The interaction between core and crust produces a variety of surface morphologies (Fink & Griffiths 1990, Kilburn 1990), cooling regimes (Crisp & Baloga 1990) and modes of flow advance (Kilburn & Lopes 1988, Borgia & Linneman 1990, Whitehead

263

& Helfrich 1991). Such variations are especially evident comparing the main categories of subaerial basaltic lava, *pahoehoe* and *aa* (Dutton 1884, Macdonald 1953). Pahoehoe lavas have smooth crusts which break locally across a flow front and allow advance by the extension of numerous small tongues. Aa lavas, in contrast, have uneven surfaces, usually hidden beneath loose debris; crustal failure is widespread and fronts tend to advance as single units.

While detailed textural differences between the lava types (Macdonald 1967, Kilburn 1990) reflect differences in surface condition before chilling (Kilburn 1990), the gross difference in crustal morphology (at length scales of decimetres or more) can be explained in terms of crustal growth under stress (Kilburn 1990). A smooth surface forms when crustal resistance dominates the imposed stress; otherwise, the crust autobrecciates *during* cooling (as opposed to breakage when already formed), producing aa.

Here it is argued that the contrasting styles of advance between pahoehoe and aa also result from differences in crustal growth and disruption. The main assumption is that less widespread crustal disruption implies greater crustal restraint. Curiously, this simple assertion has been largely ignored for the last 25 years, after Robson (1967) and Shaw et al. (1968) presented evidence that crystallizing lavas are non-Newtonian fluids. Since then, attention has been focused on describing (a) how lava is non-Newtonian (e.g. Shaw 1969, Gauthier 1973, Pinkerton & Sparks 1978, McBirney & Murase 1984, Ryerson et al. 1988) and (b) how the non-Newtonian properties of lava cores influence flow behaviour (e.g. Hulme 1974, Sparks et al. 1976, Moore et al. 1978, Fink & Zimbleman 1990, Ch. 9 this volume).

However, there is no reason to suppose that crusts are always of minor influence (e.g. Iverson 1990). Indeed, simple observations suggest the opposite. The formation of secondary boccas, for example, requires escape of internal lava through a hole in the flow crust. This implies that core lava had previously been retarded by the chilled carapace of the flow and, hence, that flow dimensions had been conditioned by the presence of a crust.

This chapter utilizes a limiting requirement for crustal disruption identified in the lava flow model of Kilburn & Lopes (1991). It is postulated that aa lavas normally satisfy this requirement, but that pahoehoe lavas do not and so are more strongly influenced by crustal resistance. The model is briefly reviewed and applied to field data. It accounts well for established empirical relations between (a) flow length and mean discharge rate of aa lavas (Walker 1973, Booth et al. 1975) and (b) discharge rate and the formation of aa or pahoehoe (Booth et al. 1975, Pinkerton & Sparks 1976, Rowland & Walker 1990). The results also imply that a simple timescale ratio involving core deformation and crustal healing provides a basis for identifying different regimes of flow emplacement.

Lava crusts

The lava crust is important because of its high tensile strength. In the crystallization interval (nominally from 1,200 to 950°C for basalts), the lava compressive strength increases from 0 to ~10^8 Pa, thereafter remaining at about 10^8 Pa during cooling to at least room temperature (Murrell & Chakravarty 1973; the symbol "~" denotes "of the order of"). The implied tensile strengths are about 10 times smaller (Jaeger & Cook 1979), that is, from 0 to ~10^7 Pa, although maximum values may be smaller due to structural heterogeneity. Even when a lava is an incandescent golden orange (approximately 900–1,000°C; Table 10.1), therefore, its strength may be several orders of magnitude greater than that of the lava interior; it is a common misconception that only blackened crust possesses a significant tensile strength.

While cooling promotes crustal growth, surface deformation due to flow favours crustal rupture (e.g. see Ch. 3). How crusts evolve and when they first offer significant resistance depends on the relative rates of flow deformation and of surface chilling. If the flow deforms too quickly, crustal fracture is widespread, chilling is too slow to heal newly formed cracks, and flow advance is limited by core rheology; on the other hand, if the flow deforms too slowly, chilling is able to create a continuous crust and this may eventually become the more important limiting factor.

For the lava crust to have a secondary influence on flow behaviour, it must deform

Table 10.1 Lava properties

(a) Temperatures and surface colours (Macdonald 1972, Bullard 1976)

Colour	Temperature (K (°C))
White	>1,423 (>1,150)
Golden yellow	1,363 (1,090)
Orange	1,173 (900)
Bright cherry red	973 (700)
Dull red	873 (600)
Lowest visible red	748 (475)

(b) Physical properties of basaltic lavas (Murase & McBirney 1973, Horai 1991)

Property	Nominal values
Initial temperature, θ_0	1,350–1,400 K (1,077–1,127°C)
Density, ρ	2,600 kg m^{-3}
Thermal conductivity*, K	1.3 J m^{-1} s^{-1} K^{-1}
Specific heat capacity, c_p	1,150 J kg^{-1} K^{-1}
Thermal diffusivity, k	4.2 x 10^{-7} m^2 s^{-1}
Surface emissivity, ε	1
Stefan–Boltzmann constant, σ	5.67 × 10^{-8} J m^{-2} s^{-1} K^{-4}
Chilling timescale, t_{ch}	160–200 s

*Thermal conductivity for approximately 10–15 vol% vesicularity.

at least as quickly as (a) the flow interior, and (b) the rate at which cooling heals broken surfaces (Kilburn & Lopes 1991). In other words, the timescale of crustal deformation (t_{cr}) must be smaller than the timescales of flow deformation (t_{def}) and of crustal healing (t_h). If these conditions are not met, crustal restraint significantly affects flow motion. The condition at which the lava crust first becomes important thus occurs when the three timescales t_{def}, t_{cr} and t_h are comparable (Kilburn & Lopes 1991).

Lava cooling and crustal growth

Crustal growth begins as soon as lava is exposed at the surface. Thermal energy is lost by reducing the surface temperature of the lava and by increasing the thickness of the cooled layer. The surface temperature drops at a rate controlled by radiation to the atmosphere and is initially the dominant response to heat loss. After a critical chilling time (t_{ch}), the rate of surface temperature decrease is small and chilled-layer thickening, controlled by conduction through the lava, becomes the more important response. During chilling, the surface temperature drops below the lava solidus; before $t = t_{ch}$, therefore, the tensile strength of the crust may approach its maximum value.

First-order modelling (Appendix) suggests that t_{ch} is approximately given by

$$t_{ch} = K\rho c_p/(\varepsilon\sigma\theta_0^3)^2, \tag{10.1}$$

where K, ρ, c_p and θ_0 correspond to the thermal conductivity, density, specific heat capacity and initial temperature of the lava, ε is the surface emissivity and σ is the Stefan–Boltzmann radiation constant. Representative values of lava properties (Table 10.1) indicate that t_{ch} for basalt is typically between 160 and 200 s (Crisp & Baloga 1990, Kilburn & Lopes 1991).

Because of crustal rupture, long-term flow cooling through the surface depends on conduction across cool, unbroken crust and on radiation from hotter, newly exposed surfaces (Crisp & Baloga 1990, Kilburn & Lopes 1991). Considering the time needed for each mechanism to remove the initial heat content from a lava batch, of depth h, two reference timescales may be defined (Crisp & Baloga 1990):

(a) conduction (or Graetz) timescale,

$$t_{Gz} = h^2/k, \tag{10.2a}$$

(b) radiation timescale,

$$t_R = \rho c_p h/\varepsilon\sigma\theta_0^3, \tag{10.2b}$$

where k is the thermal diffusivity of the lava ($k = K/\rho c_p$).

Inspection of Equations 10.1 & 2 then yields

$$t_R^2 = t_{Gz} t_{Ch} \tag{10.3}$$

a timescale relation useful for later descriptions of flow emplacement.

Aa flow durations have magnitudes similar to t_R but much smaller (by two to three orders of magnitude) than t_{Gz} (Crisp & Baloga 1990, Kilburn & Lopes 1991). This suggests that radiation-limited heat loss may govern flow lengthening (Crisp & Baloga 1990), although detailed interpretations remain ambiguous (Kilburn & Lopes 1991).

Aa flow advance

Aa flows typically move forward as single units, metres to kilometres wide and tens of centimetres to tens of metres thick (Pinkerton & Sparks 1976, Macdonald 1967, Kilburn & Lopes 1988, 1991, Ch. 3 this volume); some fronts (especially those with larger width:depth ratios) may show subdivisions into two or more lobes. Emplacement times for single flows are normally ~ 10^6 s or less (Kilburn & Lopes 1991). Although frontal velocities vary with time, they may remain almost steady for much of the advance of a flow (Kilburn & Lopes 1991). Crustal disruption is widespread (Macdonald 1953, 1972, Ch. 3 this volume); at any given moment, however, the crust may maintain continuity across the flow, adjacent zones of autobrecciation being surrounded by unbroken crust. When effusion continues after a front has halted, a new flow may grow following localized failure in the existing flow's margin of the flow (Pinkerton & Sparks 1976, Cristofolini 1984, Kilburn & Lopes 1988, 1991). Importantly, such failure is not relevant to how an aa front advances.

The aa model

Advance is governed by the frontal zone of a flow. This zone develops at the head of the channel system of a lava (where marginal levées are well established) and can be divided into two intergradational regions (Kilburn & Lopes 1991): the snout frontal zone (SFZ, or snout) and the rear frontal zone (RFZ).

At the leading edge of the flow, the surface of the SFZ drops rapidly downstream, forming a sloping front. Measured parallel with the ground, the length of the SFZ is typically similar to the mean thickness of the RFZ. Lava entering the snout spreads laterally, fixing the initial width of the flow. The back of the SFZ grades into the much longer RFZ, characterized by (a) very slow widening, (b) lack of distinct marginal levées and (c) a mean thickness (averaged over the RFZ width) which increases weakly in the downstream direction.

Mean advance rates of the front are considered to be controlled by the RFZ (Kilburn & Lopes 1991), where flow is treated as uniform (same mean velocity across each transverse section), steady (constant mean velocity with time) and one-dimensional (since width is much greater than depth and thickening and spreading rates are much slower than the downstream velocity). The crusts are assumed to be much thinner than the core, which is taken to be rheologically homogeneous and approximately Newtonian (the limiting condition for a presumed pseudoplastic fluid at low rates of shear deformation). Corresponding Reynolds numbers ($Re = uh/\nu$, where ν = kinematic viscosity = viscosity/density) are less than unity, implying that core motion is laminar and that inertial forces are negligible. Gravity must therefore drive the RFZ forward.

When crustal restraint in the downstream direction is small compared to core resistance, the mean velocity (u) in the RFZ is governed by a balance between lava weight and the viscous forces of the core, leading to (after Jeffreys (1925))

$$u = (\rho g h^2 \sin a)/3\mu, \tag{10.4}$$

where ρ and μ are the density and viscosity of lava, of mean thickness h, moving under gravity g over a slope of angle a.

Most widening is assumed to occur in the SFZ, where a balance between the outward-driving hydrostatic force and core resistance yields (Kilburn & Lopes 1991)

$$w = (\rho g h^2/\mu)t_w, \tag{10.5}$$

where t_w, the mean timescale of widening, is the inverse of the mean rate of shear deformation laterally.

Widening continues until it is slow enough for crustal healing to dominate crustal rupture. Using the critical timescale balance above, this condition occurs when t_w is comparable to t_h, the timescale of crustal healing. t_h is the thickening time a new crust needs before it rapidly slows the rate of widening. Anticipating structural complexity as a crust evolves, Kilburn & Lopes (1991) inferred typical ranges for t_w assuming $t_w = t_h$ and then estimating (a) minimum values of t_h from cooling arguments and (b) maximum values of t_w from field data.

The minimum value of t_h is based on the time for a surface to cool below its solidus (when its tensile strength tends to a maximum) and is estimated at ~30 s. At the other extreme, for widening in the SFZ alone, the maximum time for widening is that needed for the snout to pass a fixed position. With a length of order h and mean velocity of order u, the maximum widening time is of the order of h/u, implying that $t_w \sim h/u$ or less. Data from Etnean aa lavas indicate that $h/u < 10^3$ s. Nominal limits for t_w are thus 30 and 1,000 s and so t_w may be comparable to the chilling timescale t_{ch} for much of the lengthening of a flow.

For comparable advance rates during emplacement, the flow length L at time t (from the beginning of emplacement) is of the order of ut or, from Equations 10.4 & 5,

$$L \sim [(w \sin a)/3](t/t_w). \tag{10.6}$$

From Equations 10.2 & 3, the timescale ratio t/t_w can be transformed as follows: $t/t_w = (t_R/t_{ch})(t/t_R)(t_{ch}/t_w) = (t_{Gz}/t_R)(t/t_R)(t_{ch}/t_w) = (h^2/kt)(t/t_R)^2(t_{ch}/t_w)$. For the combination $(t/t_R)^2(t_{ch}/t_w) \sim 1$ (see below), $t/t_w \sim h^2/kt$ and Equation 10.6 becomes

$$L = C(wh^2 \sin a)/3kt \tag{10.7a}$$

or, rearranging,

$$t = C[(w/L)h^2 \sin a]/3k, \tag{10.7b}$$

where C is a constant.

Using maximum flow width, average flow thickness and mean ground slope to represent w, h and sin a, Kilburn & Lopes (1991) found that, given the uncertainties of field measurements, Equation 10.7b reasonably describes aa (and blocky) flow field behaviour for $C = 1 \pm 0.5$.

Implications and limits of the aa model

Each of the relations utilized to arrive at Equations 10.7 are order-of-magnitude approximations. As discussed in Kilburn & Lopes (1991), the errors on these approximations seem either to be small or to counterbalance each other; in particular, the term $(t/t_R)^2(t_{ch}/t_w)$ appears to remain close to unity, increases in t with respect to t_R being balanced by increases in t_w with respect to t_{ch}. Such counterbalancing may be fortuitous or a real physical effect and, although the second interpretation is preferred, the first cannot be excluded without further study of how flow parameters (e.g. u, w, h and t_w) change during emplacement.

The widening condition (Eq. 10.5) requires that spreading stops rapidly when $h/u \geqslant t_h$. The term h/u is also the mean timescale of deformation for flow downstream. At first sight, therefore, the condition for no further widening seems to imply that flow advance must also cease. This is clearly unreasonable, since flows continue lengthening after snouts have established their widths. Two possible explanations are (a) that the healing timescale for downstream flow (i.e. for crusting across the snout face) is longer than that for widening (i.e. for surface crusting towards the rear of the SFZ), or (b) that, although both are comparable to h/u, the timescale for widening is slightly longer than that for flow advance. In each case, all the timescales may still be of the same order of magnitude; this not only highlights a limitation in the sensitivity of the current analysis, but also suggests that aa fronts may tend to a critical state in which only small differences in applied forces decide whether or not motion can occur.

Equations 10.7 are most appropriate for emplacement of single flows. They may be extended to flow fields, consisting of several flows, only if (a) all main flows have similar dimensions and emplacement times and (b) the propagation of new flows leads predominantly to flow field widening (Kilburn & Lopes 1991). The important feature here is that an aa flow can be expected to halt when it satisfies Equations 10.7, further effusion causing flow thickening and widening or the propagation of a new flow.

Equations 10.7 are also of interest because they lack terms explicitly involving gravity, lava rheology and temperature. These three factors are accounted for implicitly through their influence on flow dimensions. Such a result is especially convenient, since temperature and rheological measurements on anything but the outer layers are often unobtainable during eruption. The form of Equations 10.7 are also well suited to remote sensing studies of emplaced aa lavas (on Earth or other planets; e.g. Lopes & Kilburn 1990, Ch. 4 this volume). When applied to extraterrestrial or submarine flows, however, it is important to remember that Equations 10.7 have been determined for subaerial flows on Earth, and modifications may be necessary to account for different ambient conditions.

Application 1: length and discharge rate of Etnean aa lavas

Multiplying both sides of Equation 10.7a by L and setting $hwL/t = Q$, the mean discharge rate of the flow, leads to

$$L = (C/3k)^{1/2}(h \sin a)^{1/2}Q^{1/2}. \tag{10.8}$$

Comparing a set of lava flows, if $h \sin a$ covers a much smaller range than Q, then Equation 10.8 would anticipate a positive dependence of flow length on discharge rate. As first emphasized by Walker (1973), empirical studies do indeed show that L and Q tend to increase together.

A good example is provided by Etna's aa lavas, for which $h \sin a$ lies between 0.1 and 1 m (Walker 1967, Robson 1967), while measured values of Q range from 0.001 to 100 m^3 s^{-1} (Table 10.2). For $0.1 < h \sin a < 1$ and $0.5 < C < 1.5$, Equation 10.8 yields limiting ranges of L as:

(a) minimum length,

$$L = 0.2Q^{1/2}, \tag{10.9a}$$

(b) maximum length,

$$L = 1.1Q^{1/2}, \tag{10.9b}$$

for L in kilometres and Q in cubic metres per second.

Equations 10.9 are shown in Figure 10.1 together with field measurements for single (both initial and successive) Etnean flows (Table 10.2). Most of the data are independent from those used by Kilburn & Lopes (1991) and so also provide an additional test for the governing relations (Equations 10.7).

The limits agree well with the spread of data. This not only supports the validity of the aa model, but also has two immediate implications for Etna's lavas. The first results from the fact that lava discharge rates generally decrease with time. Equation 10.9b used with estimates of Q made at the beginning of eruption (i.e. when Q is largest) should thus provide an upper limit to the potential length of a flow. The second follows because the data in Figure 10.1 involve both the initial flows of an eruption and flows propagated after earlier ones have halted. Time variations in eruption conditions, notably of changes in discharge rate, are thus likely to have been very different among the measured flows. Apparently, therefore, such variations do not have a controlling influence on *final* flow length.

Application 2: advance of pahoehoe and aa lavas

According to dominant surface morphology, subaerial basaltic lavas can be classified as pahoehoe or aa (Dutton 1884, Macdonald 1953): at length scales of decimetres to metres, pahoehoe lavas are characterized by smooth surfaces, while aa surfaces are highly irregular and commonly support a layer of broken debris. Although morphologi-

Table 10.2. Lengths and discharge rates of single aa flows on Etna

Date	Length L, (km)	Mean discharge rate Q, (m^{3}s^{-1})	Source
(a) Initial flows (fed from start of eruption)			
1908	3.1	13.3	1
1949	4.5	16.7	1
1950–1	5.1	25.5	2
1979	6.0	66.0	1
1980a	2.9	52.0	3
1980b	2.4	56.0	3
1981	8.3	100.0	1, 4
1983	4.4	48.0	5
(b) Successive flows (after initial flows; fed from main vent or secondary bocca)			
1634–8	3.1	10.0	6
1634–8	4.9	19.0	6
1974	1.7	1.8	7
1975	0.03	0.006	8
1975	0.04	0.001	8
1975	0.04	0.0025	8
1975	0.05	0.0125	8
1975	0.08	0.01	8
1975	0.08	0.015	8
1975	0.08	0.02	8
1975	0.1	0.01	8
1975	0.1	0.06	8
1975	0.125	0.03	8
1975	0.2	0.013	8
1991–2	1.5	1.0	9

Lengths measured from main vent or, where relevant, from secondary feeding bocca.
Typical errors estimated at: ± 5% on L, ± 20% on Q.
Sources. 1, Lopes (1985); 2, Cumin (1954); 3, Personal field data; 4, Guest et al. (1987); 5, Frazzetta & Romano (1984); 6, Corsaro & Cristofolini (1989); 7, Guest et al. (1974); 8, Booth et al. (1975); 9, Istituto Internazionale di Vulcanologia, Catania, Italy (personal communication).

cal, the classification defines a fundamental difference in lava behaviour (e.g. Macdonald 1953, Peterson & Tilling 1980, Kilburn 1981, Rowland & Walker 1987, Kilburn 1990, Rowland & Walker 1990) for, while both aa and pahoehoe may be emplaced initially as sheets, the two lava types rapidly evolve distinctive modes of advance.

The clearest quantitative data distinguishing pahoehoe and aa concern the flow discharge rate Q. Higher values of Q are associated with aa morphology, the critical discharge rate (Q^*) to be exceeded being estimated at 0.001–0.002 m^3 s^{-1} for small flows on Etna (Booth et al. 1975, Pinkerton & Sparks 1976) and at 5–10 m^3 s^{-1} for large flows on Hawaii (Rowland & Walker 1990). It is argued here that the discharge rate control reflects the critical timescale condition for lava crusting.

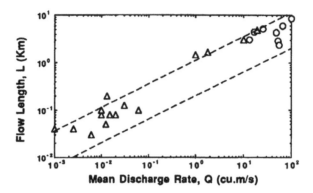

Figure 10.1   Variation of flow length with mean discharge rate for single flows on Etna (data in Table 10.2). Circles denote initial flows. Triangles represent successive flows. The dashed lines show the limits given by Equations 10.9.

Pahoehoe flow advance

Pahoehoe flow fronts cover dimensions similar to those for aa lavas (Macdonald 1953, 1967, 1972, Wentworth & MacDonald 1953). They advance, however, by propagating (or *budding*) smaller tongues, metres or less in cross-sectional dimensions (Macdonald 1953, 1967, Walker 1971, Swanson 1973, Rowland & Walker 1990, Hon et al. 1993, Chapter 5 this volume). Many tongues can be active simultaneously, budding either directly from different locations at the main front, or from breaches in the crusts of earlier tongues. Crusts are normally glassy (Jaggar 1930, Macdonald 1953, 1967, Swanson 1973) and, when revealed by drainage of internal lava, may be only a few centimetres thick (minimum thicknesses appear to be about 1–3 cm (Macdonald 1953, Swanson 1973)). For a connected sequence of tongues, budding commonly occurs at intervals of $< 10^4$ seconds (Baldwin 1953, Rowland & Walker 1990); in comparison, advance times for a main front may reach $\sim 10^7$ seconds (Macdonald 1953, Rowland & Walker 1990). Unlike aa lavas, therefore, localized crustal failure is essential to the advance of the main pahoehoe front. Moreover, for periods at least comparable with the lifetime of a tongue, the progress of a main pahoehoe front is likely to be less steady than for aa flows.

Flow advance and crustal disruption

The style of pahoehoe advance contrasts markedly with that described earlier for aa. Obviously different are the modes of crustal disruption. Widespread crustal failure allows the near-steady advance of aa fronts, while localized failure promotes the intermittent advance of pahoehoe tongues. These states correspond to when advance is limited by core rheology or by crustal resistance.

One requirement for pahoehoe emplacement, therefore, is that the lava discharge rate Q is too slow to satisfy the limiting condition for aa crustal rupture (discussed after Equation 10.5). In other words, Q^*, the maximum Q for pahoehoe growth, can be

C. R. J. Kilburn

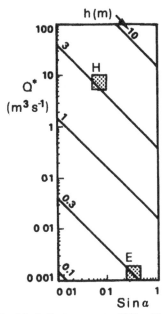

Figure 10.2 Variation of critical discharge rate (Q^*) with underlying slope (sin a) according to Equation 10.12. The sloping lines correspond to the labelled critical depths (h). The shaded areas show ranges of Q^* and sin a measured on Etna (E) and Hawaii (H). The model values of h are 0.2–0.4 m for Etna and 2–4 m for Hawaii; corresponding field estimates are 0.25–0.35 m and 2–6 m. See Table 10.3 for sources of data.

estimated from the limiting condition for a crust to inhibit aa snout widening, i.e. $t_w/t_h \sim 1$.

The snout widening timescale, t_w is described by $t_w \sim h/u = wh^2/Q$. From Equations 10.4 & 5

$$t_w/t_h \sim 3h^3/(Qt_h \sin a) \qquad (10.10)$$

and so, in the limit ($t_w/t_h \sim 1$, $Q = Q^*$)

$$Q^* \sim 3h^3/(t_h \sin a). \qquad (10.11)$$

To estimate t_h for pahoehoe, it is assumed that budding pahoehoe crusts must achieve the observed minimum thicknesses of 1–3 cm. Thinner crusts are associated with either initial sheet flow or formation of tiny lava bulbs, neither of which is relevant to the sustained advance of budding pahoehoe. The assumed crustal thickness coincides with the 1–2 cm expected during initial surface chilling (estimated as $(4kt_{ch})^{1/2}$). Accordingly, t_h is taken to be $\sim t_{ch}$ and Equation 10.11 becomes

$$Q^* \sim 3h^3/(t_{ch}\sin a). \qquad (10.12)$$

Equation 10.12 relates the maximum discharge rate at which budding pahoehoe can form to the mean thickness of the main flow front and underlying slope. This relation can be readily tested using measured values of Q^* and to estimate critical flow front thicknesses (Fig. 10.2). For both Etnean and Hawaiian lavas, the model and field

273

estimates of h compare extremely well (Table 10.3). Given that the thicknesses range over an order of magnitude, and the associated values of Q^* over almost four orders of magnitude, the consistent h estimates strongly support the crustal restraint model.

Noting from the velocity Equation 10.4 that $h/u \sim 3v/(gh \sin a)$, where the kinematic viscosity v is μ/ρ, the limiting condition $h/u \sim t_{ch}$ also gives

$$h \sim 3v/(gt_{ch} \sin a), \tag{10.13}$$

which, substituting for h in Equation 10.12, leads to

$$Q^* \sim (v/g)^3[3/(t_{ch} \sin) a)]^4. \tag{10.14}$$

Larger values of Q^* are thus required for lavas with greater kinematic viscosities and for shallower slopes. Crudely, the more viscous a lava and the shallower the slope, the

Table 10.3 Measured maximum discharge rates for pahoehoe lavas

	Etna	Hawaii
(a) Field data		
Maximum discharge rate, Q^* (m^3s^{-1})	0.001–0.002	5–10
Ground inclination, a (°)	14–25	3–6
Sin a	0.24–0.42	0.05–0.10
Critical front depth, h (m)	0.25–0.35	2–6
(b) Model data		
Critical front depth, h (m)	0.2–0.4	2–4
Lava kinematic viscosity, v (m^2s^{-1})	30–105	62–265
Chilling timescale, t_{ch} (s)	160–200	160–200

Field estimates: Etna, 1975 lavas (box E in Fig. 10.2):

Q^* Mean discharge rate per flow (Booth et al 1975, Pinkerton & Sparks 1976).
h Channelled flows develop aa morphology (Pinkerton & Sparks 1976) when fronts reach 10–30 m from vent (Booth et al. 1975). Measured channel depths at vent are about 0.20–0.25 m (Sparks et al. 1976). Aa fronts reach metre depths when 1 km or more from vent (Pinkerton & Sparks 1976). Assuming linear thickening of front with distance, an initial depth of 0.25 m and a thickness of 3 m after 1 km, maximum front thicknesses at 30 m from vent estimated at 0.33 m, rounded-up to 0.35 m.
a Values from Sparks et al. 1976.

Field estimates: Hawaii, historical flows from Mauna Loa and Kilauea (box H in Fig. 10.2)

Q^* Mean discharge rate per flow (Rowland & Walker 1990).
h Estimated from mean flow thicknesses for pahoehoe and aa lavas bounding the values of Q^* quoted in Rowland & Walker 1990. Measured flows are: (a) from Mauna Loa: 1843, 1859, 1880–81, 1916, 1950; (b) from Kilauea: 1955, 1972–74; 1983–86.
a Principal range for listed flows.

Depths and slopes estimated from data in Lipman & Swensen 1984, Lopes 1985, Lockwood et al. 1988, Rowland & Walker 1990, US Geological Survey 1986, and Wolfe et al. 1988.

faster a flow must advance (hence larger Q^*) to deform more quickly than the crust can heal. For the Etnean and Hawaiian data (Table 10.3), Equation 10.14 implies that v is comparable at $\sim 10^2$ m^2 s^{-1}. Apparently, the main reason why Hawaiian lavas show critical discharge rates and thicknesses greater than the Etnean examples is because the Hawaiian slopes are shallower.

Conclusions

The mode of lava advance depends on the importance of crustal resistance. Crusts can normally be disrupted. Effective crustal resistance is thus determined by how quickly broken crust can reheal. When rehealing is slow, emplacement is governed by the lava core and advance is almost steady. When rehealing is fast, emplacement is dominated by crustal resistance and advance rates oscillate with time.

Qualitatively, this association resembles a single dynamic system (a lava) which, because of non-linear changes in governing conditions with time (core and crust interaction), possesses more than one equilibrium state (modes of flow advance; for analogies, see May (1976), Berry et al. (1987), Shaw (1987) and Glass & Mackey (1988)). Quantitatively, a criterion distinguishing core- and crustal-dominated regimes is the value of the timescale ratio $t_w/t_h \sim h/ut_h$. The lava core dominates when t_w/t_h is less than about 1 and the crust dominates when t_w/t_h is larger than 1. The transition occurs when $t_w/t_h \sim 1$.

One feature distinguishing pahoehoe and aa lavas is their mode of advance. Another is the condition of surface lava upon cooling, which determines their characteristic textures (Kilburn 1990). This second condition is reflected by budding pahoehoe having values of t_h of the order of t_{ch} (Eq. 10.1). The difference between pahoehoe and aa thus coincides with a change from a crustal (pahoehoe) to a core (aa) dominated regime *when the crustal healing timescale $\sim t_{ch}$*. This condition, in turn, explains why, all else being equal, aa lavas are associated with higher discharge rates than pahoehoe.

The value of t_h, however, is not fixed at t_{ch} and so both pahoehoe and aa may show crust- or core-dominated behaviour. Pahoehoe sheet flow, for example, may be core dominated, while aa lavas may show short-term oscillations about a longer-term mean trend (see Ch. 3). Indeed, it can be argued that secondary bocca formation in aa lavas indicates they have entered the crustal-dominated regime.

Among aa lavas, the requirement that snout spreading is limited by crustal rehealing is a key part of the Kilburn & Lopes (1991) flow model. This model accounts for the observation on Etna that greater flow lengths are favoured by higher effusion rates and offers the hope of realistically forecasting maximum flow lengths from measurements of initial discharge rate. Moreover, when groundslope and flow front thickness are also known, it should be possible to determine if an uncrusted flow will evolve as aa or pahoehoe.

The results presented here are based on first-order modelling and further studies are needed to examine the underlying assumptions in detail. In particular, field observations are required on (a) crustal structure, (b) modes and rates of crustal disruption and rehealing, (c) rates of snout widening and advance, and (d) velocities and thicknesses in the frontal zone and how these change with time and with position in a flow. In addition to testing and refining the current model, such data will also help better identify emplacement regimes in terms of timescale and other process ratios.

Acknowledgements

Thanks are due to Joy Crisp for discussing earlier versions of this chapter. Part of this work was carried out during tenure of a National Research Council research associateship at the Jet Propulsion Laboratory, California Institute of Technology, under contract with the National Aeronautics and Space Administration.

C. R. J. Kilburn

Appendix. Lava chilling

Assuming a linear temperature gradient from the flow surface inwards, the thermal energy per unit surface area (E) lost by the crust at a time t after chilling begins is

$$E \sim \rho c_p[\theta_0 - (\theta_0 + \theta_s)/2]\delta$$
$$\sim \rho c_p\theta_0[1 - (\theta_s/\theta_0)]\delta/2, \tag{A10.1}$$

where ρ, c_p and θ_0 are the lava density, specific heat capacity and initial temperature (all temperatures absolute), and θ_s and δ are the surface temperature and crustal thickness at time t.

This energy is lost by surface radiation. Assuming a linear decrease in the rate of radiated energy loss with time then, by time t,

$$E \sim (\varepsilon\sigma t/2) [(\theta_0^4 - \theta_a^4) - (\theta_s^4 - \theta_a^4)]$$
$$\sim (\varepsilon\sigma\theta_0^4 t/2) [1 - (\theta_s/\theta_0)^4], \tag{A10.2}$$

where ε is the surface emissivity, σ is the Stefan–Boltzmann radiation constant and θ_a is the ambient temperature.

Equating Equations A10.1 & 2 leads to

$$t \sim (\rho c_p/\varepsilon\sigma\theta_0^3)\delta\varphi, \tag{A10.3}$$

where $\varphi = [1-(\theta_s/\theta_0)]/[1-(\theta_s/\theta_0)^4]$.

For conductive crustal thickening, $\delta \sim (4Kt/\rho c_p)^{1/2}$, where K is the lava thermal conductivity. Substituting this expression for δ in Equation A10.3 gives

$$t \sim 4\varphi^2(K\rho c_p/\varepsilon\sigma\theta_0^3)^2. \tag{A10.4}$$

When $t = t_{ch}$, θ_s tends to a constant value (actually decreasing slowly) and the rate of thermal energy loss is limited by conduction through the lava and not by surface radiation, as at higher temperatures.

Field observations show that lava darkens rapidly and continuously from an incandescent state. Since colour reflects temperature, it seems that near-constant temperatures are attained only after surface darkening. θ_s must thus lie between θ_a (say 300 K) and the maximum temperature for a dark surface (about 740 K, Table 10.1). Setting θ_0 to 1,400 K, φ ranges between 0.5 and 1 for θ_s between 740 and 300 K, respectively; corresponding limits for t_{ch} are $K\rho c_p/(\varepsilon\sigma\theta_0^3)^2$ and $4K\rho c_p/(\varepsilon\sigma\theta_0^3)^2$. Since θ_s is expected to be nearer 740 K than 300 K, t_{ch} is expected to be $\sim K\rho c_p/(\varepsilon\sigma\theta_0^3)^2$, as given by Equation 10.1 in the main text.

References

Baldwin, E. D. 1953. Notes on the 1880–81 lava flow from Mauna Loa. *The Volcano Letter* **520**, 1–3.

Berry, M. V., I. C. Percival, N. O. Weiss (eds) 1987. *Dynamical chaos. Proceedings of the Royal Society of London*. Princeton: Princeton University Press.

Booth, B., R. S. J. Sparks, G. P. L. Walker 1975. Notes on the eruption of Mount Etna in April 1975. UK *Research on Mt Etna 1974, Royal Society*, 69–71.

Borgia, A. & S. R. Linneman 1990. On the mechanisms of lava flow emplacement and volcano growth: Arenal, Costa Rica. In *IAVCEI Proceedings in Volcanology*. Vol. 2, *Lava flows and domes: emplacement mechanisms and hazard implications*, J. H. Fink (ed.), 208–43. Berlin: Springer.

Bullard, F. M. 1976. *Volcanoes of the Earth*. Austin: University of Texas Press.

Corsaro, R. A. & R. Cristofolini 1989. Campi di lava composta e regimi eruttivi dell'Etna. *Bolletino della Accademia Gioenia di Catania* **22**, 335–56.

Crisp, J. A. & S. M. Baloga 1990. A model for lava flows with two thermal components. *Journal of Geophysical Research* **95**, 1,255–70.

Cristofolini, R. 1984. L'eruzione etnea del 1983. *Atti della Accademia Gioenia di Catania* **160**, 39–78.

Cumin, G. 1954. L'eruzione laterale etnea del novembre 1950 - dicembre 1951. *Bulletin Volcanologique* **15**, 1–70.

Dutton, C. E. 1884. Hawaiian volcanoes. US *Geological Survey 4th Annual Report*, 75–219.

Fink, J. H. & R. W. Griffiths 1990. Radial spreading of viscous gravity currents. *Journal of Fluid Mechanics* **221**, 485–501.

Fink, J. H. & J. Zimbleman 1990. Longitudinal variations in rheological properties of lavas: Puu Oo basalt flows, Kilauea volcano, Hawaii. In *IAVCEI Proceedings in Volcanology*. Vol. 2, *Lava flows and domes: emplacement mechanisms and hazard implications*, J. H. Fink (ed.), 157–73. Berlin: Springer.

Frazzetta, G. & R. Romano 1984. The 1983 Etna eruption: event chronology and morphological evolution of the lava flow. *Bulletin Volcanologique* **47**, 1079–96.

Gauthier, F. 1973. Field and laboratory studies of the rheology of Mount Etna lava. *Philosophical Transactions of the Royal Society, London* **A274**, 83–98.

Glass, L. & M. C. Mackey 1988. *From clocks to chaos. The rhythms of life*. Princeton: Princeton University Press.

Guest, J. E., A. T. Huntingdon, G. Wadge, J. L. Brander, B. Booth, S. Carter, A. M. Duncan 1974. Recent eruptions of Mount Etna. *Nature* **250**, 385–7.

Guest, J. E., C. R. J. Kilburn, H. Pinkerton, A. M. Duncan 1987. The evolution of lava flow fields: observations of the 1981 and 1983 eruptions of Mount Etna, Sicily. *Bulletin of Volcanology* **49**, 527–540.

Hon, K., Kauahikaua, J., McKay, K. 1993. Emplacement and inflation of pahoehoe sheet flows – observations and measurements of active Hawaiian lava flows. *Bulletin of Volcanology* (in press).

Horai, K. 1991. Thermal conductivity of Hawaiian basalt: a new interpretation of Robertson and Peck's data. *Journal of Geophysical Research* **96**, 4125–32.

Hulme, G. 1974. The interpretation of lava flow morphology. *Geophysical Journal of the Royal Astronomical Society* **39**, 361–83.

Iverson, R. M. 1990. Lava domes modeled as brittle shells that enclose pressurized magma, with application to Mount St. Helens. In *IAVCEI Proceedings in Volcanology*. Vol. 2, *Lava flows and domes: emplacement mechanisms and hazard implications*, J. H. Fink (ed.), 47–69. Berlin: Springer.

Jaeger, J. C. & N. G. W. Cook 1979. *Fundamentals of rock mechanics*, 3rd edn. London: Chapman & Hall.

Jaggar, T. A. 1930. Distinction between pahoehoe and aa or block lava. *The Volcano Letter* **281**, 1–3.

Jeffreys, H. 1925. Flow of water in an inclined channel of rectangular section. *London, Edinburgh and Dublin Philosophical Magazine and Journal of Science* **49**, 793–807.

Kilburn, C. R. J. 1981. Pahoehoe and aa lavas: a discussion and continuation of the model by Peterson and Tilling. *Journal of Volcanology and Geothermal Research* **11**, 373–89.

Geological Society of America, Bulletin **84**, 615–626.

Tanguy, J.-C. & D. Biquand 1967. Quelques proprietés physiques du magma actuel de l'Etna. *Academie des Sciences à, Paris Comptes Rendus* **264**, 699–702.

US Geological Survey 1986. *Hawaii volcanoes 1:100,000 topographic map.*

Walker, G. P. L. 1967. Thickness and viscosity of Etnean lavas. *Nature* **213**, 484–5.

Walker, G. P. L. 1971. Compound and simple lava fields. *Bulletin Volcanologique* **35**, 579–590.

Walker, G. P. L. 1973. Lengths of lava flows. *Philosophical Transactions of the Royal Society, London* **A274**, 107–18.

Wentworth, C. K. & G. A. Macdonald 1953. Structures and forms of basaltic rocks in Hawaii. *US Geological Survey, Bulletin* **994**, 1–98.

Whitehead, J. A. & K. R. Helfrich 1991. Instability of a flow with temperature-dependent viscosity: a model of magma dynamics. *Journal of Geophysical Research* **96**, 4145–55.

Williams, H. & A. R. McBirney 1979. *Volcanology.* San Francisco: Freeman, Cooper.

Wolfe, E. W., C. A. Neal, N. G. Banks, T. J. Duggan 1988. Geologic observations and chronology of eruptive events. In *The Puu Oo eruption of Kilauea volcano, Hawaii: episodes 1 through 20, January 3, 1983, through June 8, 1984,* E. W. Wolfe (ed.). US Geological Survey Professional Paper 1463, 1–97.

CHAPTER ELEVEN

Thermal feedback mechanisms and their potential influence on the emplacement of lavas

Harry C. Hardee

Abstract

Heating by viscous dissipation occurs in all flowing fluids. When the fluid viscosity is also strongly temperature-dependent, thermal runaway may occur, in which viscous heating reduces fluid viscosity but triggers faster shearing and results in still higher rates of heat generation. In the crystallization interval, magmas have strongly temperature-dependent rheologies and the effects of thermal feedback may impose limits on eruptive behavior. Examples discussed here are limiting conduit sizes for effusive eruption, limiting lava thicknesses for sustained flow, and temperature changes during emplacement of lavas and of ash or mud flows.

Introduction

The thermofluid analysis of lava flows must consider both the flow geometry and the thermofluid properties of the flowing lava. The flow geometry may be treated as a one-dimensional flow layer for large surface flows, or it may be treated as conduit flow for localized flow regions. Conduit flow can be either open or closed; for instance, flow in a lava tube is analyzed as flow in a closed conduit, while flow in a lava channel is analyzed as open-conduit (or channel) flow. Such flow geometries are fully discussed in elementary fluid mechanics texts.

Eruption of magma can produce liquid flows (lavas) or solid flows (ash flows). Although liquid flows frequently contain some entrained solids, the physical laws governing flow characteristics depend on whether the flow acts primarily as a liquid or a fluidized powder. In the case of lavas, flow can be treated as that of a viscous non-Newtonian fluid (Gauthier 1973, Hulme 1974, Kushiro et al. 1976, Hardee & Dunn 1981, Dragoni et al. 1986). The lavas usually consist of materials at subliquidus, but greater than solidus, temperatures, although liquidus temperatures may occasionally be approached (Macdonald 1963); at temperatures greater than the liquidus, they appear to act as simple Newtonian fluids (Gauthier 1973, Kushiro et al. 1976, Hardee & Dunn 1981).

Solid flows, in contrast, consist of powder-like material such as ash at subsolidus temperatures. These can be analyzed using theories developed for powders and fluidized beds (Soo 1982, Cheremisinoff & Cheremisinoff 1984).

Heating by viscous dissipation occurs in all flowing fluids, although the effect is usually small in common liquids. This type of heating is a result of the conversion of some of the flow energy into heat through internal dissipation of fluid motion by viscous effects. Viscous heating is larger in fluids like lava, rather than simpler liquids like water, because of the higher viscosity of the lava and its potential non-Newtonian behavior. Viscous dissipation in lavas may result in a temperature rise of only a few degrees centigrade but even such a modest rise may be very important in eruptive phenomena. In particular, heat losses from lavas are minimized by the insulating effects of chilled crust. Thus, only a small amount of viscous dissipation heating is required to maintain high lava temperatures.

In addition to the direct effect that viscous dissipation heating can have on flow temperature, the related mechanism of *thermal feedback* is important for establishing eruptive conditions and, in some cases, may be important to lavas. Thermal feedback can occur if the lava viscosity is highly temperature dependent. In this case, viscous heating not only causes the lava temperature to rise by a few degrees, but also significantly reduces its viscosity (possibly by orders of magnitude). The reduced viscosity results in faster velocities and increased viscous dissipation until the process begins to run away. In feeding conduits, such a runaway condition can also be important to such eruption characteristics as extrusion rate, eruption temperature and, possibly, cyclic eruption frequency (Shaw 1969, Fujii & Uyeda 1974, Hardee & Larsen 1977, Nelson 1981, Hardee 1986, 1987).

The study of thermal runaway caused by temperature-dependent internal heat generation goes back many years. One of the first investigators was Frank-Kamenetskii (1939), followed by many others, including Chambre (1952) and Thomas (1958). Many of these investigators used an Arrhenius heat generation term in the energy equation. Gruntfest (1963) and Shaw (1969) noted that heat generation and thermal feedback could have important implications in geological problems, particularly those involving magma and lava. Fujii & Uyeda (1974) and Hardee & Larson (1977) both showed that thermal runaway can occur during magma ascent through conduits.

Thermal feedback can also occur in surface lava flows, although, with respect to magma ascent, the degree of feedback is more restricted by lower driving forces (due to gravity) and, sometimes, by mass rates of supply. The next two sections will examine the conditions for thermal runaway both during magma ascent and subsequent emplacement as lava flows. The final section will briefly consider implications for particulate systems, such as ash flows.

Viscous energy dissipation and thermal feedback in conduits

The rate of viscous energy dissipation per unit volume of fluid for one-dimensional flow through a slot-like conduit is given by $\mu(du/dy)^2$ (Gebhart 1961), where μ is the magma viscosity and du/dy is the velocity gradient across the conduit width (the cross-flow extent of the conduit perpendicular to its width is considered to be infinite). As Gebhart notes, the net result of this kind of term in the energy equation is equivalent to the presence of a distributed source of energy in the fluid. Thus, a thermal energy balance per unit volme of magma for one-dimensional flow yields

$$\rho c_p u(dT/dx) = -k(d^2T/dy^2) + \mu(du/dy)^2, \tag{11.1}$$

where ρ, c_p, μ and k correspond to the magma density, specific heat capacity, viscosity and thermal conductivity, T is the temperature, u is the velocity in the x direction and y measures the position in the cross-stream direction.

Equation 11.1 describes the rate of change of thermal energy (left-hand side) due to conductive heat transfer (first term on the right-hand side) and viscous heat generation (second term on the right-hand side). The additional energy supplied by the viscous dissipation term can here offset some of the loss of energy through the thermal boundary layers along the conduit walls and so help perpetuate flow. If the viscosity is strongly temperature-dependent, then thermal feedback can occur and may lead to sudden large temperature rises in the magma, or thermal runaway.

Shaw (1972) has shown that the temperature dependence of viscosity for Newtonian magma can be approximated by an Arrhenius relation of the form

$$\mu = a \exp(b/T), \tag{11.2}$$

where T is the absolute temperature of the magma and a and b are constants. The constant a is related to Avogadro's number, Planck's constant and the mole volume of the magma. The constant b is equal to the molar free energy of activation divided by the universal gas constant. In other words, a is primarily related to the absolute magma viscosity, while b is primarily related to the temperature variation of viscosity.

As discussed by Bird et al. (1960), Equation 11.2 is valid for pseudoplastic fluids in general (as well as Newtonian fluids) as long as the ratio $SV/RT \ll 1$, where S is the local shear stress, V is the volume of a mole of fluid and R is the gas law constant. This approximation is satisfied for most subliquidus lavas, particularly those of basaltic compositions. Even when the approximation is not reasonable, Bird et al. (1960) show that the major shear stress temperature dependence for pseudoplastic liquids will be an exponential inverse temperature dependence of the the form of Equation 11.2.

Theoretical analyses for magma flow in conduits (Fujii & Uyeda 1974, Hardee & Larson 1977, Hardee 1986, 1987) indicate that eruption temperature is controlled by viscous dissipation effects and the pressure driving flow, usually the gravitational potential. The critical temperature rise (θ_c) after which thermal runaway begins is a function only of the constant b and wall temperature (T_w). The associated critical magma temperature T_c at the center of the magma conduit is then defined as:

$$T_c = \theta_c + T_w. \tag{11.3}$$

Here T_w is the wall temperature or effective solid temperature (i.e. the temperature above which motion occurs), usually considered (e.g. Shaw et al. 1968) to be the temperature corresponding to 50% crystallinity (i.e. higher than the magma solidus) and θ_c is the excess temperature above this value at which thermal feedback effects predominate.

To a good approximation (Hardee 1986)

$$\theta_c = T_w^2/(b + T_w). \tag{11.4}$$

For Hawaiian basalt, $T_w = 1{,}343$ K and the viscosity can be approximated by

$$\mu = 10^{-6} \exp(26{,}170/T) \tag{11.5}$$

(thus $a = 10^{-6}$ Pa s and $b = 26{,}170$ K). Substituting these values into Equation 11.4 gives θ_c a value of 76.8 K. The corresponding value for T_c is thus $1{,}343 + 76.8$, or about 1,420 K (1147°C).

This temperature is well below the liquidus (at about 1,220°C) for Hawaiian basalt (Hardee & Dunn 1981). Although the concept of thermal runaway suggests that the liquid temperature would rise dramatically at this condition, the temperature tends to be limited close to the critical value T_c because (a) of the inability of the supply to maintain the increased mass flow rates and required pressure gradient, and (b) the flow rate tends to be buffered by melting of the conduit walls and resultant absorption of heat by phase change.

Calculations for andesitic and rhyolitic magmas give similar results. Viscosity data obtained by Murase & McBirney (1973) for Mount Hood andesite can be fitted to the equation

$$\mu = 1.122 \times 10^{-8} \exp(38{,}330/T). \tag{11.6}$$

Using 38,330 K for b and assuming a melt temperature of 1,223 K (950°C) leads to a runaway excess temperature (θ_c) of 42 K. In the case of rhyolitic magmas, Piwinskii & Weed (1980) obtained the viscosity relation

$$\mu = 0.857 \times 10^{-6} \exp(36{,}194/T). \tag{11.7}$$

Assuming a melt temperature of 873 K (600°C), and setting $b = 36{,}194$ K, θ_c for rhyolite is 22 K.

While the value of the runaway excess temperature varies with composition as the viscosity function varies, the above values of θ_c are typical.

The preceding analysis assumes that an apparent viscosity exists for subliquidus magma. A more rigorous analysis should treat lava as a pseudoplastic fluid, using a rheological model similar to that developed by Bird et al. (1960). The viscous energy dissipation term in the energy equation (Equation 11.1) would be defined in terms of shear stress and velocity gradient rather than viscosity. The shear stress result from the Bird et al. (1960) general liquid model would then be used in the heat generation term. It turns out that the rheological term replacing viscosity in Equation 11.1 is still primarily a function of $\exp(b/T)$, where b is the same as in the simpler Equation 11.2. Indeed, Bird et al. (1960) show that the constant b depends only on the free energies of activation and the universal gas constant, regardless of whether the liquid is Newtonian or not. The condition for thermal runaway for this general, non-Newtonian case remains similar to Equation 11.4.

The value of b is determined from laboratory measurements of lava viscosity as a function of temperature. Magma and freshly erupted lava may have volatile contents different from those of remelted samples. The volatile content, particularly of H_2O, is known to have a significant effect on viscosity (Shaw 1972, Yoder 1982). However, as can be seen from the viscosity/H_2O data in Shaw (1963), it does not have a strong effect on the value of b or the runaway excess temperature θ_c: laboratory measurements should thus yield reliable values of b and θ_c.

Critical conduit diameters

The energy and momentum equations for flow can be combined with the results from the thermal feedback analysis to determine the critical diameters (D_c) of feeding conduits, that is, the minimum conduit diameter that can support and maintain flow. D_c is a function of θ_c, as well as other factors (Hardee 1986, 1987):

$$D_c = [32/(P/L)]^{1/2}(9k\mu\theta_c/8)^{1/4}. \tag{11.8}$$

where k is the magma thermal conductivity and P/L is the pressure gradient in the direction of flow.

For the frequently studied case of vertical conduit flow driven by gravitational buoyancy, $P/L (= g(\rho - \rho_{country})$, where $\rho_{country}$ is the density of the country rock surrounding the conduit) is nominally 1,000 Pa m^{-1}. Assuming typical magma properties (Hardee 1986), Equation 11.8 gives critical conduit diameters of 1.9 m for basalts, 10.6 m for andesites and 336 m for rhyolites. These values, which agree with calculations by Fujii & Uyeda (1974), predict the minimum sizes of conduits that can remain open because of heat generated by thermal feedback. When such conduits become smaller, heat losses to the walls dominate and allow the conduit magma to solidify.

Thermal feedback in lava flows

The calculations on conduit diameter can be extended to tubes or open channels in lava flows. The pressure gradient P/L is now $\rho g H/L$, where H/L is the slope in the direction of flow. For a slope of 1/30 (angle of about 2°), P/L is approximately 900 Pa m^{-1}, implying from Equation 11.8 critical tube or channel diameters similar to those calculated above for conduits (when $P/L = 1,000$ Pa m^{-1}). On 10 and 0.6° slopes, the corresponding diameters would be about half and twice those for a 2° slope.

In the case of lavas advancing as wide sheets (width much greater than depth, even if bounded by lateral margins), the effects of thermal feedback can be estimated assuming approximately Couette flow (Gebhart 1961). This yields for a crusted flow with the same constant basal and surface temperature T_w (see first part of Gebhart's solution)

$$T - T_w = (h^4/12\,\mu k)(P/L)^2[y/h - (y/h)^4], \tag{11.9}$$

where h is the flow depth and y is the perpendicular distance from the flow surface (i.e. y is 0 at the flow surface and h at the flow base); it is also assumed that the flow velocity increases from zero at the flow base to a maximum at its top surface.

Introducing T_b as the bulk lava temperature, found by averaging the temperature across the flow, it is found that

$$T_b - T_w = \frac{3}{10}(h^4/12\,\mu k)(P/L)^2 \tag{11.10}$$

and, at the onset of thermal runaway, $T_b - T_w = \theta_c$. Substituting this result into Equation 11.10 yields the critical flow thickness hc to be

$$h_c = [32/(P/L)]^{1/2}(5k\mu\theta_c/128)^{1/4}, \tag{11.11}$$

a result very similar to that obtained for the critical diameter in conduit flow (Equation 11.8).

Setting $P/L = \rho gH/L$, Equation 11.11 becomes

$$h_c = [32/(\rho gH/L)]^{1/2}(5k\mu\theta_c/128)^{1/4}. \tag{11.12}$$

For a 2° slope ($H/L = 1/100$), Equation 11.12 gives a critical flow thickness of 2 m. Once again (because h_c varies with the square root of $1/(P/L)$), critical depths are about half and twice this value for 10 and 0.6° slopes respectively.

Flows thinner than h_c should solidify comparatively rapidly since there is no mechanism to make up for heat lost through the flow surface and base. One can argue that thicker flows will spread in width, if unrestrained, until their depths approach h_c. This suggests that a thermal feedback mechanism may cause flows to develop thicknesses near h_c.

During Kilauea's 1983 eruption, for example, a steady lava flow, 2 m thick, developed on a 1.4° slope ($H/L = 0.024$) during the afternoon of 26 February. Using a thermocouple, the temperature at the vent was measured as 1,118°C. Measurements made less than an hour later 2.5 km downstream gave values of 1,119°C (Hardee 1983). While the apparent 1°C increase is probably within statistical error, the measurements were carefully made with the same accurately calibrated and electronically compensated digital temperature indicator. The interesting point is that the flowing lava encountered no noticeable temperature reduction in flowing 2.5 km and may even have gained in temperature. The possibility of a small temperature gain is not unreasonable and, indeed, Shaw & Swanson (1970) note that viscous dissipation heating could in some cases actually produce superheating.

Thermal feedback in solid flows

Solid flows, such as ash flows, can be analyzed using theories developed for the flow of two-phase gas/solid systems, flow of powders, and fluidized beds (Eckert & Drake 1959, Grace 1982, Soo 1982, Cheremisinoff & Cheremisinoff 1984). There are some interesting effects that can occur with these kinds of flows. Volcanic ash, for example, is an excellent thermal insulator. Indeed, some of the best commercial insulation uses compacted powders that are chemically similar to ash. During ash flow, therefore, any internal heating, such as exothermic reactions or frictional heating due to flow, are enhanced by the outer ash insulation.

This situation can be treated as an insulated body with internal heat generation and, theoretically, it can lead to very high internal temperatures, even if the heating mechanism is small. An example is the heating by gravitational energy during emplacement of a relatively cool ash flow. For an ash body at ambient temperature flowing downslope, a simple energy balance indicates that gravitational energy will be converted to heat and kinetic energy. The kinetic energy gain is a small factor and can be ignored. The temperature rise of the ash (ΔT) is then approximately

$$\Delta T = gH/c_p, \tag{11.13}$$

where g is gravitational acceleration, H is the vertical drop in elevation of the flow, and c_p is the specific heat capacity of ash ($2{,}000$ J kg^{-1} K^{-1}). This result is roughly equivalent to a 5 K rise in temperature for every kilometer of descent.

Such an effect may have been important during the 1985 eruption of Ruiz volcano in Colombia. A mud/ash flow, of ice melted by the 1985 activity and ash from an earlier eruption, apparently heated up from ambient temperatures during its descent from the volcano into the valley below (Decker 1982, personal communication). The summit of Ruiz stands at 5,200 m and so the ash flow had dropped several kilometers before entering the valley. This could have produced a 10–20 K temperature rise if all the potential energy was converted to heat by viscous effects. Under some conditions, a temperature rise by such an amount is sufficient to produce second- and third-degree burns on people and animals trapped in the flow (Hardee 1977), even if they were not completely buried.

Conclusions

Thermal feedback induced by viscous heating may profoundly influence how flowing magmatic systems behave. For effusive phenomena in particular, thermal feedback may limit the widths of magma conduits and the thicknesses of lava flows. Viscous heating may also maintain or even increase temperatures in lavas and ash or mud flows. In the case of lavas, the maintained temperatures may allow lavas to advance much further than anticipated by first-order thermal analyses. A goal for future lava measurements in the field is to establish the effectiveness of viscous heating, especially near the flow base and channel/levée margins where rates of shearing are highest.

References

Bird, R. B., W. E. Stewart, E. N. Lightfoot 1960. *Transport phenomena*. New York: John Wiley.

Chambre, P. L. 1952. On the solution of the Poisson–Boltzmann equation with application to the theory of thermal explosions. *Journal of Chemical Physics* **20**, 1795–7.

Cheremisinoff, N. P. & P. N. Cheremisinoff 1984. *Hydrodynamics of gas–solids fluidization*. Houston: Gulf.

Dragoni, M., M. Bonafede, E. Boschi 1986. Downslope flow of a bingham liquid: implications for lava flows. *Journal of Volcanology and Geothermal Research* **30**, 305–25.

Eckert, E. R. G. & R. M. Drake 1959. *Heat and mass transfer*. New York: McGraw-Hill.

Frank-Kamenetskii, D. A. 1939. Diffusion and heat exchange in chemical kinetics. *Acta Physicochimica (USSR)* **10**, 365.

Fujii, N. & S. Uyeda 1974. Thermal instabilities during flow of magma in conduits. *Journal of Geophysical Research* **79**, 3367–9.

Gauthier, F. 1973. Field and laboratory studies of the rheology of Mount Etna lava. *Philosophical Transactions of the Royal Society, London* **A274**, 83–98.

Gebhart, B. 1961. *Heat transfer*. New York: McGraw-Hill.

Grace, J. R. 1982. Fluidization. In *Handbook of multiphase systems*, G. Hetsroni (ed.), Ch. 8. New York: Hemisphere.

Gruntfest, I. J. 1963. Thermal feedback in liquid flow: plane shear at constant stress. *Transactions of the Society of Rheology* **7**, 195–207.

Hardee, H. C. 1977. A simple conduction model for skin burns resulting from exposure to chemical fireballs. *Fire Research* **1**, 199–205.

Hardee, H. C. 1983. Heat transfer mechanisms of the 1983 Kilauea lava flow. *Science* **222**, 47–8.

Hardee, H. C. 1986. Replenishment rates of crustal magma and their bearing on potential sources of thermal energy. *Journal of Volcanology and Geothermal Research* **28**, 275–96.

Hardee, H. C. 1987. *Heat and mass transport in the east-rift-zone magma conduit of Kilauea volcano*. US Geological Survey Professional Paper 1350, 1471–86.

Hardee, H. C. & J. C. Dunn 1981. Convective heat transfer in magmas near the liquidus. *Journal of Volcanology and Geothermal Research* **10**, 195–207.

Hardee, H. C. & D. W. Larson 1977. Viscous dissipation effects in magma conduits. *Journal of Volcanology and Geothermal Research* **2**, 299–308.

Hulme, G. 1974. The interpretation of lava flow morphology. *Geophysical Journal of the Royal Astronomical Society* **39**, 361–83.

Kushiro, I., H. S. Yoder, B. O. Mysen 1976. Viscosities of basalt and andesite melts at high pressures. *Journal of Geophysical Research* **81**, 6351–6.

Macdonald, G. A. 1963. Physical properties of erupting Hawaiian magmas. *Geological Society of America, Bulletin* **74**, 1071–8.

Murase, T. & A. R. McBirney 1973. Properties of some common igneous rocks and their melts at high temperatures. *Geological Society of America, Bulletin* **84**, 3563–92.

Nelson, S. A. 1981. The possible role of thermal feedback in the eruption of siliceous magmas. *Journal of Volcanology and Geothermal Research* **11**, 127–37.

Piwinskii, A. J. & H. C. Weed 1980. Dynamic viscosity of some silicate melts to 1688°C under atmospheric pressure. *Thermochimica Acta* **37**, 189–95.

Shaw, H. R. 1963. Obsidian–H_2O viscosities at 1000 and 2000 bars in the temperature range 700 to 900°C. *Journal of Geophysical Research* **68**, 6337–43.

Shaw, H. R. 1969. Rheology of basalt in the melting range. *Journal of Petrology* **10**, 510–35.

Shaw, H. R. 1972. Viscosities of magmatic silicate liquids: an empirical method of prediction. *American Journal of Science* **272**, 870–93.

Shaw, H. R. & D. A. Swanson 1970. Eruption and flow rates of flood basalts. In *Proceedings of the Second*

Columbia River Basalt Symposium, H. Gilmour & D. Stradling (eds), 271–99. East Washington State University.

Shaw, H. R., T. L. Wright, D. L. Peck, R. Okamura 1968. The viscosity of basaltic magma: an analysis of field measuremets in Makaopuhi lava lake, Hawaii. *Journal of Geophysical Research* **266**, 225–64.

Soo, S. L. 1982. Gas–solid systems. In *Handbook of multiphase systems*, G. Hetsroni (ed.), Ch. 3. New York: Hemisphere.

Thomas, P. H. 1958. On the thermal conduction equation for self-heating materials with surface cooling. *Transactions of the Faraday Society* **54**, 60–65.

Yoder, H. S. Jr 1982. Experimental methods for determination of transport properties of magma. *Physics and Chemistry of the Earth* **13/14**, 375–408.

CHAPTER TWELVE

Cellular automata methods for modelling lava flows: simulation of the 1986–1987 eruption, Mount Etna, Sicily

D. Barca, G. M. Crisci, S. Di Gregorio, F. Nicoletta

'A lava

Ora la sciara niura e pitrusa
'nfoca a lu suli tutta la jurnata,
nuddu aceddu ca passa si cci posa,
non crisci 'n filu d'erba acquazzinata

The lava

Now dead, the black and stony flow,
The whole day in the sun inflames.
No passing bird alights there,
No dampened blade of grass can grow.

Francesco Guglielmino

Abstract

This chapter describes a two-dimensional, cellular automata model for simulating lava flows and its application to the 1986–87 flow field on Mount Etna, Sicily. A lava is viewed as a dynamic system whose growth can be followed through local interactions at discrete time and space intervals. The space intervals are represented by cells. Each cell is characterized by specific values of selected physical parameters, including flow altitude,

291

thickness and lava temperature. Lava rheology is considered indirectly through its effect on lava thickness. The boundary values constraining a simulation are those describing the underlying topography, lava discharge rate, eruption temperature and rheology, and solidus temperature and rheology. The cellular automata model has been tested against growth data for Etna's 1986–87 flow field. Even though the data set is heterogeneous, the model and real flow fields show strikingly similar growth patterns. The close similarity highlights the flexibility of the cellular automata approach, especially the ability to monitor growth at small time intervals and, by adjusting surface topography as emplacement continues, to describe the formation of numerous flows.

Introduction

An important goal in volcanology is the ability to simulate eruptive phenomena to enable reliable forecasting of their potential hazard. Such phenomena, including lava flows, show complex behaviour and their governing equations cannot be solved analytically without making substantial simplifications. Detailed simulations thus rely on numerical solutions to the governing equations, a technique ideally suited to modern computing methods.

Traditional flow simulations are based on the differential conservation equations (see Ch. 13) and require knowledge of how the physical state of a system (e.g. lava viscosity and density) and enviromental conditions (e.g. surface topography) change with time and position. In the case of lava flows, particular complexities arise because of irregular ground topography and because lavas, while solidifying during emplacement, may range rheologically from approximately Newtonian liquids to brittle solids. As a result, it is extremely arduous to solve the governing flow equations (e.g. the Navier–Stokes equations for viscous fluids).

To overcome these difficulties, several lines of approach have been tried for simulating lava flow growth. Condarelli et al. (1975) and R. S. J. Sparks (personal communication) considered scale models of an area, combining analogue and digital computing techniques. Such models are strongly dependent on the non-trivial reconstruction of surface morphology and (to our knowledge) have not been further developed. Crisci et al. (1982, 1986) and Barca et al. (1987a,b) designed a three-dimensional cellular automata model and then simplified it to a two dimensional version (Barca et al. 1988a,b). This new approach, in which changes in time are considered as discrete, not continuous, quantities, overcomes many of the complexities associated with differential equations and allows such features as multiple flow development to be simulated. Using an approach similar to cellular automata, Ishihara et al. (1988, 1990) started from the Navier–Stokes equations and deduced formulations for discrete space and time intervals; their method, however, cannot be applied to multiple flows or to flows which are extruded intermittently. Finally, Young & Wadge (1990) have presented a clever simulation, faster than those cited above, but applicable only to simple flow fronts.

This chapter reviews the two-dimensional automata simulation for lava flows of Barca et al. (1988a,b) and shows the results obtained for Etna's 1986–87 flow field.

Cellular automata

Cellular automata (CA) are a powerful tool for modelling natural phenomena. Conceived in the 1950s to investigate self-reproduction (von Neumann 1966), CA have been used mainly for studying the parallel computing and the formal properties of model systems (Burks 1970, Lindenmayer 1971). However, with the rapid advances in computational resources during the 1980s, CA have become increasingly utilized for more general computer simulations (Toffoli & Margolus 1987); examples in the Earth sciences include studies of earthquake genesis (Rundle 1989) and landslide behaviour (Barca et al. 1987c). The applied aspects of CA modelling have also been widely investigated from a theoretical viewpoint (Vichniac 1984, Margolus et al. 1986).

CA capture the peculiar character of systems (e.g. a lava flow) that evolve through local interactions of their constituent parts (Burks 1970, Wolfram 1984), while guaranteeing computational universality (Thatcher 1970). To model local behaviour, the system is partitioned into cells of uniform size. Cells are distinguished by their values for the parameters chosen to describe the physical state of a systems (e.g. local lava temperature and thickness). Once initial values are assigned at the start of the simulation, the values of the parameters in each cell may vary as a result of (a) interaction between neighbouring cells, and (b) changes in imposed conditions which affect all cells simultaneously. Such changes are considered for the system as a whole after discrete time intervals.

The CA approach thus models a system in terms of discrete volume (i.e. a cell) and time intervals, unlike differential equation modelling, which examines continuous variations in conditions with time and position (Toffoli 1984). As a result, CA modelling can be much simpler than using differential equations when variations are highly non-linear, while the reverse is true for linear or weakly non-linear systems (e.g. a pendulum).

In our case, the CA simulation treats a lava surface as a collection of uniform cells. Flow is allowed to occur from one cell to an adjacent cell only if (a) the lava surface is lower in the adjacent cell, and (b) the lava has not solidified. A key feature of the simulation is that it describes the changing physical state of a lava in terms of variation in flow thickness and temperature. The methods by which this is achieved are described qualitatively in the next section. A more formal presentation, with the algorithms used, is given in the Appendix.

Design of the simulation

By breaking a flow down into cells, the task of the CA simulation is to describe the flow of lava between adjacent cells given the boundary constraints affecting the lava flow as a whole. Cell interaction is described by "local parameters", while the boundary conditions are described by "global parameters". These are treated separately in the next two sections.

Local parameters: the cells of a lava flow

The flow surface is treated as a mesh of square cells. Each cell thus has four neighbours, labelled "north", "south", "east" and "west".

The parameters chosen to characterize the state of the lava in a given cell are its altitude, lava thickness and temperature, and the four lava fluxes, or outflows, to adjacent cells. Other physical properties, such as lava rheology, are evaluated indirectly through their influence on the chosen parameters. After each time interval, the parameter values are changed as follows.

(a) The cell altitude remains unchanged until lava begins to solidify. Since a single value is used to characterize flow temperature (see below), the altitude of a cell is increased by the local flow thickness once the lava temperature drops below the solidus. At the same time, the lava thickness of the cell is reset to zero.

(b) Until solidification, the cell thickness increases by the difference between the net inflow and outflow through a cell. Assuming a constant-density lava, mass conservation gives that lava accumulating in a cell is the difference between the mean volume input and output during the given time interval. Since the surface dimensions of a cell are specified, the amount of lava accumulated can be described in terms of changes in lava thickness.

The outflow from a cell (or, alternatively, part of the inflow into an adjacent cell) depends on the hydrostatic pressure gradients across the cell due to differences in lava thicknesses compared with neighbouring cells, and on the rheology of the lava in the cell. The effect of variations in lava thickness are accounted for by minimizing the total height difference (cell altitude plus lava thickness) between a cell and its four neighbours after each time interval.

During solidification, the rheological resistance of a lava is strongly dependent on temperature, the resistance increasing as the temperature decreases. Because of complexities inherent in specifying lava rheology and its variation with temperature (e.g. McBirney & Murase 1984, Ch. 9 this volume), the authors have chosen to model rheological resistance in terms of an "adherence parameter" v, which represents the amount of lava (expressed as a thickness, as discussed above) that cannot flow out of a cell because of rheological resistance. v is assumed to vary with temperature (T) according to a simple inverse-exponential relation, $v = a\,e^{-bT}$, where a and b are constants describing lava rheology. Results from earlier simulations of non-Etnean flows (Barca et al. 1987a,b, 1988a,b) and preliminary studies of Etnean flows other than the main 1986–87 flow field suggest that the Etnean examples are reasonably represented by $v = 0.7$ m for $T = 1{,}373$ K and $v = 7$ m for $T = 1{,}123$ K (the nominal solidus). For these limiting conditions, the constants a and b become 2.17×10^5 m and 0.0092 K^{-1}, respectively.

(c) Changes in cell temperature are modelled as a two-step process to describe changes due to lava motion through a cell and to thermal energy loss from the cell surface. It is also assumed that a flow can be treated as if thermally well mixed, so that each cell has the same temperature throughout. Although such a cooling model may not always be realistic (Kilburn & Lopes 1991), it does yield a characteristic cooling timescale comparable to the typical emplacement times of aa flows (Crisp & Baloga 1990, Kilburn & Lopes 1991, Ch. 10 this volume).

To account for heat transported by lava motion, the first step takes the average temperature of lava passing through a cell and assigns that value to the cell at the end of each time interval. The second step then estimates the temperature drop due to thermal energy losses at the surface, assuming that losses through other sides of the cell (by other than flow motion) are small in comparison. For lava much hotter than ambient temperatures, the rate of energy loss is approximated by (Park & Iversen 1984)

$$\rho c V \frac{dT}{dt} = -\varepsilon \sigma A T^4,$$

where ρ is the lava density, c the specific heat, V the volume, σ the Stefan–Boltzmann constant $(5.67 \times 10^{-8} \text{ W m}^{-2} \text{ K}^{-4})$, T the temperature, A the surface area of the cell and ε is the surface emissivity $(0 \leqslant \varepsilon \leqslant 1)$.

Integrating the flux equation from t_1 to t_2, we obtain the new temperature of lava in the cell after a time interval Δt $(\Delta t = t_2 - t_1)$ as

$$T_2 = T_1[1 + (3T_1^3 \varepsilon A \, \Delta t)/(\rho c V)]^{-1/3} = T_1[1 + (T_1^3 p A/V)]^{-1/3},$$

where the "cooling parameter" $p = 3\varepsilon\sigma \, \Delta t/\rho c$. The combination $\varepsilon\sigma/\rho c$ describes the physical properties of the lava, while the time interval Δt depends on the length of a cell's side. For the simulation, p was determined empirically starting from values in Harrison & Rooth (1976) and following the model of Dragoni (1989); the nominal value obtained was $4.7 \times 10^{-14} \text{ m}^3 \text{ K}^{-3}$.

Global parameters: initial and boundary constraints

The previous procedures describe the local flow of lava from one cell to the next. Such interaction, however, is also constrained by conditions limiting the bulk evolution of a flow field. For the CA simulation, the relevant initial and boundary constraints are as follows:

(a) lava flow temperature at the vent, T_V, and at solidification, T_S;
(b) lava adherence, v, at the vent and at solidification;
(c) cooling parameter, p;
(d) lava discharge rate;
(e) groundslope, topography and location of vents.

To these natural constraints must be added the computational limit imposed by assigning the surface length, L, of a cell.

295

Implementation and methodology

The model was implemented on a Macintosh II fx computer. The SCIARA program was written in Think Pascal and the results displayed on-screen in cartographic form.

The program flow chart is shown in Figure 12.1. The assigned physical and chemical parameters remain fixed in "read global parameters", while topographic changes induced by flow solidification are automatically introduced in "σ". Discontinuous features of an eruption, such as the opening of new vents and associated changes in effusion rate, can be simulated by interrupting the program and changing the values of relevant parameters. The first step in the simulation is to assign values to the global parameters (Naylor 1966, Emshoff & Sisson 1970). The simulation then describes how the flow field lengthens, widens and thickens with time.

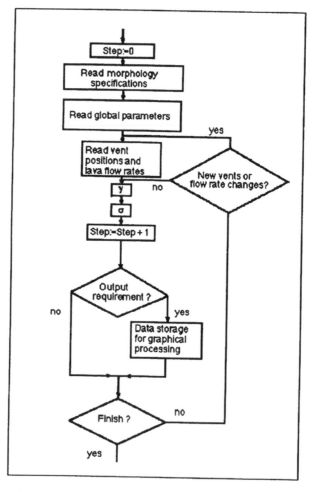

Figure 12.1 Flow chart of the SCIARA program. The function y adds the emitted lava to the thickness of the lava source cells. The function σ updates the cell parameters.

296

Simulation of the 1986–87 Etnean lava flow field

The 1986–87 eruption produced a complex flow field typical of long-duration effusions on Etna (see Ch. 3). This eruption was chosen to test the CA simulation because sufficient data are available for both the chronology of flow-field emplacement and the pre-eruption surface morphology. The eruption chronology (Caltabiano et al. 1987) is summarized below (see also Table 12.1 & Fig. 12.2).

Table 12.1   Evolution of Etna's 1986–87 flow field (all times local)

1986	30 October, 3 a.m.	Lava flow begins at 2,900 m a.s.l.
	30 October, 11 a.m.	Lava reaches 1,750 m a.s.l.
	30 October, 1 p.m.	Two vents open at 2,200 m a.s.l.
	30 October, 2 p.m.	The vent at 2,900 m a.s.l. ceases activity
	31 October, 8 a.m.	Flows from the two 2,200 m vents reach 1,700 m a.s.l.
	17 November	A vent opens at 2,600 m a.s.l.
	17 November	A vent opens at 2,350 m a.s.l. (Monte Rittmann)
	19 November	Flows reach 1,300 m a.s.l.
	25 November	The two 2,200 m vents stop activity
1987	25 January	The 2,600 m vent stops activity
	27 February	The last vent (Monte Rittmann) stops erupting; for the past month the eruption rate has been extremely low

The main effusion began early on 30 October 1986 from vents aligned along a fissure between 2,900 and 2,500 m a.s.l. (above sea level) on the western wall of the Valle del Bove (Fig. 12.2). Within 8 hours the lava had advanced 2.5 km, descending to 1,750 m a.s.l.

In the same interval, another vent opened at higher altitude, slowly feeding a small lava heading south-east. By early afternoon, a second fissure had opened between 2,250 and 2,200 m a.s.l., to the north-west of Monte Simone. New flows issued from two of the four active vents on the fissure. Passing around Monte Simone, the flows joined to feed a single front heading east-south-east.

Supplied by the vents at 2,250–2,200 m (active until 25 November) and 2,350 m, the flow field evolved as a complex fan, almost 2 km wide, of overlapping and adjacent flows. Initially forming against the northern wall of the Valle del Bove, the flow field expanded southwards with time, another branch finally forming to the north-east of Monte Centenari.

Most of the flows halted above the level of Rocca Musarra (at 1,500 m a.s.l.). The longest flow, however, reached 1,300 m a.s.l. on 19 November, about 5 km from its source.

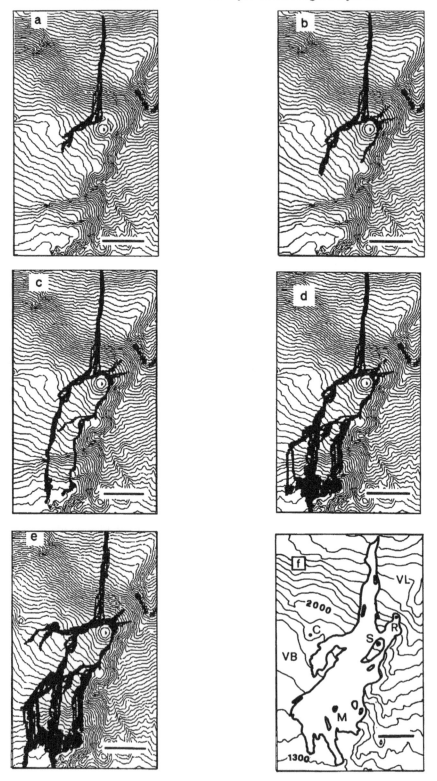

Simulation parameters

Although the eruption was well observed (Caltabiano et al. 1987), gaps occur in the available data set because of bad weather (typical for the winter season) and poor accessibility to all parts of the flow field. Nevertheless the data are sufficient to describe the main changes in vent location and in flow field lengthening and widening.

Less well constrained are variations in discharge rate with time. To accommodate the average effusion rate of 5.5 m^3s^{-1}, the event was simulated keeping discharge rates between mainly 2 and 12 m^3s^{-1}. Only for two brief periods were larger values assumed: during the first hours of eruption (22 m^3s^{-1}) and on the 18th day of activity (17 November) when four vents were feeding the flow field (22 m^3s^{-1}).

The pre-eruption topography was obtained from 1:10,000 base maps of the area (Regione Siciliana 1988). Cell dimensions of 10 m were chosen to satisfy the requirements that (a) a cell is much smaller than the flow width and, possibly, comparable to (or smaller than) the flow depth, and (b) the total number of cells remains manageable for the computational resources used. The initial lava temperature was set at 1,373 K (1,100°C) and the cooling coefficient p taken as 4.7×10^{-14} m^3 K^{-3}.

Finally, the initial and final adhesion coefficients for the viscosity parameter were given empirical values of 0.7 and 7 m respectively. These values were based on results from earlier simulations of rhyolite lavas (Barca et al. 1987a,b, 1988a,b) and of minor flows emplaced during Etna's 1986–87 eruption, but which did not form part of the main flow field.

Figure 12.2 (Facing page) Simulation of Etna's 1986–87 flow field. North is to the right. The contour interval is 200 m, except for part f, in which it is 1,000 m. Darker lava is cooler than lighter lava. The scale bar represents 1 km. Location names are given in part f. (**a**) Lava flow 8 hours after the start of eruption. Only one vent (about 2,900 m a.s.l.) is active with a high eruption rate of 22 m^3 s^{-1}. The lava deviates around Monte Simone. The front has cooled by about 85°C. (**b**) The situation 15.3 hours after eruption starts. The highest vent (2,900 m a.s.l.) is no longer active. Flow is fed from two other vents at about 2,250 and 2,200 m a.s.l. and an eruption rate of 22 m^3 s^{-1} is obtained. The lava breaks into two branches: one passing to the north of Monte Simone, the other joining the flow from the 2,900 m vent. Note the contact between the cooled (black) and hotter (paler) new lava. The left and right fronts have cooled by about 115°C and 45°C, respectively. (**c**) 1 November (day 3). The two vents are still active, with a low eruption rate of 4 m^3 s^{-1}. Lava channels can just be distinguished along the uppermost kilometre of flow (below vents on the right). (**d**) 15 November (day 17). Four vents are simultaneously active (at 2,600, 2,350, 2,250 and 2,200 m a.s.l.). Having reached its maximum length by 8 November, the flow field starts to migrate southwards along the Valle del Bove and to fill the spaces between earlier flows. The eruption rate is 9 m^3 s^{-1}. (**e**) Day 35. The final result of the simulation. Note the new branch formed towards Monte Centenari (to the left). The eruption rate is 6 m^3 s^{-1}. Compare with part f. (**f**) Actual final outline of the 1986–87 flow field (after Caltabiano et al. 1987). Etna's summit region is about 1 km west (to top) of the 2,900 m vent. VB, Valle del Bove; VL, Valle del Leone; C, Monte Centenari (1,830 m a.s.l.); M, Rocca Musarra (1,500 m); R, Monte Rittmann (2,350 m); S, Monte Simone (2,070 m).

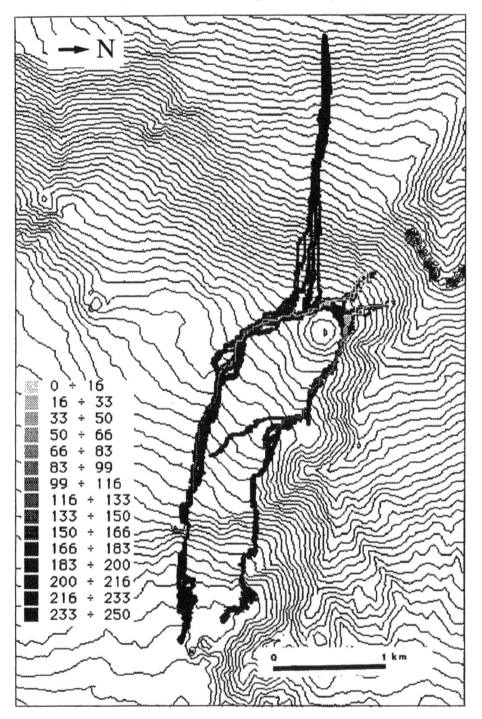

Figure 12.3 Enlarged version of Figure 12.2c for day 3. North is to the right. The grey scale shows the amount of lava cooling in degrees Centigrade. Note the well developed channel structure in flows below the lower vents.

Simulation results

The pre-eruption surface morphology and final outline of the 1986–87 flow field (Caltabiano et al. 1987) are shown in Figure 12.2f. The results of the simulation are summarized in Figures 12.2a–e. Figure 12.3 is an enlarged version of Figure 12.2c and highlights the detail of information which can be obtained by the simulation.

After the first 8 hours of eruption (Fig. 12.2a) the flow has reached 1,700 m a.s.l., fed at 22 m^3 s^{-1} by a vent at 2,900 m a.s.l. The flow deviates about Monte Simone, confirming the sensitivity of the model to morphological variations. Fifteen hours after the start of eruption (Fig. 12.2b) the flow field is being fed by two vents at 2,250 and 2,200 m a.s.l. Flows from these vents branch around Monte Simone, one subsequently joining the earlier lava from the 2,900 m vent.

Figures 12.2c & d show conditions 3 and 17 days after the start of eruption. The 2,250–2,200 m vents continue to feed the flow field which, inhibited by solidification, lengthens only to the 1,500 m contour level, beginning also to branch and widen southwards. Another branch develops by day 22, moving to the north-east of Monte Centenari. The essential planimetric form of the flow field is established by day 35 (Fig. 12.2e). The furthest flow front has reached 1,300 m a.s.l., while the others remain above 1,500 m.

Conclusions

The simulation satisfactorily reproduces the main planimetric evolution of Etna's 1986–87 flow field. Although only the first 35 days of eruption have been considered, the following activity hardly modified the flow-field outline, being characterized by the slow effusion of small flows which thickened the flow field and filled in spaces between existing branches.

The good agreement between the simulated and actual flow fields illustrates the flexibility inherent in CA modelling. In particular, the ability to alter cell altitudes can account for changes in surface topography with time (due to lava cooling and flow superposition).

Changing lava rheology can also be reasonably described by varying the adhesion coefficients of the model. Although empirically determined, the same range of adhesion coefficients (0.7–7 m) may be applicable to Etnean lavas in general: the current simulation may thus have immediate value as a forecasting tool on Etna.

The present simulation, however, remains an initial model of lava emplacement using cellular automata. Future improvements are expected by refining, in particular, the treatments of flow cooling and lava rheology. Greater computing power would also make practicable the use of smaller cells, so allowing the study of smaller-scale flow processes. Indeed, a new implementation of this simulation is currently being prepared, using a transputer net to increase computation power by a factor of 10.

Acknowledgments.

We would like to thank GIAST and GNV for support and encouragement. Thanks are also due to M. P. Bernasconi and C. R. J. Kilburn for useful discussions of earlier versions of this manuscript. This research was supported by CNR (No. 9004096 CT 12).

Appendix. Cellular automata algorithms

A homogeneous CA can be considered as a d-dimensional Euclidean space, the cellular space, partitioned into cells of uniform size, each one embedding an identical Moore elementary automaton (e.a.). Input for each e.a. is given by the states of the e.a. in the neighbouring cells, where neighbourhood conditions are determined by a pattern invariant in time and constant over the cells. At the time $t = 0$, e.a. are in arbitrary states and the CA evolves, changing the state of all e.a. at discrete times, according to the transition function of the e.a.

Formally a CA is a quadruple $A=(E^d, S, X, \sigma)$, E^d is the set of cells identified by the points with integer coordinates in a d-dimensional Euclidean space (i.e. partitioned with a square–cubic–hypercubic tessellation), S is the finite set of state of the e.a. X, the neighbourhood index, is a finite set of d-dimensional vectors, which defines the set $V(X, i)$ of neighbours of cell $i = \{i_1, i_2, ..., i_d\}$ as follows: let $X = \{\xi_1, \xi_2,..., \xi_m\}$ with $m = \#X$, then $V(X, i) = \{i + \xi_1, i + \xi_2,..., i + \xi_m\}$. $\sigma: S^m \rightarrow S$ is the deterministic transition function of the e.a. $C = \{c|c: E^d \rightarrow S\}$ is the set of possible state assignments to A and will be called the set of configurations. $c(i)$ is the state of cell i.

When appropriate, such a formal definition may be easily extended to different types of space, e.g. Riemannian space, or different tessellations, e.g. hexagonal or triangular tessellations in two-dimensional space.

Alava flow
The previous formal definition can be modified for our particular purpose, when applied to $A_{lava flow}$; the specification of a CA for lava flow is

$$A_{lava flow} = (R_2, L, S, X, \sigma, \gamma g)$$

where $R_2 = \{(x, y)|x, y \in N, 0 \leqslant x \leqslant l_x, 0 \leqslant y \leqslant l_y\}$ is the set of points with integer co-ordinates in the finite region where the phenomenon evolves. N is the set of natural numbers, and $L \subset R_2$ specifies the lava source cells. The finite set S of states of the e.a. are $S = S_a \times S_h \times S_T \times S_f^4$, where S_a is the altitude of the cell, S_h is a parameter correlated to the lava thickness in the cell, S_T is a parameter correlated to the lava temperature of the cell and S_f is a parameter correlated to the lava flow from a cell to its four neighbours.

The set X identifies the geometrical pattern of cells which influences the change in cell state. These cells are the cell itself and the "north", "south", "east" and "west" neighbouring cells, corresponding to

$$X = \{(0,0), (0,1), (0,-1), (1,0), (-1,0)\}.$$

Finally, $\sigma: S^5 \rightarrow S$ is the deterministic state transition for the cells in R_2, while $\gamma: S_h \times N \rightarrow S_h$ specifies the lava emitted from the source cell at time t. In this case the set of natural numbers N identifies the time intervals of the CA.

At the beginning of the simulation, we specify the states of the cells in R_2, defining the initial configuration of the CA. At each next step the function σ is applied to all cells in R_2, and the function γ corrects the substate S_h for cells in L. The configuration changes and the evolution of the $A_{lava\ flow}$ is obtained.

Main characters of σ

The main mechanisms of the transition function concern the computation of the possible "lava outflows" from the single cells towards the cells with common sides (Fig. 12.4) and updating the cell "temperature"; they are illustrated by the following Pascal-like algorithms, which compute the values of the cell substates at time $t+1$ according to the values of neighbouring cells' states at time t. Indexes 0, 1, 2, 3, 4 are used for the cell itself and neighbours "north", "east", "west", "south" respectively.

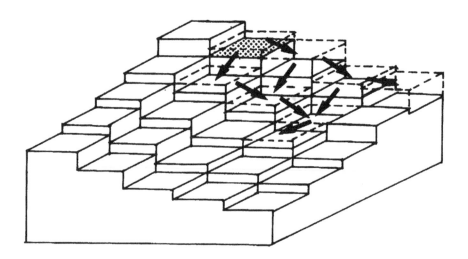

Figure 12.4 Representation of lava flow. The cell with stippled base shows the point of emission. Dashed cells show lava thicknesses. Arrows indicate flow directions.

S_a

The cell altitude "alt" changes only if the cell temperature "T" is equal or less than the solidification temperature of lava "Tsol". In this case we add to the altitude the lava thickness "h" of the cell.

procedure new_altitude;

    ....

```
begin
   if T≤ Tsol
      then new_alt := alt+h
      else new_alt := alt
end;
```

S_h

The new cell thickness "new_h" at time $t+1$ is given by the thickness "h" at time t plus the contribution determined by lava inflows "fi[i]" from the neighbour cell i minus the contribution determined by lava outflows "fe[i]" from the cell toward the neighbour cell i at time t. Note that fi[i] and fe[i] are expressed as thicknesses (see main text). The thickness averaging procedure is dealt with separately in S_{f_4} below.

procedure new_thickness;

```
...............
begin
  new_h := h;
  for i := 1 to 4 do
    new_h := new_h + fi[i] - fe[i]
end;
```

S_T

The new temperature "new_T" is obtained in two steps. The first one computes the average temperature "av_temp" determined by mixing lava "l[i]" of different temperatures "T[i]" of the cell and of its neighbours.

procedure average_temperature;

```
....................
begin
  lava_sum := l[0];
  for i := 1 to 4 do
    lava_sum := lava_sum + l[i];
  temp_lava_sum := l[0] * T[0];
  for i := 1 to 4 do
    temp_lava_sum := temp_lava_sum + l[i] * T[i];
  av_temp := temp_lava_sum / lava_sum
end;
```

The second step computes the decrease of the average temperature due to radiation (Park & Iversen 1984, Dragoni 1989); "V" is the lava volume in the cell, "S" is the radiating surface, "p" is a phenomenological parameter expressing physical properties of the particular lava type, "new_T" is the new temperature, "cubic_root" is the homonymous function.

procedure new_temperature;

```
....................
begin
  new_temp := av_temp/cubic_root(1+(av_temp*av_temp*av_temp*S*p/V))
end;
```

S_f^4

The outflow from a cell is calculated according to the algorithm given below, which minimizes the differences of heights between neighbouring cells.

The starting values at time t of the heights "z[i]" are the altitude plus lava thickness, for neighbour cells north, south, east and west, and altitude plus "viscosity" for the cell itself; l is the lava amount to be distributed, i.e. lava thickness minus lava "viscosity", described by the temperature-dependent adherence parameter v (see main text).

Cells where lava cannot flow in are identified considering the average height given by

$$av\_height = (1 + \sum_{i=0}^{4} z[i])/5.$$

"z[i]" greater than "av_height" indicates that lava cannot flow towards neighbour i, so i must be eliminated from the distribution and "v_height" computations.

This computation is iterated with the remaining cells, calculating the new "av_height" and eliminating cells with "z[i]" greater than "av_height" until there are no more cells to be eliminated; the quantity "av_height" – "z[i]" for the remaining neighbour cells then represents the lava inflow to the neighbour cell i.

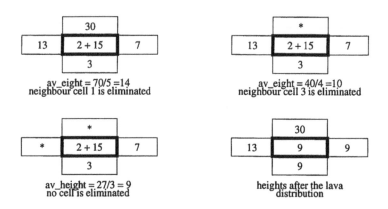

Figure 12.5 Example of steps in thickness-minimizing routine. Sequence is from top left, right, to bottom left, right.

Figure 12.5 shows the minimization procedure for the following example:

$$z[0] = 2,\ z[1] = 30,\ z[2] = 7,\ z[3] = 13,\ z[4] = 3;\ l = 15.$$

The associated Pascal-like algorithm is given below; "eliminated[i]" is a logical variable true when cell i is eliminated; "z_sum" specifies the sum of the heights plus the quantity of lava to be distributed; "new_control" is a logical variable true when a new "av_height" must be calculated; "fe[i]" is the outflow towards the neighbour cell i, "count" counts the remaining cells.

```
procedure lava_distribution
.......................
begin
   for  i := 0 to 4 do
      eliminated[i] := false;
   repeat new_control := false;
      z_sum := l;
      count := 0;
      for i := 0 to 4 do
         if not eliminated[i]
         then
            begin z_sum := z_sum + z[i];
               count := count + 1
         end;
      av_height:= z_sum / count;
      for i := 0 to 4 do
         if (z[i] > av_height) and (not eliminated[i])
         then
            begin new_control := true
               eliminated[i] := true
         end
      until not new_control;
   for i := 1 to 4 do
      if eliminated[i]
      then fe[i] := 0
      else fe[i] := av_height – z[i];
end;
```

References

Barca, D., G. M. Crisci, S. Di Gregorio, F. P. Nicoletta 1987a. Lava flow simulation by cellular automata: Pantelleria's examples. *Proceedings of the Applied Modelling and Simulation International Conference, Cairo, 1987* **4A**, 9–15.

Barca, D., G. M. Crisci, S. Di Gregorio, F. P. Nicoletta, M. T. Pareschi 1987b. Gli automi cellulari e la simulazione delle colate laviche. *Bolletino del Gruppo Nazionale per la Vulcanologia, 1987* 89–96.

Barca, D., S. Di Gregorio, F. P. Nicoletta, M. Sorriso Valvo 1987c. Flow type landslide modeling by cellular automata: Pantelleria's examples. *Proceedings of the Applied Modeling and Simulation International Conference, Cairo, 1987* 4A, 3–9.

Barca, D., G. M. Crisci, S. Di Gregorio, S. Marabini, F. P. Nicoletta 1988a. Nuovo modello cellulare per flussi lavici: colate di Pantelleria. *Bolletino del Gruppo Nazionale per la Vulcanologia, 1988*, 41–51.

Barca, D., G. M. Crisci, S. Di Gregorio, S. Marabini, F. P. Nicoletta 1988b. Lava flow simulation by cellular automata and Pantelleria's example. *Proceedings of the Kagoshima International Conference on Volcanoes, Kagoshima, Japan*, 475–8.

Burks, A. W. (ed.) 1970. *Essays on cellular automata*. Urbana: University of Illinois Press.

Caltabiano, T., S. Calvari, R. Romano 1987. Rapporto sull'attività eruttiva dell'Etna nel periodo Gennaio 1986–Febbraio 1987. *Bolletino del Gruppo Nazionale per la Vulcanologia, 1987*, 215–31.

Condarelli, D., G. M. Lechi, E. Lo Giudice, A. A. Tonelli 1975. Studio di fattibilità di un modello fisico-matematico dell'Etna. *Rivista Mineraria Siciliana* **5**, 27–40.

Crisci, G. M., S. Di Gregorio, G. Ranieri 1982. A cellular space model of basaltic lava flow. *Proceedings of the Applied Modeling and Simulation International Conference, Paris, 1982* 11, 65–7.

Crisci, G. M., S. Di Gregorio, O. Pindaro, G. Ranieri 1986. Lava flow simulation by a discrete cellular model: first implementation. *International Journal of Modelling and Simulation* **6**, 137–40.

Crisp, J. A. & S. M. Baloga 1990. A model for lava flows with two thermal components. *Journal of Geophysical Research* **95**, 1255–70.

Dragoni, M. 1989. A dynamical model of lava flows cooling by radiation. *Bulletin of Volcanology* **51**, 88–95.

Emshoff, J. R. & R. L. Sisson 1970. *Design and use of computer simulation models*. New York: Macmillan.

Harrison, C. G. A. & C. Rooth 1976. The dynamics of flowing lavas. In *Volcanoes and tectonospheres*, H. Aoki & S. Iizuka (eds), 103–13. Tokyo: Tokai University Press.

Ishihara, K., M. Iguchi, K. Kamo 1988. Numerical simulation of lava flows at Sakurajima. *Proceedings of the Kagoshima International Conference on Volcanoes, Kagoshima, Japan*, 479–82.

Ishihara, K., M. Iguchi, K. Kamo 1990. Numerical simulation of lava flows on some volcanoes in Japan. In *IAVCEI Proceedings in Volcanology*. Vol. 2, *Lava flows and domes: emplacement mechanisms and hazard implications*, J. H. Fink (ed.), 174–207. Berlin: Springer.

Kilburn, C. R. J. & R. M. C. Lopes 1991. General patterns of flow field growth: aa and blocky lavas. *Journal of Geophysical Research* **96**, 19,721–32.

Lindenmayer, A. A. 1971. Cellular Automata, formal languages and development systems. *IV International Congress for Logic, Methodology and Philosophy of Science, Bucarest*, 3–15.

McBirney, A. R. & T. Murase 1984. Rheological properties of magmas. *Annual Review of Earth and Planetary Sciences* **12**, 337–57.

Margolus, N., T. Toffoli, G. Vichniac 1986. Cellular Automata supercomputers for fluid-dynamics modelling. *Physics Review Letters* **56**, 1694–6.

Naylor, Th. H. 1966. *Computer Simulations Techniques*. New York: Wiley.

Park, S. & J. D. Iversen 1984. Dynamics of lava flow: thickness growth characteristics of steady two-dimensional flow. *Geophysical Research Letters* **11**, 641–4.

Regione Siciliana 1988. *Carte topografiche scala 1:10000*, Maps 625010 & 625050.

Rundle, J. 1989. Earthquakes, self-organization and scaling. *Physics World* November, 22–3.

Thatcher, J. W. 1970. Universality in the von Neumann cellular model. In *Essays on Cellular Automata*,

A. W. Burk (ed.), 132–86. Urbana: University of Illinois Press.

Toffoli, T. 1984. Cellular Automata as an alternative to (rather than an approximation of) differential equations in modeling physics. *Physica* **10D**, 117–27.

Toffoli, T. & N. Margolus 1987. *Cellular automata machines*. Massachusetts: MIT Press.

Vichniac, G. 1984. Simulating physics with cellular automata. *Physica* **10D**, 96–115.

von Neumann, J. 1966. *Theory of self reproducing automata*. Urbana: University of Illinois Press.

Wolfram, S. 1984. Computation theory of cellular automata. *Communications in Mathematical Physics* **96**, 15–57.

Young, P. & G. Wadge 1990. Flowfront: simulation of a lava flow. *Computers and Geosciences* **16**, 1171–91.

CHAPTER THIRTEEN

A short introduction to continuum mechanics

Søren-Aksel Sørensen

Chiedi al rio perchè gemente
dalla balza ov'ebbe vita
corre al mar che a sè l'invita,
e nel mar sen va a morir:
ti dirà che lo strascina
un poter che non sa dir.

Ask the river why, lamenting
from the rock from which it sprang
it runs down to the sea which beckons it on,
and goes to its death in the ocean:
it will tell you it is dragged there
by a power it cannot explain.

G. Donizetti

Abstract

Current lava flow models rely on highly simplified forms of the equations of continuum mechanics. Although simplifications make calculations easier, they often also obscure fundamental theoretical assumptions. This chapter discusses how the general flow equations are derived from the concept of treating collections of particles as continuous media. It aims to provide a general introduction to the principles of continuum mechanics, rather than to show specific applications to lava flows.

Introduction

Disasters have played an important role in the history of science. The ultimate ambition of humankind has always been to harness the destructive forces of nature and employ them in its own interest. Natural science is founded on the belief that this can be accomplished by accumulating ordered and systematic knowledge about the world. In the latter part of the 17th century Newton published his *Principia Mathematica* and revolutionized scientific thinking. As always, the new ideas caused a wave of optimism to pass through the scientific community, and the prevailing mood was that the last threshold had been conquered. Soon it would be possible to create mechanical models of the entire natural world based on the Newtonian approach. However, for Newtonian mechanics to be successful it needed to explain the behaviour of fluids. Modern society has a fascination for electronics, and we tend to forget how the physics of fluids dominates our transport system and power generation facilities. In addition they play a central role in all major natural disasters. Without a mechanical model for fluids, modern science could not have evolved.

The problems encountered by the Newtonian approach were associated with the existence of flow resistance or viscosity. Fluid friction had already received considerable attention. The main attributes were described by Leonardo da Vinci and the phenomenon had been investigated experimentally by Rene Descartes and a host of other eminent scientists. An expression for the viscous force had been derived by Newton in his *Principia Mathematica*. However, because Newtonian mechanics is so closely associated with the particle concept it had proved impossible to create realistic models that could account for viscosity. The breakthrough was made by the Swiss mathematician Leonhard Euler. Euler was a mathematical virtuoso and one of the true giants of science, and pioneered modern mathematical analysis as well as the mechanics of continua. A continuum is an abstract form of matter that reflects the statistical behaviour of large quantities of interacting particles. Euler employed this concept with great success, and managed to derive a set of partial differential equations describing the motion of continua. These equations would determine the behaviour of a fluid under the action of a given force provided they could be integrated. The task of solving Euler's equations was undertaken by the French mathematician Louis de Lagrange. He showed that a solution could be found for one of two cases: either when the flow contained no rotation (potential flow) or when the parameters describing a flow did not change with time (steady flow). The integral that Lagrange derived was one of the fundamental equations of fluid dynamics, and is known today as Bernoulli's equation. The integration of Euler's equations in their general form is still a problem waiting to be solved, but significant progress has been made in fields like aerodynamics and hydrodynamics where the basic equations take on a particularly simple form.

The introduction of continuum mechanics to volcanology has been hampered by the fictitious tenet that volcanic systems are too complex to be reliably modelled. Such an attitude has changed over the last decade and mathematical modelling is now more common, especially in studies of explosive eruptions and of the withdrawal and ascent of magma from higher-level reservoirs.

This chapter aims to present the basic elements of continuum mechanics to readers familiar with elementary calculus, but with little interest in the rigorous mathematical background to the subject. Unfortunately, it is not realistic to abandon mathematics completely and use a purely qualitative approach. Mechanics is a physical description of a phenomenon. It uses physical concepts (such as mass or force) to predict the behaviour of a system in terms which can be verified by experiment. However, the language used to report the state of a mechanical system and describe how its various physical properties are related is provided by kinematics, and the kinematic concept of motion is purely mathematical (kinematics is often considered a branch of geometry rather than physics). An understanding of kinematics is fundamental to the study of mechanics, and this makes it impractical to dispense completely with mathematical notation. All the relevant equations have therefore been included here for completeness. The equations, however, are not essential to understanding the text, and readers who agree with the Scottish biochemist John Haldane that a law such as

$$\frac{\partial x}{\partial t} = \nu \left(\frac{\partial^2 x}{\partial y^2} \right)$$

is less simple than "it oozes" should not be put off by the amount of mathematical notation.

Lavas are normally taken to be incompressible and to vary rheologically from simple viscous fluids to elastic solids as their temperatures drop through the crystallization interval (see Ch. 9.). As a result, this chapter first deals with the basic properties of continua and then examines some of the simplifications made to study lavas. No attempt has been made to address all lava models, emphasis being placed instead on the principles involved in simplifying the general governing equations. In particular the energy equations are not directly considered.

The first part of the chapter develops the equations of motion in a form that is applicable to all continua. The second part concentrates on the constitutive equations. These equations express the rheological properties of continua, i.e. the properties that make the various types of solids and fluids behave differently. The discussion covers both the ideal fluid approximation and the Reiner–Rivlin approximation that can be applied to a wide range of fluids, including simple Newtonian viscous flows. Indeed, from the latter approximation, we will derive the linearized Newtonian momentum equation known as the Navier–Stokes equation, as well as its non-Newtonian counterpart. Final sections will then look at how the equations of motion can be (a) determined for elastic solids (hypoelastic materials) and (b) simplified if a fluid is incompressible. Numerous texts are now available introducing fluid dynamics and mechanics at different levels and so, in place of text references, a selected bibliography is given at the end of the chapter.

Properties of continua

It has already been mentioned that Newtonian mechanics runs into problems when applied to systems which contain large numbers of particles. First, we need to solve a large number of simultaneous equations, but that is purely an administrative problem which could be solved given enough time and resources. The real problem is much more fundamental. A specific model is not defined by the equations of motion but by its initial conditions. In order to solve a specific model we would therefore need to establish the initial condition for the system, and that would provide us with the impossible task of establishing the position, mass and velocity of every single particle at a given instant in time.

The effect is not just theoretical. We can experience it by simply picking up a handful of dry sand and compare the feel of it to that of a handful of gravel. Mechanically, there is little difference between the two systems. They have very similar mass and are both made from a collection of roundish particles each of which obeys the laws of Newtonian mechanics. The only obvious difference is that the sand consists of smaller and therefore more particles. This may lead us to expect the two systems to behave in a similar fashion, but experience quickly tells us that this is not the case. While the gravel is retained easily in the hand, the sand will continuously shift under our fingers, and the harder we try to grasp it, the more difficult the task seems. What happens is that our nervous system is able to collect information from particles in the vicinity of the controlling fingers. Interaction with the remaining particles is indirect and consequently unreliable. Because the gravel has fewer particles, a higher proportion is under our control, and we may eventually manage to find an equilibrium state. However, when we are dealing with the sand the task becomes impossible and the feeling we recognize as that of a fluid is simply the feeling of losing control.

Information problems in many-body systems are not restricted to mechanics. Democratic governments face similar problems when they interact with the population. In small communities it is possible to take the views and welfare of each individual into account when decisions are made, but as the population grows this is no longer practical. The authorities are forced instead to adopt a statistical approach and make decisions which are beneficial to the average citizen. Based on information gathered through censuses and opinion polls they define "the average person", a fictitious character who is 59% female, has 0.67 spouses and 2.15 children. This subtle change is mathematically quite revolutionary. It allows us to replace real people with a substance with attributes which can be described by continuous single-valued functions, i.e. functions we can treat using normal calculus. A substance of this kind is called a *continuum* and the variables describing its properties are called *fields*.

Fluid mechanics uses the same methodology. Instead of considering the position, mass and velocity of individual particles, the volume containing the many-body system is partitioned into a grid of small cells called fluid elements and the average values of the attributes of particles inside the volume element are adopted as field variables. Fluid elements are volumes rather than particle families, and do not claim any knowledge

S.-A. Sørensen

about the identity of individual members. Particles are thus free to migrate between neighbouring elements. A fluid element is sometimes referred to as a fluid particle, but this term can be misleading. While a particle is defined as an infinitesimally small mathematical point with mass, a fluid element is a finite volume, small enough for its size to be insignificant compared to the dimensions of the continuum, but still large enough to contain so many particles that the impact on the value of field variables from particle migration between neighbouring elements is negligible. A fluid element must also be significantly larger than the mean free path of a particle. For a normal gas, a cube with $10\,\mu m$ sides, which contains around 10^{10} molecules, would easily satisfy these criteria. A fluid element can, however, be considerably larger. An example of a system with large fluid elements is an avalanche. To a distant observer, the avalanche will appear as a continuum, a waterfall running down the mountain slope, but people trapped inside the avalanche will not find themselves surrounded by a smooth continuous substance but, rather, by a hazardous mixture of stones and small boulders, each with its own very painful identity.

The physical attributes of the particles are now replaced by the corresponding field values. This can be done formally using statistical mechanics, but for the present purpose simple mean values will suffice. The *density* ρ (the mass per unit volume) is thus defined as

$$\rho = \frac{1}{\delta V} \sum_{i \in \delta V} m_i,$$ (13.1)

where m_i is the mass of the ith particle in a fluid element with volume δV, and the sum is over all particles in the fluid element. Similarly, the velocity components (v_x, v_y, v_z) are replaced by the means

$$\bar{v}_x = \frac{1}{\delta V} \sum_{i \in \delta V} (v_x)_i,$$

$$\bar{v}_y = \frac{1}{\delta V} \sum_{i \in \delta V} (v_y)_i,$$ (13.2)

$$\bar{v}_z = \frac{1}{\delta V} \sum_{i \in \delta V} (v_x)_i,$$

and the *momentum* components are defined as the product of the density and mean velocity.

Because velocities are vectors, the component values can be positive or negative and may cancel each other out. The velocity of the fluid element may therefore be insignificant, even though the individual particles themselves have large velocities. This effect has to be considered when the energy of a fluid element is defined. The energy of a particle is its kinetic energy $(E_i = \frac{1}{2}m_i v_i^2)$. If this is replaced by the kinetic energy of the fluid element,

$$E = \frac{1}{2}\rho \left(\frac{1}{\delta V} \sum_{i \in \delta V} v_i \right)^2,$$

315

energy could clearly be lost in the process, and one of physics' fundamental principles will have been breached. The magnitude of the error can be determined by considering the velocity of a particle (v_i) as the sum of the fluid element velocity ($\bar{v}$), and a deviation (v_i^*). The averaging process will then take the form

$$E = \frac{1}{2}\rho\frac{1}{\delta V} \sum_{i \in \delta V} v_i^* = \frac{1}{2}\rho\frac{1}{\delta V} \sum_{i \in \delta V} (\bar{v}^2 + v_i^{*2} + 2\bar{v}v_i^*)$$

(13.3)

$$= \frac{1}{2}\rho\bar{v}^2 + \frac{1}{2}\rho\frac{1}{\delta V} \sum_{i \in \delta V} v_i^{*2} + 0 .$$

The energy of a fluid element is now composed of two terms. The first is the kinetic energy of a fluid element. In addition, there is a second term which measures the amount of kinetic energy lost if the velocity of the fluid element is adopted as the typical velocity. By introducing the continuum we have thus been forced to introduce a new concept – internal energy or heat – to allow for the fact that a fluid element at rest can still possess kinetic energy because it contains moving particles.

The concept of internal energy has been forced upon us because continuum mechanics is based on approximate descriptions, and this is perhaps the appropriate place to remind the reader that the laws we will derive are based on a statistical approximation and are therefore not fundamental laws of physics in the same way as Newton's laws in mechanics or Maxwell's equations in the theory of electromagnetism.

Kinematics

Before we proceed to derive the equations of motion, let us look at how fields are used to describe continua. This will also provide the opportunity to introduce co-ordinate transformations and the concept of tensors. Tensors represent something of a dilemma. Many practitioners will tell you that they saw the use of tensors as the main obstacle to their understanding of the subject. Tensors nevertheless provide us with a simple and very powerful tool for solving problems in continuum mechanics. A large section of the mathematical community will argue that fluid mechanics is little more than an exercise in tensor calculus. Here, we will limit the use of tensors as much as possible with a few exceptions. The spin and deformation tensors which will be introduced below can actually aid understanding because of their physical connotations.

The first problem facing us when attempting to describe a lava flow is the selection of a reference frame. Our description has no significance if it cannot be verified by an independent observer. All observations must therefore be converted to a standard reference frame. Continuum mechanics employs two different reference frames. The Eulerian approach describes changes in the field variables (density, velocity, etc.) at fixed positions (x, y, z) in space. A field variable M is therefore a function of both the observation point and time, $M = M(x, y, z, t)$. The Lagrangian description, also devised by Euler, follows instead the history of individual fluid elements as they progress, $M = M(x(t), y(t), z(t), t)$, in much the same way as we would do with particles in Newtonian mechanics. The equations of motion are normally simpler under the Eulerian description and that approach is therefore commonly used. However, in certain cases the Lagrangian description can simplify the equations considerably.

Tensors

The first thing to secure is that the positions can be expressed in both reference frames. We therefore have to establish a transformation between the respective co-ordinate systems. In Figure 13.1a the two reference systems are aligned and the transformation can be performed by a simple translation:

$$x^* = x + l_x$$
$$y^* = y + l_y$$

(13.4)

However, in most cases a compensation for translation will lead to the situation in Figure 13.1b, where the two co-ordinate systems have the same origin but are rotated by an angle a with respect to each other. The relation between the two co-ordinate systems is in this case

$$x^* = x \cos a - y \sin a$$
$$y^* = x \sin a + y \cos a.$$

(13.5)

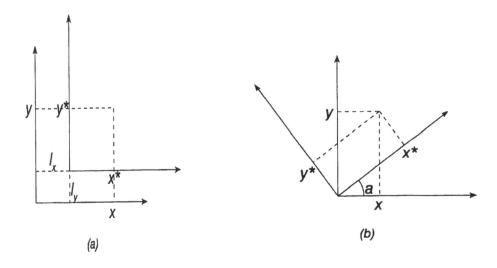

Figure 13.1 Co-ordinate transformations.

If the Equations 13.5 are rewritten in the form

$$\begin{bmatrix} x^* \\ y^* \end{bmatrix} = \begin{bmatrix} \cos a & -\sin a \\ \sin a & \cos a \end{bmatrix} \begin{bmatrix} x \\ y \end{bmatrix} \qquad (13.6)$$

we see that the co-ordinates of a point in the rotated system (x^*, y^*) can be obtained by multiplying the original position vector (x, y) with a transformation tensor. Transformations in the opposite direction are simply performed by using the inverse tensor.

Equation 13.6 is valid only for a 2-dimensional Cartesian co-ordinate system. However, the principle remains valid for any co-ordinate system, although the number of elements in the tensor and their exact form may be different. It is therefore advantageous to replace Equation 13.6 by a general expression like

$$x_i^* = \sum_{j=1}^{3} T_{ij} \, x_j, \qquad (13.7)$$

where a generic tensor T_{ij} has been introduced and subscripts have been used to identify the co-ordinates. In this particular case the co-ordinates would be $x_1 = x$ and $x_2 = y$, while the tensor would have the elements $T_{11} = T_{22} = \cos a$ and $T_{21} = -T_{12} = \sin a$.

Two classes of tensors merit a closer look because they have distinct geometric properties. One is the class of symmetric tensors, i.e. all tensors which are symmetric with respect to their diagonal ($T_{ij} = T_{ji}$). The other class contains the antisymmetric tensors which are symmetric except that the two halves have opposite sign ($T_{ij} = -T_{ji}$). Figure 13.2 shows the result of applying a symmetric and an antisymmetric transformation tensor to a square. Observe that the antisymmetric tensor affects the angular information but leaves distances unchanged (not surprising as the rotation tensor in Equation 13.6 is antisymmetric). The symmetric tensor, on the other hand, leaves the angular information unchanged, but affects the linear dimensions, causing the square to be deformed. These properties are common to all tensors belonging to the two classes, and we have found a very useful rule: *If we can identify an antisymmetric tensor in the dynamic equations, then we know that solid body rotation is involved. If we discover a symmetric tensor, the dynamic process involves deformation.*

Material derivatives

It is not only the co-ordinates that need to be converted. If the two reference frames are moving with respect to each other, the rates of change in the observed variables (their derivatives) will be quite different. The Eulerian derivatives which record local (i.e. fixed in space) changes in the field variables cannot be used in the Lagrangian description. In order to follow a fluid element it is necessary to replace the Eulerian derivatives by a set of derivatives suitable for an observer moving with the flow. Such derivatives are known as *material derivatives*. To determine what these derivatives look like, let us consider the density $\rho = \rho(x(t), y(t), z(t))$ of a (Lagrangian) fluid element

318

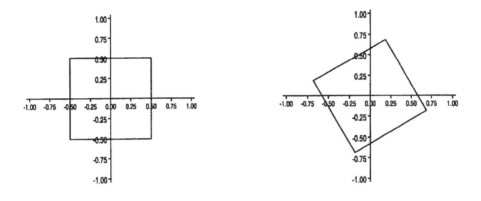

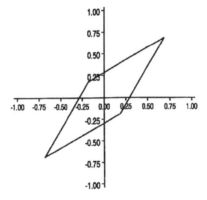

Figure 13.2 Effect of applying symmetric (top right) and antisymmetric (bottom) tensors to a square (top left).

moving with the velocity $v = v(v_x, v_y, v_z)$. During a small time interval dt, the fluid element moves from its original position $x(t_0) = (x, y, z)$ to $x(t_0 + dt) = (x+v_x\, dt, y+v_y\, dt, z+v_z\, dt)$, and the resulting change in ρ is as follows:

$$\rho = \frac{\partial \rho}{\partial x}\, dx + \frac{\partial \rho}{\partial y}\, dy + \frac{\partial \rho}{\partial z}\, dz + \frac{\partial \rho}{\partial t}\, dt \tag{13.8}$$

or

$$d\rho = \frac{\partial \rho}{\partial x}\, v_x\, dt + \frac{\partial \rho}{\partial y}\, v_y\, dt + \frac{\partial \rho}{\partial z}\, v_z\, dt + \frac{\partial \rho}{\partial t}\, dt. \tag{13.9}$$

The material derivative, which we normally denote by the capital D is then

$$\frac{D\rho}{Dt} = \lim_{dt \to 0}\left(\frac{d\rho}{dt}\right) = \frac{\partial \rho}{\partial t} + \left[v_x\frac{\partial \rho}{\partial x} + v_y\frac{\partial \rho}{\partial y} + v_z\frac{\partial \rho}{\partial z}\right], \tag{3.10}$$

where the terms in square brackets represent the convective change due to the movements of the fluid particle. If we were considering velocity, the expression Dv/Dt becomes the acceleration. It will, however, relate to any quantity, and we therefore refer to it as the material rate of change rather than acceleration.

Instantaneous motion

In this section we will consider the motion of two neighbouring fluid elements which, at the instant t, are found at positions with co-ordinates (x, y, z) and $(x+dx, y+dy, z+dz)$ respectively. After the interval dt one particle will have moved to the position

$$(x + v_x(x,t)dt, \ y + v_y(y,t)dt, \ z + v_z(z,t)dt) \tag{13.11}$$

while the other has moved to

$$(x + dx + v_x(x + dx,t) \ dt, \ y + dy + v_y(y + dy,t) \ dt, \ z + v_z(z + dz,t) \ dt). \tag{13.12}$$

Expressions like $v_x(x+dx, t)$ do not mean much to us. However, because the two points are very close to each other, we can assume that the velocity field is linear over the short distance. In that case, the velocity at $(x+dx)$ is simply the velocity at x plus the local velocity gradient $(\partial v_x/\partial x, \partial v_y/\partial y, \partial v_z/\partial z)$ multiplied by the distance to the neighbouring point (dx, dy, dz). The x co-ordinate of the particle is thus

$$x + dx + v_x(x + dx,t) \ dt = x + v_x dt + dt \left(\frac{\partial v_x}{\partial x}dx + \frac{\partial v_x}{\partial y}dy + \frac{\partial v_x}{\partial z}dz \right) \tag{13.13}$$

The right-hand side of Equation 13.13 can be simplified by adding the zero value term

$$\frac{1}{2}dt \left[\left(\frac{\partial v_x}{\partial x} - \frac{\partial v_x}{\partial x} \right)dx, \ \left(\frac{\partial v_y}{\partial x} - \frac{\partial v_y}{\partial x} \right)dy, \ \left(\frac{\partial v_z}{\partial x} - \frac{\partial v_z}{\partial x} \right)dz \right] \tag{13.14}$$

and reorganized to obtain

$$x + dx + v_x(x + dx,t) = x + dx + \left[v_x \ dt \right]$$

$$+ \frac{1}{2}dt \left[\left(\frac{\partial v_x}{\partial x} - \frac{\partial v_x}{\partial x} \right)dx + \left(\frac{\partial v_x}{\partial y} - \frac{\partial v_y}{\partial x} \right)dy + \left(\frac{\partial v_x}{\partial z} - \frac{\partial v_z}{\partial x} \right)dz \right] \tag{13.15}$$

$$+ \frac{1}{2}dt \left[\left(\frac{\partial v_x}{\partial x} + \frac{\partial v_x}{\partial x} \right)dx + \left(\frac{\partial v_x}{\partial y} + \frac{\partial v_y}{\partial x} \right)dy + \left(\frac{\partial v_x}{\partial z} + \frac{\partial v_z}{\partial x} \right)dz \right],$$

with similar expressions for the other two components. At a first glance this does not appear to have simplified matters much, quite the contrary. However, looking back to the section on tensors, we can see that Equation 13.15 has become a dynamic expression. The relative change of position between neighbouring elements is now the results of three separate factors. The first term can readily be identified as a translation. The other two terms are less obvious, but closer examination shows one to be the result of applying the antisymmetric co-ordinate transformation tensor

$$S_{ij} = \frac{1}{2} \left(\frac{\partial v_j}{\partial x_i} - \frac{\partial v_i}{\partial x_j} \right) \tag{13.16}$$

to the distance vector between the two elements. It can therefore be identified as the result of a solid body rotation around the point (x,y,z) with the local angular velocity vector

$$w = \left(\frac{1}{2}\left(\frac{\partial v_z}{\partial y} - \frac{\partial v_y}{\partial z}\right), \frac{1}{2}\left(\frac{\partial v_x}{\partial z} - \frac{\partial v_z}{\partial x}\right), \frac{1}{2}\left(\frac{\partial v_y}{\partial x} - \frac{\partial v_x}{\partial y}\right) \right). \tag{13.17}$$

The tensor S_{ij} is commonly known as the *spin tensor*, and the field w as the *vorticity* of the velocity field.

The last term in Equation 13.15 is also the result of a tensor transformation. The tensor

$$D_{ij} = \frac{1}{2}\left(\frac{\partial v_j}{\partial x_i} + \frac{\partial v_i}{\partial x_j}\right) \tag{13.18}$$

is very similar in form to the spin tensor, but this time it is symmetric and consequently represents deformation. The tensor D_{ij} is called the *rate of deformation tensor*. It measures the instantaneous change in the distance between the two fluid particles due to variations in the velocity field. The diagonal terms (D_{ii}) denote the rate of compression (or extension) while the non-diagonal components are the shear rates. If the rate of deformation tensor vanishes, the material will move as a rigid body at that particular moment. The instantaneous motion of fluid elements can therefore be divided into three distinct parts: a translation, a solid body rotation and a deformation,

$$x^* = x + v \, dt + Sx + Dx.$$

The instantaneous velocity field can be visualized by drawing a collection of arrows representing the field vector at selected points. Another common method is to use *field lines*. A field line is an oriented curve indicating the direction of the field vector at any point. The field vectors are thus tangential to the field lines at any point. Field lines can be used to visualize any vector field. If the field is the velocity field we call the field lines *streamlines*, while the field lines of the vorticity w are known as vortex lines.

The instantaneous snapshots of the flow provided by both streamlines and velocity vectors greatly help our understanding of a flow. It is, however, often impossible to obtain the instantaneous velocity field from experiments. In most cases we are confined to making our observations at certain locations, and we are usually unable to make more than a very limited number of simultaneous readings. In aerodynamics, where we mainly are interested in a very thin boundary layer near the surface of a solid body, we can observe the instantaneous velocity field by attaching thin cotton threads in a grid across the surface and photographing them. More frequently, a flow is visualized by inserting a series of trace markers at regular intervals at a fixed point and recording their subsequent positions. The type of marker used depends on the material we study. For gases we will normally use smoke and for transparent liquids we can release dye or bubbles in the flow. Lava flows are notoriously hard to observe. The bulk of the flow is inaccessible to us and observations are restricted to placing markers on the surface of the flow. The curves outlined by this procedure are called *streaklines*, and in their very nature they show a history of the velocity field at the time the marker passed various locations rather than the instantaneous picture provided by the streamlines. A third type of visualization is through *pathlines* which outline the history of selected fluid particles. They are generated by placing a marker into the flow and subsequently plotting its position at constant intervals. Both of the two latter methods of visualization involve an element of dynamics and will therefore be different from the streamlines. Only in the rare cases when the flow is steady will the three types of line coincide.

Equations of motion

The previous section concentrated on the instantaneous state of affairs in a continuum. The next task is to derive the equations of motion, i.e. a set of equations which will predict how a flow will develop in the future. For that we will use the fundamental principle of physics that basic quantities like matter and energy are conserved. Therefore, provided the values of these variables are known at a certain time t_0, the task is reduced to a simple case of book-keeping:

$$\text{assets}_{now} = \text{assets}_{before} + (\text{revenues} - \text{expenses}). \tag{13.20}$$

At first sight this does not appear to have helped much. The expression allows us to calculate the current assets but not to predict how they change in the future. However, if we assume that the rates of revenue and expense remain constant for a short period δt, we can predict the future assets:

$$\text{assets}_{t-\delta t} = \text{assets}_{t} + \delta t \left(\frac{\partial(\text{revenues})}{\partial t} - \frac{\partial(\text{expenses})}{\partial t} \right), \tag{13.21}$$

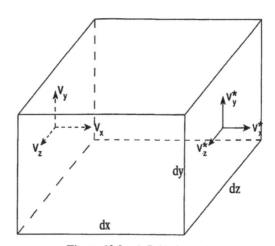

Figure 13.3   A fluid element.

Let us now use the methodology to derive equations of motion for a continuum. Consider the small Eulerian volume element shown in Figure 13.3. To simplify the problem the element is a small parallelepiped with its sides parallel to the axes of a Cartesian co-ordinate system (x, y, z) and volume $dV = dx\, dy\, dz$. The continuum is described by five variables (density (ρ), three velocity components (v_x, v_y, v_z), and the internal energy (ε)), and an equation of motion must be derived for each of them. None of these variables is conserved, and it is therefore necessary first to introduce their conserved equivalents: mass (m), momentum (mv) and total (internal + kinetic) energy (E).

The first variable is mass $(m = \rho\, dV)$. Mass is exchanged between volume elements as particles migrate across their contact surface. Figure 13.4 shows the left-hand contact surface (S_A) of the volume element in Figure 13.3. It is perpendicular to the x axis and has the surface area $dS = dy\, dz$. As we attempt to determine the amount of matter transported across this interface, we see the mathematical benefit of the continuum approximation. Instead of individual molecules, we have on the left-hand side of the wall homogeneous matter with density ρ^L and velocity (v_x^L, v_y^L, v_z^L), and on the right-hand side homogeneous matter with density ρ^R and velocity (v_x^R, v_y^R, v_z^R). During a small time interval dt the matter contained in the volume $v_x^L\, dt\, dy\, dz$ will enter the element, provided v_x^L is positive, and similarly the matter in the volume $v_x^R\, dt\, dy\, dz$ will cross the interface in the opposite direction if v_x^R is negative. The total transfer of mass resulting from this process is thus

$$\delta m = \delta m^L + \delta m^R \tag{13.22}$$

where

$$\delta m^L = \begin{cases} \rho^L v_x^L\, dt\, dy\, dz & v_x^L > 0 \\ 0 & \text{otherwise} \end{cases} \tag{13.23}$$

and

$$\delta m^R = \begin{cases} \rho^R v_x^R\, dt\, dy\, dz & v_x^R < 0 \\ 0 & \text{otherwise.} \end{cases} \tag{13.24}$$

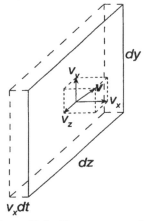

Figure 13.4   Flux across a surface.

Introducing mean values of density and velocity, the mass flux – the amount crossing a unit area per unit time – is

$$\frac{\delta m^{(S_A)}}{dt\, dy\, dz} = \rho^L v_x^L + \rho^R v_x^R$$

$$= \frac{1}{2}(\rho^L + \rho^R)\frac{1}{2}(v_x^L + v_x^R) + \frac{1}{2}(\rho^L - \rho^R)\frac{1}{2}(v_x^L - v_x^R) \qquad (13.25)$$

$$= \rho^{S_A} v_x^{S_A} + O(x^2),$$

where the second term is small enough to be discarded. The matter crossing the interface will not only transfer mass. It will also carry the momentum and energy associated with the source. The flux of these quantities can be determined in a similar manner to give

$$\text{mass flux} = \rho^{S_A} v_x^{S_A},$$

$$x \text{ momentum flux} = v_x^{S_A}\rho^{S_A} v_x^{S_A},$$

$$y \text{ momentum flux} = v_y^{S_A}\rho^{S_A} v_x^{S_A}, \qquad (13.26)$$

$$z \text{ momentum flux} = v_z^{S_A}\rho^{S_A} v_x^{S_A},$$

$$\text{energy flux} = E^{S_A}\rho^{S_A} v_x^{S_A}.$$

The fluxes considered above originated purely from the transport of matter across the plane S_A. This is the only way matter can be transported, but other physical quantities can be transported by different means, and the flux caused by matter carried across a plane is therefore called the convective flux (from the Latin *vehere*, meaning to carry).

Returning to the volume element in Figure13. 3, it is now a simple matter to construct the equations of motion. The changes in density can be calculated directly as the sum of mass transfers across all surfaces. In the x direction, the flux across the left interface (S_A) may be different from the flux across the right interface (S_B). The result is a net change in the local mass equal to

$$dm_x = (\rho^{S_B} v_x^{S_B} - \rho^{S_A} v_x^{S_A})\, dy\, dz\, dt, \qquad (13.27)$$

with similar contributions coming from the other two directions. The resulting change in density is

$$\rho_{t-dt} = \frac{1}{dV}[\rho t\, dv - dt(dm_x - dm_y - dm_z)], \qquad (13.28)$$

which, in the limit as $dt \to 0$ becomes

$$\frac{\partial\rho}{\partial t} + \frac{\partial(\rho v_x)}{\partial x} + \frac{\partial(\rho v_y)}{\partial y} + \frac{\partial(\rho v_z)}{\partial z} = 0 \qquad (13.29)$$

This equation – the *continuity equation* – expresses the fact that changes in the density inside the small volume are caused only by variations in the mass flux at its boundaries, i.e. mass is conserved.

Not all conserved field variables are changed only through convection. Just as for a normal particle in Newtonian mechanics, a fluid particle can gain momentum and kinetic energy through the influence of normal body forces. For lavas, the only relevant body force is that due to gravity and is proportional to the mass ($m = \rho\,dV$). In addition, internal energy can be transported by conduction and radiation. Thus although the equation of motion for any conserved quantity (H) has the same basic form as Equation 13.29), it may be necessary to include a term to represent these additional sources of change:

$$\frac{\partial H}{\partial t} + \frac{\partial(\rho H v_x)}{\partial x} + \frac{\partial(\rho H v_y)}{\partial y} + \frac{\partial(\rho H v_z)}{\partial z} = \text{non-convective sources } (H) \quad (13.30)$$

The body force is not the only force we need to take into consideration. Just as the statistical definition of continua forced us to introduce the concept of internal energy, it now compels us to consider an additional force term – the contact force (also sometimes referred to as a surface force). This new force term originates from the definition of a fluid particle. When local mean values were adopted as the field variables it was stressed that, although the number of particles in a fluid element remain constant, it is not necessarily the same particles which make up a specific fluid element at all times. An exchange of particles across the interface between neighbouring fluid elements is perfectly admissible. While the local density is unchanged as long as the number of particles leaving a fluid element equals the number entering, it is unlikely that the mean velocity of the particles leaving the fluid element is exactly the same as the mean velocity of those entering (the probability of choosing identical members from two populations by random selection is extremely small). Particle exchanges will thus continuously cause small changes to the local values of momentum and energy even when the convective flux is zero. In Newtonian mechanics, changes in momentum are the result of a force acting on the particle, and we therefore attribute these changes to a fictional force $\mathbf{dF} = (F_x, F_y, F_z)$, proportional to the surface area (dS) and exerted on the material inside a fluid element by the matter outside (this is known as the Euler–Cauchy stress principle).

In continuum mechanics the force is conventionally measured as positive in the direction away from the fluid element ($T = F/dS$ is known as the surface traction vector), but the tradition in geology is to reverse the sign and consider compression as positive. The force vector $\mathbf{dF}$ is not necessarily perpendicular to the surface (dS), and the surface traction vector T normally consists of a component perpendicular to dS (normal stress) and two components tangential to dS (shearing stresses). By considering planes perpendicular to the three co-ordinate axes we find that the surface forces can be expressed by the stress tensor T:

$$
\begin{aligned}
F_x &= T_{xx}\,dS_x + T_{xy}\,dS_y + T_{xz}\,dS_z\,, \\
F_y &= T_{yx}\,dS_x + T_{yy}\,dS_y + T_{yz}\,dS_z\,, \\
F_z &= T_{zx}\,dS_x + T_{zy}\,dS_y + T_{zz}\,dS_z\,.
\end{aligned}
\quad (13.31)
$$

A short introduction to continuum mechanics

Both contact and body forces act on the fluid elements and the rate of change of linear momentum must be equal to the sum of the two. Likewise, the total couple exerted by the body and contact forces must be equal to the rate of change of angular momentum within the fluid element, and this requires the stress tensor to be symmetric ($T_{ij} = T_{ji}$).

It is now possible to evaluate the source terms in Equation 13.30 and generate a set of conservation equations consisting of the continuity equation showing the conservation of mass:

$$\frac{\partial \rho}{\partial t} + \frac{\partial}{\partial x}(\rho v_x) + \frac{\partial}{\partial y}(\rho v_y) + \frac{\partial}{\partial z}(\rho v_z) = 0, \tag{13.32}$$

three equations for the conservation of momentum:

$$\frac{\partial}{\partial t}(\rho v_x) + \frac{\partial}{\partial x}(v_x \rho v_x) + \frac{\partial}{\partial y}(v_x \rho v_y) + \frac{\partial}{\partial z}(v_x \rho v_z) = \rho F_x + \frac{\partial T_{xx}}{\partial x} + \frac{\partial T_{yx}}{\partial y} + \frac{\partial T_{zx}}{\partial z},$$

$$\frac{\partial}{\partial t}(\rho v_y) + \frac{\partial}{\partial x}(v_y \rho v_x) + \frac{\partial}{\partial y}(v_y \rho v_y) + \frac{\partial}{\partial z}(v_y \rho v_z) = \rho F_y + \frac{\partial T_{xy}}{\partial x} + \frac{\partial T_{yy}}{\partial y} + \frac{\partial T_{zy}}{\partial z}, \tag{13.33}$$

$$\frac{\partial}{\partial t}(\rho v_z) + \frac{\partial}{\partial x}(v_z \rho v_x) + \frac{\partial}{\partial y}(v_z \rho v_y) + \frac{\partial}{\partial z}(v_z \rho v_z) = \rho F_z + \frac{\partial T_{xz}}{\partial x} + \frac{\partial T_{yz}}{\partial y} + \frac{\partial T_{zz}}{\partial z},$$

and an equation for the conservation of total specific energy,

$$\frac{\partial \rho E}{\partial t} + \frac{\partial}{\partial x}(E\rho v_x) + \frac{\partial}{\partial y}(E\rho v_y) + \frac{\partial}{\partial z}(E\rho v_z) = T_{xx}D_{xx} + T_{xy}D_{xy} + T_{xz}D_{xz}$$

$$+ T_{yx}D_{yx} + Hx_{yy}D_{yy} + T_{yz}D_{yz} + T_{zx}D_{zx} + T_{zy}D_{zy} + T_{zz}D_{zz} - \left(\frac{\partial Q_x}{\partial x} + \frac{\partial Q_y}{\partial y} + \frac{\partial Q_z}{\partial z}\right) \tag{13.34}$$

where D_{ij} is the rate of deformation tensor, and the heat flux Q_i is a measure of the rate at which energy is transported by radiation or other non-mechanical means. This family of equations is valid for any type of continuum, including gases, liquids, solids and plasmas. However, we are still not in the position to solve them as we have only five equations but a total of 12 unknowns, namely the density, the three velocity components, the specific energy, the heat flux and six stress components (reduced from nine because the tensor is symmetric).

Rheology

In the previous section we managed to determine five of the 12 equations of motion and, in the process, we have exhausted the information which can be gained from our definition of continua. In order to determine the last seven equations we must make use of information concerning the physical properties of the specimen matter under investigation. In the following we will simplify matters considerably by not including the energy equation in the discussion. That leaves the problem of determining the six stress components T_{ij} from the mechanical properties of the specimen. This relationship is specified by a tensor equation known as the constitutive equation.

The study of constitutive equations is often considered as a separate branch of continuum mechanics called rheology from the Greek *rheos* (flow) and *logos* (to discourse). The object of rheology is to study the factors which cause the various continua to behave differently. We will discuss here three approximations starting with the assumption that the stress tensor is isotropic. We then consider the case where fluids can sustain shear stresses while in motion, and lastly deal with materials which are able to remember their previous deformation history.

Ideal fluids

The simplest constitutive equation is that where the continuum is isotropic, i.e. stress has no perception of direction. The mechanical properties for such a substance are the same in all directions. It is also homogeneous because the form of the constitutive equation remains the same everywhere in the fluid. The form of the isotropic stress tensor

$$T^{\#} = \begin{bmatrix} \frac{1}{3}(T_{xx} + T_{yy} + T_{zz}) & 0 & 0 \\ 0 & \frac{1}{3}(T_{xx} + T_{yy} + T_{zz}) & 0 \\ 0 & 0 & \frac{1}{3}(T_{xx} + T_{yy} + T_{zz}) \end{bmatrix} \tag{13.35}$$

is normally rewritten as

$$T^{\#} = \begin{bmatrix} -P(\rho, \Psi) & 0 & 0 \\ 0 & -P(\rho, \Psi) & 0 \\ 0 & 0 & -P(\rho, \Psi) \end{bmatrix}, \tag{13.36}$$

where the tensor sum has been replaced by a scalar P (the pressure) which is a function of density (ρ) and temperature (Ψ).

A continuum with a stress tensor of the form in Equation 13.36 is called an ideal fluid. The assumption which allows us to disregard all non-diagonal terms is that the fluid is unable to sustain shear stresses. This assumption is well founded when the fluid is at rest. The very fact that shear stresses cannot be sustained is normally used as the definition of a fluid. It is much harder to justify the assumption for moving fluids as all real fluids will sustain shear stresses during motion. It is these shear stresses which are responsible for the drag forces we experience when solid objects are moved through the fluid. Ideal fluids are therefore inviscid. Although lavas clearly cannot be modelled by the ideal approximation, there are some fluid systems for which the shear stresses are sufficiently small compared to their isotropic counterparts that shearing can be neglected (examples are some pyroclastic flows where shear stresses are reduced through continuous outgassing; this allows the flow to reach very high velocities). Indeed, because the isotropic tensor is so fundamental to continuum mechanics, it is convenient to consider any stress tensor as consisting of two parts; the isotropic stress (T^*) and the stress deviator $(T^* = T - T^*)$.

Under the ideal fluid approximation the momentum equations take the form

$$\frac{\partial \rho v_x}{\partial t} + \frac{\partial v_x \rho v_x}{\partial x} + \frac{\partial v_x \rho v_y}{\partial y} + \frac{\partial v_x \rho v_z}{\partial z} = \rho F_x - \frac{\partial P}{\partial x},$$

$$\frac{\partial \rho v_y}{\partial t} + \frac{\partial v_y \rho v_x}{\partial x} + \frac{\partial v_y \rho v_y}{\partial y} + \frac{\partial v_y \rho v_z}{\partial z} = \rho F_y - \frac{\partial P}{\partial y}, \qquad (13.37)$$

$$\frac{\partial \rho v_z}{\partial t} + \frac{\partial v_z \rho v_x}{\partial x} + \frac{\partial v_z \rho v_y}{\partial y} + \frac{\partial v_z \rho v_z}{\partial z} = \rho F_z - \frac{\partial P}{\partial z}.$$

The constitutive equation does not specify the functional form of the pressure explicitly and the ideal fluid approximation can be applied for a wide variety of equations of state. The most common form is that for the perfect gas,

$$P = \frac{R \rho \Psi}{\mu}, \qquad (13.38)$$

where μ is the mean molecular weight of the fluid and R the universal gas constant. Alternatively we could use the van der Waal approximation

$$\left(P + \frac{\rho^2}{\mu^2} a \right)\left(1 - \frac{\rho}{\mu} b \right) = \frac{R \rho \Psi}{\mu}, \qquad (13.39)$$

where a and b are positive constants. A special case is found at the extremes where the thermal transport either is so efficient that gas is isothermal or so poor that we can consider the gas adiabatic $(P = a\rho^\gamma; a, \gamma = \text{constants})$. Under these circumstances the pressure can be expressed as a function only of the local density $(P = P(\rho))$. Fluids of this type are called *barotropic*.

Reiner–Rivlin fluids

In this section we will consider fluids which are able to sustain shear stresses during motion. Experience shows that drag increases as we try to move an object faster through a fluid. The obvious first approximation would therefore be to consider that the shear stresses were related to the velocity. Unfortunately, such a relation is not possible as it would provide us with a situation where the stresses (and thus the drag) depended on the choice of reference frame. However, if we instead assume that the stresses are functions of the current velocity gradient with a stress deviator of the general form

$$T_{xy}^* = F\left(\frac{\partial v_y}{\partial x}, \rho, \Psi \right),$$

the stresses will be decoupled from any specific reference frame.

If the velocity gradient is restructured as

$$\frac{\partial v_y}{\partial x} = \frac{1}{2}\left(\frac{\partial v_y}{\partial x} + \frac{\partial v_x}{\partial y} \right) + \frac{1}{2}\left(\frac{\partial v_y}{\partial x} - \frac{\partial v_x}{\partial y} \right) \qquad (13.40)$$

S.-A. Sørensen

the stress becomes a function $F(\mathbf{D}, \mathbf{S}, \rho, \Psi)$ of the rate of deformation tensor $\mathbf{D}$ and the spin tensor $\mathbf{S}$. It has, however, already been established that the stress tensor itself is symmetric. The stress deviator can therefore only incorporate symmetric tensors and scalars. This has two implications. First, it is not possible for the function F to include the antisymmetric spin tensor. This is not surprising as we would not expect shear stresses to depend on rigid body rotation. Secondly, it is now possible to use the Cayley–Hamilton theorem and simplify the expression for the stress. We will not here consider the details here; the interested reader is referred to one of the many standard textbooks on tensor analysis. For the current purpose, it will be accepted without proof that the stress can be expressed as

$$\mathbf{T}^* = a\mathbf{I} + \beta\mathbf{D} + \gamma\mathbf{D}^2, \tag{13.41}$$

where

$$\mathbf{I} = \begin{bmatrix} 1 & 0 & 0 \\ 0 & 1 & 0 \\ 0 & 0 & 1 \end{bmatrix} \tag{13.42}$$

and a, β and γ are functions of density, temperature and the tensor invariants

$$I_1 = D_{xx} + D_{yy} + D_{zz},$$
$$I_2 = D_{yy}D_{zz} - D_{yz}D_{zy} + D_{zz}D_{xx} - D_{xz}D_{zx} + D_{xx}D_{yy} - D_{xy}D_{yx}, \tag{13.43}$$
$$I_3 = D_{xx}(D_{yy}D_{zz} - D_{yz}D_{zy}) + D_{xy}(D_{yz}D_{zx} - D_{yx}D_{zz}) + D_{xz}(D_{yx}D_{zy} - D_{yy}D_{zx}).$$

Continua with constitutive equations of the form of Equations 13.43 are known as Reiner–Rivlin fluids. For the special case

$$a = \lambda(\rho, \Psi)(D_{xx} + D_{yy} + D_{zz}),$$
$$\beta = 2\mu(\rho, \Psi), \tag{13.44}$$
$$\gamma = 0,$$

Equation 13.41 is reduced to a linear function of $\mathbf{D}$:

$$T^*_{ij} = \begin{cases} \lambda(D_{xx} + D_{yy} + D_{zz}) + 2\mu D_{ii} & i = j \\ 2\mu D_{ij} & i \neq j. \end{cases} \tag{13.45}$$

These fluids are known as *Newtonian fluids*. The elements along the diagonal in Equation 13.45 are commonly rearranged as

$$T^*_{ii} = (\lambda + \tfrac{2}{3}\mu)(D_{xx} + D_{yy} + D_{zz}) + 2\mu D_{ii} - \tfrac{2}{3}\mu(D_{xx} + D_{yy} + D_{zz}), \tag{13.46}$$

where the bulk viscosity $(\lambda + \tfrac{2}{3}\mu)$ is normally assumed to vanish, leaving the constitutive equation

$$
T_{ij} = \begin{cases} -P + 2\mu\left(\dfrac{\partial v_i}{\partial x_i}\right) - \dfrac{2}{3}\mu\left(\dfrac{\partial v_x}{\partial x} + \dfrac{\partial v_y}{\partial y} + \dfrac{\partial v_z}{\partial z}\right) & i = j \\[4mm] \mu\left(\dfrac{\partial v_j}{\partial x_i} + \dfrac{\partial v_i}{\partial x_j}\right) & i \neq j \end{cases}
\tag{13.47}
$$

and the momentum Equation 13.33 reduces to the Navier–Stokes equation

$$
\frac{\partial}{\partial t}(\rho v_x) + \frac{\partial}{\partial x}(v_x \rho v_x) + \frac{\partial}{\partial y}(v_x \rho v_y) + \frac{\partial}{\partial z}(v_x \rho v_z) =
$$

$$
\rho F_x - \frac{\partial P}{\partial x} + \rho\left(\frac{\partial D_{xx}}{\partial x} + \frac{\partial D_{yx}}{\partial y} + \frac{\partial D_{zx}}{\partial z}\right) + \frac{1}{3}\mu\frac{\partial}{\partial x}\left(\frac{\partial v_x}{\partial x} + \frac{\partial v_y}{\partial y} + \frac{\partial v_z}{\partial z}\right),
$$

$$
\frac{\partial}{\partial t}(\rho v_y) + \frac{\partial}{\partial x}(v_y \rho v_x) + \frac{\partial}{\partial y}(v_y \rho v_y) + \frac{\partial}{\partial z}(v_y \rho v_z) =
$$

$$
\rho F_y - \frac{\partial P}{\partial y} + \rho\left(\frac{\partial D_{xy}}{\partial x} + \frac{\partial D_{yy}}{\partial y} + \frac{\partial D_{zy}}{\partial z}\right) + \frac{1}{3}\mu\frac{\partial}{\partial y}\left(\frac{\partial v_x}{\partial x} + \frac{\partial v_y}{\partial y} + \frac{\partial v_z}{\partial z}\right),
$$

$$
\frac{\partial}{\partial t}(\rho v_z) + \frac{\partial}{\partial x}(v_z \rho v_x) + \frac{\partial}{\partial y}(v_z \rho v_y) + \frac{\partial}{\partial z}(v_z \rho v_z) =
$$

$$
\rho F_z - \frac{\partial P}{\partial z} + \rho\left(\frac{\partial D_{xz}}{\partial x} + \frac{\partial D_{yz}}{\partial y} + \frac{\partial D_{zz}}{\partial z}\right) + \frac{1}{3}\mu\frac{\partial}{\partial y}\left(\frac{\partial v_x}{\partial x} + \frac{\partial v_y}{\partial y} + \frac{\partial v_z}{\partial z}\right).
$$

$$\tag{13.48}$$

Hypoelastic fluids

The Reiner–Rivlin approximation can account for a large range of viscous fluids, including many lava flows. As a fluid cools and starts to solidify, we can, however, observe behaviour which is inconsistent with that expected from a Reiner–Rivlin fluid. The most obvious difference is that the fluid not only resists deformation but, when deformed, it will now attempt to return to its previous state. Thus, while the mechanical properties of a Reiner–Rivlin fluid depended only on the current conditions, the new material must retain information about the history of the flow, normally of some previous stress-free reference state. Materials with such properties are known as elastic. For an elastic material the stress rate T^∇ is a measure of the material rate of change of the stress components with respect to a fixed co-ordinate system. The rate of change

$$
\frac{\partial T}{\partial t} + v_x \frac{\partial T}{\partial x} + v_y \frac{\partial T}{\partial y} + v_z \frac{\partial T}{\partial z}
\tag{13.49}
$$

cannot be used for the same reason which caused us to reject a relation between stress and velocity for Reiner–Rivlin fluids: the expression is not invariant to rigid body rotation and thus not independent of the reference frame used. Instead we must write the stress rate as the invariant

$$T_{ij}^{\nabla} = \frac{\partial T_{ij}}{\partial t} + v_x\frac{\partial T_{ij}}{\partial x} + v_y\frac{\partial T_{ij}}{\partial y} + v_z\frac{\partial T_{ij}}{\partial z}$$

$$-\frac{1}{2}\left[T_{ix}\left(\frac{\partial v_j}{\partial x} - \frac{\partial v_x}{\partial x_j}\right) + T_{iy}\left(\frac{\partial v_j}{\partial y} - \frac{\partial v_y}{\partial x_j}\right) + T_{iz}\left(\frac{\partial v_j}{\partial z} - \frac{\partial v_z}{\partial x_j}\right)\right] \quad (13.50)$$

$$-\frac{1}{2}\left[T_{jx}\left(\frac{\partial v_i}{\partial x} - \frac{\partial v_x}{\partial x_i}\right) + T_{jy}\left(\frac{\partial v_i}{\partial y} - \frac{\partial v_y}{\partial x_i}\right) + T_{jz}\left(\frac{\partial v_i}{\partial z} - \frac{\partial v_z}{\partial x_i}\right)\right].$$

If we now consider the change in stress at a fluid particle situated at a point (x, y, z) as the result of a small displacement $(du_i = v_i\,dt; i = x, y, z)$ due to the local velocity v_i we find

$$dT_{ij} = \left(\frac{\partial T_{ij}}{\partial t} + v_x\frac{\partial T_{ij}}{\partial x} + v_y\frac{\partial T_{ij}}{\partial y} + v_z\frac{\partial T_{ij}}{\partial z}\right)dt$$

$$= T_{ij}^{\nabla}\,dt + \frac{dt}{2}\left[T_{ix}\left(\frac{\partial v_j}{\partial x} - \frac{\partial v_x}{\partial x_j}\right) + T_{iy}\left(\frac{\partial v_j}{\partial y} - \frac{\partial v_y}{\partial x_j}\right) + T_{iz}\left(\frac{\partial v_j}{\partial z} - \frac{\partial v_z}{\partial x_j}\right)\right]$$

$$+\frac{dt}{2}\left[T_{jx}\left(\frac{\partial v_i}{\partial x} - \frac{\partial v_x}{\partial x_i}\right) + T_{jy}\left(\frac{\partial v_i}{\partial y} - \frac{\partial v_y}{\partial x_i}\right) + T_{jz}\left(\frac{\partial v_i}{\partial z} - \frac{\partial v_z}{\partial x_i}\right)\right] \quad (13.51)$$

$$= T_{ij}^{\nabla}\,dt + \frac{dt}{2}\left[T_{ix}\left(\frac{\partial u_j}{\partial x} - \frac{\partial u_x}{\partial x_j}\right) + T_{iy}\left(\frac{\partial u_j}{\partial y} - \frac{\partial u_y}{\partial x_j}\right) + T_{iz}\left(\frac{\partial u_j}{\partial z} - \frac{\partial u_z}{\partial x_j}\right)\right]$$

$$+\frac{dt}{2}\left[T_{jx}\left(\frac{\partial u_i}{\partial x} - \frac{\partial u_x}{\partial x_i}\right) + T_{jy}\left(\frac{\partial u_i}{\partial y} - \frac{\partial u_y}{\partial x_i}\right) + T_{jz}\left(\frac{\partial u_i}{\partial z} - \frac{\partial u_z}{\partial x_i}\right)\right].$$

We can also use the relation above to replace the velocities (v_x, v_y, v_z) by the displacements (u_x, u_y, u_z) so the change of stress depends purely on the displacement and not on the speed with which the displacement takes place; the material is therefore inviscid. Furthermore, the displacement is completely reversible. Materials which maintain this type of one-to-one correspondence between the current and initial position of the individual fluid particles are called hypoelastic. A special case is the constitutive equations for an isotropic hypoelastic material which is near a stress-free state ($T_{ij} = 0$; $i,j = x,y,z$). Under these restrictions, equation Equation 13.50 reduces to

$$T_{ij}^{\nabla} = \begin{cases} \lambda(D_{xx} + D_{yy} + D_{zz}) + 2\mu D_{ii} & i = j \\ 2\mu D_{ij} & i \neq j, \end{cases} \quad (13.52)$$

or

$$T_{ij}^{\nabla} = \begin{cases} \lambda(U_{xx} + U_{yy} + U_{zz}) + 2\mu U_{ii} & i = j \\ 2\mu U_{ij} & i \neq j. \end{cases} \quad (13.53)$$

This is the generalization of Hooke's law to three dimensions. It is similar in form to the constitutive equations derived for Newtonian fluids, except that the rate of deformation tensor has been replaced by the deformation itself. The constants λ and μ are known as Lamé's constants.

Incompressible flow solutions

With the construction of constitutive equations, the equations of motion are complete. There is, however, one more hurdle to cross before the task is finished. This task, solving the equations, turns out to be the most difficult. While it is within our powers to find a solution for ideal gases, there is still no general analytical solution to the Navier–Stokes equation. Computer simulation techniques have been used extensively over the last two decades. Although each experiment can provide us only with one of the many possible solutions, the techniques have now reached a stage where they may be able to assist in solving the general problem. Until then, an analytical solution to the equations of motion can only be found under very special circumstances.

Fortunately, most lava flows fall into the category of incompressible fluids, i.e. the pressure is sensitive to very small changes in the density. An example of such a fluid is an adiabatic gas ($P = a\rho^\gamma$; $a,\gamma =$ constants) which becomes increasingly incompressible as $\gamma \to \infty$, and at the limit the density can be considered a constant $\rho \equiv \rho_0$. The continuity equation (Eq. 13.32) is then reduced to

$$\frac{\partial v_x}{\partial x} + \frac{\partial v_y}{\partial y} + \frac{\partial v_z}{\partial z} = 0 , \tag{13.54}$$

which simply states that any mass flux into a fluid element must be accompanied by an identical out-flux. Matter cannot be accumulated in any fluid element. The momentum equations reduce to

$$\frac{\partial v_x}{\partial t} + v_x\frac{\partial v_x}{\partial x} + v_y\frac{\partial v_x}{\partial y} + v_z\frac{\partial v_x}{\partial z} = F_x - \frac{\partial}{\partial x}\left(\frac{P}{\rho}\right) + v\left(\frac{\partial^2 v_x}{\partial x^2} + \frac{\partial^2 v_x}{\partial y^2} + \frac{\partial^2 v_x}{\partial z^2}\right), \tag{13.55}$$

$$\frac{\partial v_y}{\partial t} + v_x\frac{\partial v_y}{\partial x} + v_y\frac{\partial v_y}{\partial y} + v_z\frac{\partial v_y}{\partial z} = F_y - \frac{\partial}{\partial y}\left(\frac{P}{\rho}\right) + v\left(\frac{\partial^2 v_y}{\partial x^2} + \frac{\partial^2 v_y}{\partial y^2} + \frac{\partial^2 v_y}{\partial z^2}\right), \tag{13.56}$$

$$\frac{\partial v_z}{\partial t} + v_x\frac{\partial v_z}{\partial x} + v_y\frac{\partial v_z}{\partial y} + v_z\frac{\partial v_z}{\partial z} = F_z - \frac{\partial}{\partial z}\left(\frac{P}{\rho}\right) + v\left(\frac{\partial^2 v_z}{\partial x^2} + \frac{\partial^2 v_z}{\partial y^2} + \frac{\partial^2 v_z}{\partial z^2}\right), \tag{13.57}$$

where $v = \mu/\rho_0$ is the *kinematic viscosity*. Another important consequence of constant density is that the energy equation becomes decoupled from the other equations of motions. Unless we specifically need to know the temperature of the material or are considering effects like buoyancy, the energy equation has become superfluous.

The equations of motion can be simplified further if we restrict ourselves to study two-dimensional flow problems. It is then possible to replace the velocity vector v with two scalar functions. The stream function (ψ) is defined so that

$$v_x = \frac{\partial \psi}{\partial y} \quad \text{and} \quad v_y = -\frac{\partial \psi}{\partial x} \tag{13.58}$$

and the vorticity

$$w = \frac{\partial v_y}{\partial x} - \frac{\partial v_x}{\partial y},$$
(13.59)

which is Equation 13.17 for the two-dimensional case. The stream function has acquired its name because

$$d\psi = \frac{\partial \psi}{\partial x} dx + \frac{\partial \psi}{\partial y} dy = v_x \, dy - v_y \, dx$$
(13.60)

and ψ is therefore constant along lines for which $dy/dx = v_y/v_x$, i.e. the velocity field lines or streamlines.

Through the introduction of the stream function, the continuity equation takes the form

$$\frac{\partial v_x}{\partial x} + \frac{\partial v_y}{\partial y} = \frac{\partial^2 \psi}{\partial x \, \partial y} - \frac{\partial^2 \psi}{\partial y \, \partial x} = 0$$
(13.61)

and has thus been satisfied identically. In addition, we can combine Equations 13.58 & 59 to get

$$\frac{\partial^2 \psi}{\partial x^2} + \frac{\partial^2 \psi}{\partial y^2} = -w.$$
(13.62)

We now take the partial derivatives of the momentum Equations 13.55 & 56 in their two-dimensional form with respect to y and x respectively and subtract. This provides us with the Helmholtz vorticity equation

$$\frac{\partial w}{\partial t} + v_x \frac{\partial w}{\partial x} + v_y \frac{\partial w}{\partial y} = \nu \left(\frac{\partial^2 w}{\partial x^2} + \frac{\partial^2 w}{\partial y^2} \right),$$
(13.63)

where we have assumed that the local changes in the external force field are small enough to be disregarded. The equation shows that only viscosity can change the vorticity of the flow. For an inviscid fluid the dynamic problem has therefore been reduced to a single equation of motion:

$$\frac{\partial^2 \psi}{\partial x^2} + \frac{\partial^2 \psi}{\partial y^2} = 0.$$
(13.64)

However, in the general case both the stream function and vorticity equations have to be solved simultaneously.

Another case where a simplified solution can be obtained is for fluids where the vorticity vanishes, i.e. rotation-free fluids. From vector analysis we know that the rotation of a gradient is identically zero, so consequently there exists in this case a function φ (the velocity potential) such that

$$v_x = -\frac{\partial \varphi}{\partial x} \quad \text{and} \quad v_y = -\frac{\partial \varphi}{\partial y}.$$
(13.65)

The equation of motion then becomes

$$\frac{\partial^2 \varphi}{\partial x^2} + \frac{\partial^2 \varphi}{\partial y^2} = 0,$$
(13.66)

and a solution can again be obtained. It is furthermore possible to express the pressure (P/ρ_0) as a function of the velocity field as

$$\frac{\partial^2}{\partial x^2}\left(\frac{p}{\rho_0}\right) = -\left(\frac{\partial v_x}{\partial x}\right)^2 - \frac{\partial v_y}{\partial x}\frac{\partial v_x}{\partial y},$$

$$\frac{\partial^2}{\partial y^2}\left(\frac{p}{\rho_0}\right) = -\frac{\partial v_x}{\partial y}\frac{\partial v_y}{\partial x} - \left(\frac{\partial v_y}{\partial y}\right)^2,$$

(13.67)

from where the pressure can be obtained.

Lastly, let us briefly consider parallel flows where $v_y \equiv 0$ and v_x is a function purely of y and t. In this case the three equations of motion reduce to

$$\frac{\partial v_x}{\partial t} = -\frac{\partial}{\partial x}\left(\frac{p}{\rho_0}\right) + v\frac{\partial^2 v_x}{\partial y^2},$$

(13.68)

and we are again left with only one equation to solve.

The first case we will consider is when the pressure gradient

$$\frac{\partial}{\partial x}\left(\frac{p}{\rho_0}\right) = 0,$$

and the flow is in a steady state, enclosed between two parallel walls situated at $y = 0$ and $y = h$ and moving in the x direction with velocities v_0 and v_h, respectively. For this flow approximation (known as Couette flow) we find the solution to be a linear velocity profile

$$v_x = v_0 + (v_h - v_0)\frac{y}{h}.$$

(13.69)

Another common approximation is Poiseuille flow. This flow type is found when the walls are at rest and the flow has a constant pressure gradient

$$\frac{\partial}{\partial x}\left(\frac{p}{\rho_0}\right) \equiv -\lambda.$$

Assuming that the flow is at rest near each wall, we obtain a parabolic velocity profile

$$v_x = -\frac{\lambda}{2v}(y^2 - hy)$$

(13.70)

with maximum velocity

$$v_{max} = \frac{h^2\lambda}{8v}$$

(13.71)

in the centre of the flow.

Concluding comments

The flows we have dealt with in the last section are very simplistic. At first they may not appear to be of any relevance to the study of lava flows. Such an impression is correct as far as the absolute characteristics are concerned, but the impact approximate solutions can have on our understanding of relative behaviour should not be underestimated. General analytical solutions to non-Newtonian flow problems are beyond our ability at present, but some progress has been made with numerical models. Where appropriate, numerical techniques offer a set of extremely powerful tools which allows us to find possible solutions to specific flow models. At the same time they raise a new set of complex numerical problems for the investigator, but these are beyond the scope of this chapter.

Selected bibliography

This list is not exhaustive. However, in keeping with the introductory nature of this chapter, they have been chosen to provide a starting point for further study of continuum mechanics.

(a) Introductory texts which emphasize principles and the physics of continuum mechanics. (Tritton (1977) and Vardy (1990) require a knowledge of elementary calculus.)

Massey, B. S. 1983. *Mechanics of fluids*, 5th edn. Wokingham: Van Nostrand Reinhold.
Tritton, D. J. 1988. *Physical fluid dynamics*, 2nd edn. Oxford: Clarendon Press.
Vardy, A. 1990. *Fluid principles*. Maidenhead: McGraw-Hill.

(b) More advanced and standard texts on fluid mechanics.

Batchelor, G. K. 1967. *An introduction to fluid dynamics*. London: Cambridge University Press.
Bird, R. B., W. E. Stewart, E. N. Lightfoot 1960. *Transport phenomena*. New York: John Wiley.
Landau, L. M. & E. M. Lifshitz 1959. *Fluid mechanics*. New York: Pergamon Press.
Panton, R. L. 1984. *Incompressible flow*. New York: John Wiley.

(c) Introductory texts on elasticity and fracture.

Jaeger, J. C. 1969. *Elasticity, fracture and flow*, 3rd edn. London: Chapman & Hall.
Jaeger, J. C. & N. G. W. Cook 1979. *Fundamentals of rock mechanics*, 3rd edn. London: Chapman & Hall.

Part IV

MEDIATING

Preface

Volcanology is not a social science. Eruptions, however, can have a strong social impact. It is inevitable, therefore, that, as eruptive mechanisms become better understood, so the volcanological community is expected to take a more active part in advising populations at risk.

Like it or not, therefore, volcanologists must establish guidelines for communicating with non-specialists during volcanic emergencies. This idea, of course, is not new, but is still underdeveloped. The book thus concludes with a chapter on the advisory role of the volcanologist and highlights the growing responsabilities which can be expected almost certainly within the next decade.

CHAPTER FOURTEEN

Interactions between scientists, civil authorities and the public at hazardous volcanoes

Donald W. Peterson & Robert I. Tilling

Abstract

Volcanologists readily assume the challenge of applying their skills to restless volcanoes in trying to understand current behavior and determine possible future activity. Most volcanologists, however, would rather avoid the parallel challenge of conveying their findings effectively to civil officials, news reporters and the general public. But unless community and regional leaders gain an understanding of the volcanic processes and prognoses sufficient to enable them to make valid and informed decisions on emergency management, the public response to crises is likely to be ineffective, or even disastrous. Volcanologists, thus, have an obligation to help the public understand the range of possible outcomes and the uncertainties that are associated with restless volcanoes, even when members of the public are disinterested or antagonistic. Scientists can benefit from principles outlined by sociologists to help make their statements about hazards more effective. The ingredients for effective warnings include specificity, consistency, accuracy, certainty, and clarity. Most statements about restless volcanoes lack one or more of these attributes, and scientists need to find ways to help the public take the hazards seriously despite the inherent uncertainties.

Volcanic activity during the 1980s caused more human fatalities and economic loss than in any decade since the early part of the 20th century. Thus, despite great advances in scientific understanding of volcanic processes, people have not been fully effective in applying this knowledge to the benefit of society. Different eruptions throughout the world have resulted in highly varied human responses, and they have occurred at both well-studied and little-studied volcanoes and in areas where people were well prepared and poorly prepared. A review of selected examples of eruptions and of the varied public response leads to several conclusions about the factors that govern the effectiveness of relations between scientists and non-scientists:

    (a) Small, frequent eruptions at a volcano induce good communications and promote effective relations between scientists and the public.

    (b) Conversely, uncertainty over the outcome of unrest when eruptions are infrequent may cause strained and difficult relations.

    (c) Unrest at long-quiescent volcanoes is particularly difficult to diagnose; such unrest does not necessarily culminate in an eruption, but if an eruption does occur, it may be particularly violent. Either outcome poses difficult challenges to scientists, not only in their study of the volcano, but also in their public relations.

    (d) Geological studies, compilation of hazards maps, and community-wide preparation for emergencies *before* a crisis develops are important goals that will mitigate potential hazards at volcanoes in inhabited areas.

Introduction

Volcanic eruptions are among the natural phenomena that can cause serious harm to people who live near volcanoes and who must adjust accordingly. When the activity is mild to moderate, successful accommodation can generally be made; however, when the vigor of the activity exceeds expectations, loss of life and property may result. Severe losses are most common with the sudden reawakening and violent eruption of volcanoes that have been previously unrecognized or quiescent for a long time.

Throughout most of human history, people have simply accepted death and destruction from volcanic eruptions as inevitable – a manifestation of nature's wrath. But when geology began to develop into a science from the late 18th century, the study of volcanoes revealed much about their worldwide distribution, their eruptive styles and their resulting products. Eventually it was realized that systematic scientific study could be applied towards reducing the losses from volcanic eruptions.

By the early 20th century, efforts were concentrated at fledgling volcano observatories, such as those in Italy, Japan and Hawaii (see Tilling 1989a). The early volcanologists generally conveyed their findings directly to the people nearby, and relations between scientists and the public were cordial and mutually supportive. However, as investigative techniques and research methods became more sophisticated, and as other institutions became involved in volcanological studies, direct contact with the public diminished, and the results of the studies were reported chiefly to the scientific community. Not surprisingly, many of the results were too technical to be understood by non-scientists, and, except at frequently or continuously active volcanoes, many volcanologists lost sight of the historical linkage between their science and its direct service to the public. Similarly, civil officials and the public became less aware of the relevance of research on volcanoes to their safety and welfare.

In regions of long-quiescent volcanoes, most people were totally unaware of the potential volcanic hazards. Even when cognizant of the hazards, some officials preferred to ignore them, fearing that general knowledge might hamper commercial activity and growth in their jurisdiction and decrease property values. As a consequence, both the officials and the population tended to be unprepared for any emergency; misunderstanding of and antagonism towards scientists developed, emergency response was commonly deficient, and excessive losses were sometimes sustained (e.g. Saarinen & Sell 1985, Mader & Blair 1987, Voight 1990).

In contrast, at frequently active volcanoes where scientists and local officials tended to be on good terms, they understood each other's needs and problems. Where the news media as well as the general public were educated about the nature of the volcano and its potential hazards, the response to emergencies was even more effective, and when a volcanic crisis culminated in an eruption, losses tended to be minimal (e.g. Fiske 1988, Kamo 1988).

This chapter discusses the general relationships between scientists and the general public, explores some of the reasons that underlie both good and poor relations, and reviews examples of how the efficacy of public response to volcanic emergencies greatly depends on the quality of the relations between scientists, civil authorities, the general public and the news media.

Responsibilities of volcanologists during crises

When a crisis develops at a volcano, the civil officials and the general public look to scientists to assess the magnitude of the threat and to provide advice on how it can best be mitigated. If the volcano has been well studied previously and if monitoring has been in progress, scientists can generally provide reliable and useful information. If, however, the volcano has been little studied and not monitored, the amount and quality of information scientists can provide is severely limited. In either case, the scientists are expected to play a dual role: (a) they must assess the state of the volcano and try to anticipate its future behavior; and (b) they must convey this information in understandable terms to civil officials and to the general public. No one has time to do both jobs satisfactorily, and carrying them out generally requires some compromises as part of a team effort (Fiske 1984, Peterson 1988, Tilling 1989a).

Volcanologists generally relish the first of these tasks; it is, after all, an opportunity to test their ability, training and experience, and they are eager to meet the challenge. However, they may try to avoid the second task, hoping someone else will cope with it; many feel that the time and effort expended in dealing with officials and the public is a distraction from their "real work". A major reason for this attitude is that scientific stature is commonly judged chiefly by the publication of research papers; time spent with the public obviously interferes with research. Other contributing reasons are that some officials are disinterested or hostile, and journalists often distort the interviews and produce inaccurate stories. Nevertheless, the fact remains that, if the work of volcanologists is to be useful to society, it must be disseminated beyond the scientific community in an accurate and persuasive manner understandable to the public. If the information received by the public is not accurate or timely, or if it is not understood or acted upon, losses from volcanic activity are likely to be unnecessarily severe.

Table 14.1 Selected references pertinent to scientists' interactions with the public when dealing with volcanoes

Alexander (1986)	McKee et al. (1985)
Barberi et al. (1984b, 1990)	Mader & Blair (1987)
Blong (1984)	Marts (1978)
Chester et al. (1985)	Miller et al. (1981)
Crandell et al. (1984)	Murton & Shimabukuro (1974)
Decker (1986)	Perry & Greene (1983)
Fiske (1984, 1988)	Peterson (1986a,b, 1988)
Fournier d'Albe (1979)	Podesta & Olson (1988)
Hall (1990)	Saarinen & Sell (1985)
Hays & Shearer (1981)	Sheets & Grayson (1979)
Herd et al. (1986)	Smith (1985, 1986, 1988)
Hill et al. (1991)	Tilling (1985, 1989a,b)
Hodge et al. (1979)	UNDRO & UNESCO (1985)
Housner (1987)	Voight (1988, 1990)
Lowenstein (1988)	Walker (1974)
Lundgren (1988)	Williams & McBirney (1979, Ch. 15)
Macdonald (1972 (Ch. 16), 1975)	

This fundamental premise was emphasized by Peterson (1988, p. 4161) as follows:

volcanologists should regard the development of effective communications with the public just as important a challenge as that of monitoring and understanding the volcanoes. We must apply the same degree of creativity and innovation to improving public understanding of volcanic hazards as we apply to the problems of volcanic processes. Only then will our full obligation to society be satisfied.

Table 14.1 is a list of books and papers that treat the relations between scientists and the general public during or when threatened by volcanic eruptions and other natural hazards. They include discussions on how people are affected by and how they react to the crises, and some of the papers discuss the importance of the scientists' role in fostering responsible actions by officials, the news media and the general public when confronting a volcanic emergency.

Scientific studies of volcanoes: methods and motivation

Many scientific disciplines and techniques are used in the study of volcanoes, directed at the three principal aims of learning: (a) what the volcano has done in the past and why it has done so; (b) what it is doing now and why it is doing it; (c) what it is likely to do in the future, both soon and over the long term, and why it is likely to do so. Only a few volcanoes worldwide have been examined thoroughly, and, even at these, knowledge is somewhat deficient. At most volcanoes the level of surveillance and knowledge falls short of this ideal.

The history of a volcano is determined by conventional geological studies, including mapping the rock units and structure, deciphering the stratigraphy, studying the petrology and geochemistry of the rocks, analyzing the geomorphology of landforms, and determining dates of events by radiometric and other techniques. The current state of a volcano is determined through monitoring activities such as observing and photographing eruptive phenomena, recording the seismicity and the geodetic changes, gas emissions, magnetic and gravity changes, changes in groundwater level and composition, and other phenomena. Careful analysis of the monitoring and comparisons with past records, utilizing pattern recognition, are aimed at anticipating future activity during the days and weeks ahead, leading to forecasts and predictions of impending activity (e.g. Walker 1974, Fournier d'Albe 1979, Swanson et al. 1985, Decker 1986). Study and analyses of the geological data are used to make forecasts of the long-range hazards of a volcano, which comprise a useful foundation for land-use planning. Maps showing the hazards, in conjunction with other information, can be used by civil authorities to prepare zoning maps for appropriate uses for land near a volcano (e.g. Crandell et al. 1984).

To achieve these objectives volcanologists must combine a wide variety of approaches as they try to understand the volcano. Each scientist relishes the opportunity and challenge to apply his or her own specialty towards achieving the above goals. The ultimate motivation of most investigators is the personal satisfaction gained from tackling and solving a segment of the problem about how the volcano works, often as a team effort with colleagues using an array of techniques. Another major motivation is the professional recognition received upon presenting results at meetings of scientific societies and having them published. However, we suspect that very few scientists would include the opportunity to explain volcanoes to the public among their primary motivations! Even so, most volcanologists are willing to assume this task once they recognize the need, although most begin with few skills in public relations techniques. Further, regrettably, little peer recognition is gained from their efforts.

When a volcano begins to show unrest or when it actually erupts, scientists feel naturally that their primary responsibility is to study and monitor the activity in an effort to understand the current activity. They need all their ingenuity to try to decipher the activity and gauge its current impact and possible future course. Personnel, equipment and time may not be adequate to accomplish this as well as they might wish, and these immediate demands usually conflict with the urgent need for information by civil officials and other leaders responsible for emergency management of the volcanic crisis.

Scientists must realize, however, that civil and law enforcement officials have a genuine need to gain sufficient understanding to enable them to make sound decisions about the safety of the general population, including logistics on emergency response, restrictions on access, evacuation of areas, care for evacuees, search and rescue operations, medical facilities, and a host of related matters. Corporate officials and land managers need to know how the volcano might affect their operations and the safety of their work-forces. Any damage to activities and facilities such as manufacturing plants, logging operations, farms, power plants and lines, telephone lines, water supplies, railways, roads, bridges and dams can affect large segments of the population (Blong 1984). Those responsible for these functions need to know the range of possible adverse effects so they can take precautions and plan countermeasures.

In a developing volcanic crisis, the affected populace must be given the best possible information so reactions can be responsible and rational. The only feasible way to transmit information to large numbers of people is through the news media, which play an essential role in assuring that the public receives as complete and accurate an account as possible. Because scientists are the sole source of the information needed by the other segments of society, they are ethically obligated to supply the information in a form that can be understood and utilized. It is vital that they devote enough time and effort to ensure that both officials and the general public receive timely, accurate, and reliable reports on the state of the volcano. Experience has shown that this is best handled by a single designated member of the scientific team (Fiske 1984, Peterson 1988), because both the scientist spokesperson and the audience need to become well acquainted with one another to establish rapport and to understand their respective needs and methods of operating as well as their mutual limitations. If the scientific team is small, it may be appropriate for the leader to assume this responsibility. With a larger team and a more complex operation, another scientist may be appointed to represent the leader, who may share the duties. At Mount St Helens, for example, one scientist was assigned the task of interacting with civil, law enforcement and corporate authorities, and another was the scientific representative to the news media. As needed, the scientist-in-charge shared both functions.

If the hazards are immediate and obvious, such as with earthquakes being felt or actual volcanic explosions, the information from scientists is generally accepted without question and appropriate action is taken. However, if the hazards are of the long-range variety, such as at a long-quiescent but potentially active volcano, the danger may not be at all apparent to most people. In such situations, announcements about the hazards are generally not at all welcomed by the public. If the information upsets the comfortable status quo of influential people, threatens planned commercial expansion and development, or threatens to reduce tourism and land values, scientists may find themselves confronting opposition or outright hostility. In such situations, scientists are on the firing line and must immediately assimilate on-the-job lessons in public relations techniques. They must be persistent and persuasive and learn how to convey their messages effectively and convincingly, including explanations of uncertain data. It is important to anticipate the difficulties before the release of potentially unwelcome information and carefully prepare how it will be conveyed. Authorities are invariably antagonistic when they learn of adverse information second hand, as, for example, through the news media. Private briefings for officials in advance of public news announcements usually blunt such hostility. The long-term welfare of the entire population ultimately depends on the skill and effectiveness by which scientists convey their hazards announcements. It is not a responsibility that can be treated casually, but instead must rank equally in importance to the scientific work itself.

D. W. Peterson & R. I. Tilling

Components of effective warning messages

The perception of natural hazards is so readily apparent to Earth scientists that many do not realize that such hazards are not nearly so obvious to the average person. When conveying hazards or warning information, scientists must remember that they are not talking with other scientists; they must be careful to avoid scientific jargon and to use terms that are widely understood. This can be difficult at first but becomes easier with practice. They could imagine, for example, what they would say if they were explaining some scientific matter to a next-door neighbor or to an interested relative.

Social scientists have provided extremely helpful insights into how people respond to messages that warn of an impending hazard. Sorensen & Mileti (1987) have shown that response to a warning includes a series of steps: hearing the message, understanding it, believing it, personalizing it (people becoming convinced that it actually applies to *them*), and taking action. Unless a warning results in appropriate action, it is useless. If an individual or group misses one or more of these steps, action is not likely to be taken at all. Hence, when scientists are issuing warnings, they should be sure that the messages allow the intended audience to experience each of the above steps.

In addition, social scientists have recognized that the style of the message itself influences its reception. The most effective messages have the following attributes: specific, consistent, accurate, certain, and clear (Sorensen & Mileti 1987 (p. 20)). If one or more of these attributes is deficient or missing, the message is much more likely to be misunderstood, ignored or disbelieved. It is thus not surprising that volcanologists often experience difficulty in getting appropriate responses to their warnings. Usually the condition of, and knowledge about, a volcano is not precise enough to allow a warning message to have all of the above-listed attributes. When scientists realize that one or more of these attributes is deficient, they must devote extra efforts to make the warning believable. With impending volcanic activity, the missing ingredients commonly are specificity and certainty about the time, place and nature of the eruption. The message must convince people that the warning should be heeded even though it lacks these attributes; explanations in probabilistic terms are sometimes helpful.

An ideal scenario

For illustrative purposes, let us assume an idealized situation: a population that has adapted appropriately to a nearby troublesome volcano. We shall assume that in historical times the volcano has had small to moderate ash-producing explosions at frequent but irregular intervals, and that it has had stronger explosive eruptions that produced pyroclastic flows and surges at less frequent intervals. Small to moderate lahars have been generated by some of these eruptions, but lahars have also occurred during rainstorms not associated with eruptions. Lava flows have been interspersed among the explosive events. Geological studies of the volcano have revealed that during prehistoric times it had very large eruptions that produced thick, widespread pyroclastic

345

flow deposits and voluminous lahars at intervals of hundreds to thousands of years.

A well-established volcano observatory has been in operation for decades, and its staff members have systematically recorded all of the activity and carefully studied the eruptive products. Some monitoring studies are continuous, and others are conducted at regular intervals, all using the most advanced equipment and techniques. Scientists have made a complete geological study of the volcano and the surrounding region and have learned much about its setting, its history of growth and development, its magmatic evolution, and the range of its eruptive recurrence frequencies. From all of this information they have deduced the nature of the principal current hazards at the volcano and have compiled a detailed hazard-zonation map.

The local authorities use this map, supplemented with regular consultation with the scientists, to make decisions on location, types and sizes of new developments for housing and industry. The map also provides an important basis for the regular emergency drills that the civil government holds, in which all emergency-response agencies as well as the population participate.

Scientists meet regularly with civil and corporate leaders, law enforcement officials, various government agencies, and members of the news media, not only to advise them on the current state of the volcano, but also to help them better understand its history and long-term behavior. The general public is actively interested in the volcano, spurred by exhibits in local museums, newspaper articles, radio and television programs, and talks, lectures, public programs, and school classroom visits by observatory scientists. Open days are held regularly at the observatory to explain the work of the scientists and respond to questions. Members of the local and regional news media are fully informed about the state of the volcano, and they strive for accuracy in their stories.

When scientists detect an increased potential for dangerous activity, they immediately notify all appropriate parties according to emergency response plans, and the civil officials decide on matters such as restricting access and ordering evacuation. If evacuation is necessary, both the officials and the people co-operate in carrying it out according to procedures practised during the drills, with adaptations made according to any changing requirements.

If a major eruption actually occurs, this community and region are well prepared; orderly evacuation of the population is carried out according to advance plans, adjusting as necessary to the location and size of threatened areas. The scientific team, of course, is intensely occupied with the volcano, but full communications and mutual confidence are more important than ever between the scientists and the civil authorities. Hence a single spokesperson, or a small number of specified persons, are designated to deal with the news media, and one or more additional persons are appointed to handle the relations between scientists and public officials.

Such advance planning and organization ensures an optimal response to the crisis by civil authorities; they are able to make well informed and appropriate decisions, which lead to responsible and co-operative conduct by the public throughout the emergency. This system also enables the population to have confidence in decisions and assures people that they are being given full and accurate information. The myriad details of evacuation procedures, such as transportation, shelter, food, sanitation and medical care of evacuees are not within the expertise of the scientists and hence are outside the scope of this chapter. Eventually, however, when scientists determine that the crisis has ended, they inform the civil authorities, who begin the steps to restore normal functioning of the community and to direct the recovery operations.

To summarize, this idealized scenario has three underlying attributes: (a) scientists have the personnel, equipment and other resources enabling them to progress effectively towards their major goals of understanding the past through geological studies, the present through monitoring and the future utilizing the interpretations built from the past and present; (b) the scientists maintain full communications with civil authorities and other members of the community and the region, all of whom are actively interested in and responsive to the information; (c) the authorities of the community and the region respond to the scientific information by developing a land-use plan that is cognizant of the hazards, and they also develop and regularly rehearse emergency plans.

Departures from the ideal scenario

The optimal response discussed above towards coexistence with an active volcano and its occasional crises is rarely, if ever, achieved in reality. Nonetheless, the idealized scenario embodies a goal towards which communities and regions can aspire, even though deviations from the ideal occur at every level.

Few volcanoes, even in developed countries, have been studied adequately; the history, structure and eruptive recurrence of most volcanoes are poorly known. Observatories operate at relatively few volcanoes, and, even among these, many have only a small staff, limited equipment, inadequate funding and the monitoring may be minimal. In some regions, a small staff is responsible for many volcanoes as well as earthquakes and other natural hazards; insufficient time and funds do not allow them to study any of the volcanoes adequately or to establish rapport with local civil officials. Sometimes the scientists are so engrossed in their work that they regard interactions with officials, the press and the public as annoyances and distractions. Under assorted variations of these situations, significant work by the scientists may be carried out with little or no public awareness, let alone support. Worse still, if a crisis arises, the scientists may be shocked to find that their statements carry little or no credibility with the public, or they may even encounter overt hostility.

At volcanoes that remain quiescent for long periods (a generation or more), people tend to forget the hazards, and a growing population may expand into inviting but unsafe areas. If a scientific study shows that a populated and commercially developed area is at risk, those who have a financial or political stake in the area or in continuing its development inevitably resist the adverse findings. Antagonism and rancor develop, scientific credibility is questioned or attacked, and often no action is taken to alleviate the problems. Many scientists lack political skills and have little experience in public relations. Even when they demonstrate compelling evidence to support their concerns, a lack of skill in public presentation and communication may cause their reports to be overshadowed, criticized or ignored, especially if the opposition includes well-entrenched, politically astute, influential groups.

The infrequent activity and inherently uncertain behavior of some volcanoes themselves contribute to the problem. Many volcanoes remain quiescent for decades or centuries between eruptions. People forget or never even realize that the nearby mountain is actually a potentially active volcano, or they tacitly assume that the eruptions have ended permanently. With an increasing population and growing commerce and industry, development of land around the volcano may progress without regard to the potential hazards, and no contingency plans are prepared for dealing with a volcanic emergency. If volcanic unrest begins, the reaction is likely to include shock, dismay, incredulity and confusion. If a scientific team is brought in, it must make hurried assessments without adequate background studies. Under crisis conditions, even fully competent and well experienced teams will have great difficulty in making valid assessments and forecasts with such limited time and data.

Under emergency conditions, the task of educating the public about the volcano can be overwhelming and generally "too little, too late". The people and their leaders often fail to understand the nature of the problems. Overreaction may include fright or despair, whereas underreaction may include apathy or indifference. People may have unrealistic expectations of scientists' forecasting ability, and if actual events do not coincide with cautiously expressed expectations of scientists, or if scientists decline to provide specific forecasts, people may question the ability, credibility, and motives of the scientists.

If different scientific teams are working, especially if competition arises, and if they fail to communicate with each other or do not co-ordinate their statements to authorities or the news media, serious misunderstandings with the people may develop. Sometimes maverick scientists, hungry for publicity, make irresponsible statements that cause still further confusion.

Unprepared or sensation-seeking members of the news media can be a source of great confusion and misunderstanding. Few reporters have had scientific training, and some, when assigned to cover a volcanic event, have not taken the trouble to become acquainted with readily available basic facts. Other reporters see the volcano as an opportunity for lurid or sensational stories and may over-emphasize such aspects. The accounts by such reporters generally are distorted, misleading and inaccurate, and they aggravate all the other problems. In contrast, responsible reporters realize and appreciate their key role, and they take great pains to prepare accurate, fully informative stories. Fiske (1984) and Peterson (1988) have discussed these issues in some detail.

Response to volcanic crises: selected case histories

Worldwide volcanic activity during the decade of the 1980s caused more fatalities than any 10 year period since the early part of the 20th century (Fig. 14.1) (see Tilling 1989a). The eruption of Mount St Helens (USA) in 1980 opened the decade; subsequent eruptions in the volcanic belts throughout the world were comparable in size and frequency to those in prior decades (McClelland et al. 1989). Several of these eruptions, however, caused large numbers of fatalities and, along with other eruptions, substantial economic loss.

The eruptions of the 1980s displayed a wide variety of behavior and size; they occurred in localities of broadly diverse cultures, each with differing capabilities for scientific study and surveillance, which collectively resulted in various types of public response. To illustrate this diversity, we focus on a few arbitrarily selected eruptions in the 1980s, including some of the most damaging; Table 14.2 shows their different modes of volcanic activity and scientific surveillance, and Table 14.3 shows the varied types of public response. Our discussion highlights the comparability and contrasts illustrated by these tables.

For many years, American geoscientists had been urging increased study, surveillance, and monitoring of the volcanoes in the Cascade Range, but little funding or support had been generated and the volcanoes were largely unmonitored prior to 1980. Reports on the volcanic hazards of Mount St Helens, which included a forecast of an early renewal of activity (Crandell et al. 1975, Crandell & Mullineaux 1978), aroused little public attention. The unfolding volcanic events of 1980 (Table 14.2), including the uncertainties of the precursory period from late March to mid-May, the climactic eruption on May 18, and the subsequent lesser eruptions caused dramatic alterations in the perceptions of volcanic hazards by both civil officials and the general public (Table 14.3). Prior to March 1980, the concerns of scientists were largely ignored. Between March and May, although some of the signals from the volcano were ominous, the scientists could not interpret them with certainty, and the public displayed reactions

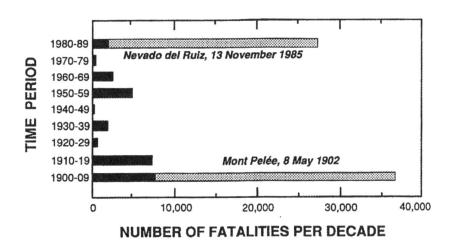

Figure 14.1. Number of fatalities caused by volcanic eruptions for each decade during the 20th century through 1989; the shaded bars indicate the fatalities from the two deadliest events of the century. (Derived from data in Decker (1986, Fig. 1), McClelland et al. (1989, Fig. 6) and Tilling (1989a, Table 1).)

Table 14.2 Summary of activity at selected volcanoes in the 1980s

Volcano with large eruption during 1980s	Background	Preliminary activity	Main event	Subsequent activity	Selected references[*]
Mount St Helens, USA (1980)	Active periods decades long at intervals of ~200–800 years. Quiescent since 1857	Earthquakes begin 20 March 1980, steam blast eruptions begin 27 March. Strong ground deformation, north flank "bulges"	18 May 1980; M-5.1 earthquake triggers gigantic landslide/debris avalanche; lateral blast released, major destruction. Also pyroclastic flows and lahars. 57 deaths	Five explosive events in May, June, July, August and October 1980. Episodic growth of lava dome in 1980-86	Lipman & Mullineaux (1981) Foxworthy & Hill (1982) Swanson et al. (1983, 1987) Tilling et al. (1990)
El Chichón, Mexico (1982)	No historical activity. Fumaroles in summit area	Earthquakes, some felt, beginning in fall of 1981	28–29 March, 30 March–3 April, and 4 April: Plinian explosions, pyroclastic flows, lahars. Approximately 2,000 deaths	Gas plume emission from lake in new crater	Havskov et al. (1983) Luhr & Varekamp (1984) Duffield et al. (1984)
Galunggung, Indonesia (1982–1983)	Destructive historical eruptions in 1822, 1894, 1918	Explosions begin 5 April 1982; no reported precursors	Many Plinian explosions May through August; with pyroclastic flows and lahars. Intermittent activity into January 83; final event a lava flow. 27 reported deaths	Fumaroles, small phreatic explosions. New lake forming in crater posing a threat when activity resumes	Katili & Sudradjat (1984a) Sudradjat & Tilling (1984) Katili (1986) Lubis et al. (1987) Tilling (1989a)
Colo, Una-Una Island, Indonesia (1983)	Destructive eruption in 1898; minor activity about 1938	Earthquakes begin in early July 1983, and progressively increase; phreatic explosions begin 18 July	Climactic Plinian eruption and pyroclastic flows on 23 July, followed by others throughout July and August. No deaths reported	Intermittent small explosions during a few months	Katili & Sudradjat (1984b)[†] McClelland et al. (1989)[†]
Nevado del Ruiz, Colombia (1985)	Several destructive historical eruptions; strongest in 1595; most recent in 1845	Beginning in November 1984 signals increase, including fumaroles, phreatic explosions, earthquakes, harmonic tremor	13 November 1985; small explosions and pyroclastic flows mixed with ice and snow, which melted, generating lahars that descended rivers. More than 25,000 deaths	Intermittent explosions and seismic swarms	Herd et al. (1986) Williams (1990a,b)
Oshima, Izu Islands, Japan (1986)	Large eruptions at intervals of ~100–200 years; most recent in 1777–92. Small eruptions more frequent – most recent in 1974	Earthquakes begin in April 1986, increases progressively, tremor begins in July and increases; strong and frequent through October and early November	15 November 1986; Strombolian activity begins, lava overflows crater. Explosions and earthquakes increase. 21 November fissures open on caldera floor, flows spread widely. Emissions and explosions end 24 November. No deaths	Gas and fume emissions, earthquakes, tremor bursts continue into 1987, then gradually decline to background	Metropolitan Tokyo (1988) Watanabe (1988) Endo et al. (1988)

Volcano with continuous activity	Background	Current activity	Selected references[*]
Kilauea, Hawaii, USA (1983–Present)	Nearly continuous lava emissions in summit caldera from 1820s to 1924; large phreato-magmatic explosion in 1924. Sporadic eruptions at summit in 1920s and 1930s; at summit and along rift zones from 1950s through 1970s, nearly continuous at Mauna Ulu (east rift zone) 1969–1974.	Continuously active since January 1983, at Puu Oo until July 1986 and at Kupaianaha since then. Lava flows have covered about 71 km² of land, and destroyed houses, roads, buildings, and significant landmarks and historic sites, and have added about 80 ha (200 acres) of new land to the island	Swanson et al. (1979) Decker et al. (1987) Tilling et al. (1989a,b) Wolfe et al. (1987, 1988)
Sakurajima, Kyushu, Japan (1955–present)	Four major explosive eruptions since about 700 AD. Largest was in 1779; most recent in 1914. It included ash explosions, large lava flows, debris flows, lahars and major subsidence of land. Moderate-sized eruption in 1946	Beginning in 1955 frequent ash explosions have been emitted from summit vents from fewer than 100 to more than 400 per year, more than 6,000 from 1955 to 1990, at average rate of 10 million m³ of ash per year. Ash lands on slopes of volcano and surrounding region, requires removal and clean-up, and causes lahars; sometimes large fragments are hurled several kilometers and cause injury and damage	Kamo (1988) Aramaki et al. (1988)

[*] Some of these volcanoes have very extensive literature; the references cited here are representative of activity of the indicated period and can also guide the reader to other references.

[†] McClelland et al. (1989) describe activity prior to 1985 of all the listed volcanoes; to avoid repetition it is cited in the table only if other references are not readily available.

Table 14.3 Summary of response to activity at selected volcanoes and of interactions between scientists and civil officials and the general public at these volcanoes

Volcano		Scientists' activities			Interaction with civil officials		Interactions with general public		Selected references
Name	Time Period	Knowledge of past, behavior	Level of Monitoring	Hazard Map?	Contacts by Scientists	Response by Officials	Scientific Outreach	Public Response	
Mount St Helens, USA	Pre-1980	Good	Little	Yes	Few, sporadic	Little interest	Little	Little interest	Miller et al. (1981) Perry & Greene (1983) Saarinen & Sell (1985) Punongbayan & Tilling (1989)
	20 March–18 May 1980		Quickly expanding	Modified to fit activity	Constant information conveyed; interpretations were indefinite	Made emergency plans and preparations; restricted public access	Frequent interviews with news media; attend public meetings	Vary from interested and cautious to confused to skeptical	
	Post-18 May		Partly destroyed, eventually excellent; volcano observatory established	Revised	Regular meetings and consultations with many agencies	Detailed and complex mitigation actions; good interagency co-operation	Intense interest by news media; lectures, programs, open days, general interest publications	Widespread interest, improved understanding	
El Chichón, Mexico	Pre-1982	Almost none	None	No	None	Not aware of hazards	None	Not aware of hazards	Tilling (1989a) Punongbayan & Tilling (1989)
	Post-April 1982	Studies made	Improved, still rudimentary		Not very effective	Slow and disorganized; hampered by remote location	Little	Passive, fatalistic	Lubis et al. (1987)
Galunggung, Indonesia	Pre-1982	Historical record only	None	Simple	None	Civil defense	None	General awareness	Tilling (1989a) Punongbayan & Tilling (1989)
	April 1982–January 1983		Quickly improved, with international assistance; observatory established		Information given to local government officials	Evacuation of vulnerable areas, housing and care of refugees, resettlement of some, mitigation actions	Agency contact with news media; government officials inform public	Voluntarily fled from area, returned whenever possible; stoical acceptance of the disruption	
	Post-1983		Monitoring reduced as activity declined, continues at rudimentary level		Information given to local government officials	Refugees either allowed to return home or resettled elsewhere; mitigation of on-going lahar hazards	Agency contact with news media; government officials inform public	Return home or resettle, depending on local effects of damage	
Colo, Una-Una Island, Indonesia	Pre-1983	Historical record	Distant seismic	Simple	None	General awareness	None	Some awareness	Katili & Sudradjat (1984b) McClelland et al. (1989)
	July 1983		On-site visit with increase in seismicity		Issued warning because of increased seismicity and visual signs	Ordered and carried out prompt evacuation of all island residents	Only through official channels	Ready response to orders in face of obvious danger	

Table 14.3 (continued)

Location	Period								References
Nevado del Ruiz, Colombia	Pre-1984	Historical Record, some on prehistoric activity	None	None	None	Little awareness	None	Little awareness	Herd et al. (1986) Voight (1988, 1990) Hall (1990, 1992) Parra & Cepeda (1990) Barberi et al. (1990)
	November 1984–November 1985		Improved, but less than needed	Prepared and given to officials	Local and foreign scientists voice concerns, recommend monitoring and emergency and contingency planning	Highly variable – deep concern to skepticism; some plans made, but communications failed during emergency	Contacts with news media; government officials convey most information	Media and public response parallel to official dichotomy; clergy and medical leaders skeptical	
	Post-1985		With international assistance a well staffed and equipped volcano observatory has been established	Refined	Systematic and regular information exchange	Contingency plans developed and improved; improved procedures developed for other volcanoes	Improved contacts and information dissemination	Improved realization of hazards; credibility of scientists accepted	
Oshima, Izu Islands, Japan	Pre-1986	Excellent understanding of historic and pre-historic record	Well-staffed and equipped volcano observatory using state-of-the-art techniques	Unpublished; available for official use	Regular contacts maintained, advised officials of increasing seismicity, April–October 1986	Multidisciplinary council evaluates all information	Media and public informed through channels	Full awareness of hazards	Metropolitan Tokyo (1988)
	November–December 1986				Information flow continues, full contacts throughout high activity and subsequent decline	Special committee convened, makes decision to evacuate, carried out promptly and efficiently; when deemed safe, residents allowed to return	Media and public fully advised of status of eruption and official decisions	Readily comply with evacuation orders; endure inconvenience stoically	
	Post-1986		Continuing to improve techniques		Resume background levels of communications and contacts	Evaluations continue	Same	Lives gradually return to normal	
Kilauea, Hawaii, USA	Early 1970s	Excellent understanding of historic and pre-historic record	Well staffed and equipped volcano observatory using state-of-the-art techniques	First issued in 1974; public information booklet in 1975	Regular contacts with National Park officials; only casual and irregular contacts with local government officials	National Park fully depends on scientists for visitor safety and making decisions; local government not interested in volcanic hazards and resists information	Information on current eruptions and hazards disseminated to news media; to general public through National Park displays	Great variation among local residents, some fully aware and concerned, others not interested	Murton & Shimabukuro (1974) Marts (1978) Mullineaux et al. (1987) Heliker (1990)
	Post-January 1983			Revised in 1987; public information booklet in 1990	Regular contacts with both National Park and local government officials; active vents on land outside Park created new government interest	Excellent working relations develop between scientists and local government officials, mutual co-operation leads to effective action during crises	Improved systems of information distribution through National Park displays, general interest publications, etc.	Long lived eruption validates scientists' credibility; citizens informed and concerned	
Sakurajima, Kyushu, Japan	As developed since 1955	Excellent understanding of historic and pre-historic record	Well staffed and equipped volcano observatories using state-of-the-art techniques	Unpublished; available for official use	Regular and systematic contacts maintained with civil officials; scientists play vital role in safety and wellbeing of the region	Entire local government and economy must constantly cope with the regular emissions of ash, and all have learned to adapt; co-operate well with scientists	Scientists inform business community, schools, etc., as well as news media on current activity and general behavior	A public museum has excellent displays; citizens must pay attention because of persistent ash; they have adapted well	Kamo (1988) Fiske (1988)

ranging from sensible caution and concern to outright skepticism over the validity of the potential danger. The cataclysmic event on 18 May was an extreme emergency that instantly forged productive co-operation between scientists and civil authorities. This co-operation was enhanced through subsequent months and years because, as the abilities of the scientists to predict eruptions improved, public confidence in the scientists increased. These improvements were fostered by regular meetings of scientists with authorities, by diligent efforts of scientists to educate and co-operate with the news media, and by the public outreach of scientists through special programs, lectures and open days. The response to the reawakening of Mount St Helens serves as a prime example of initial disarray and confusion owing to lack of preparation, followed by co-operation and effective action induced by a dire emergency and aided by quickly expanded scientific resources for volcano surveillance and interpretation.

El Chichón was an obscure, little studied volcano prior to 1982 (Table 14.2), although its hazards potential had just been recognized by geologists exploring for geothermal resources (Canul & Rocha 1981). This unpublished recognition prompted no action by either scientists or civil officials; no monitoring or other studies were initiated, and the local populace was totally unprepared for an eruption (Table 14.3). During a few days in late March and early April, a series of strong explosive eruptions caused extensive devastation and an estimated 2,000 fatalities. Official and scientific response to the disaster was slow owing to difficult logistics and organizational inertia and because no plans had been instituted to co-ordinate exchange of information between scientists and civil officials. This was the greatest volcanic disaster in the history of Mexico, and the lessons learned have now stimulated government efforts to develop more effective procedures in responding to future volcanic emergencies. El Chichón is an example of a major eruption at a long-quiescent volcano in a society that then had few scientific and economic resources to mount an effective response quickly.

With no recognized precursory activity, Galunggung erupted in April 1982, starting a series of largely explosive eruptions that continued intermittently through January 1983 (Table 14.2). Indonesia has many volcanoes, and the small but effective Volcanological Survey of Indonesia (VSI) has well-established and practised procedures for responding to volcanic emergencies (Table 14.3). Although monitoring had not been adequate to provide warnings, most of the people in the endangered areas had time to move voluntarily to safer locations, because the initial explosions, although ominous, were not highly destructive. Ultimately, about 80,000 people were evacuated and had to be housed and cared for, and by the end of the eruption about 35,000 had to be permanently relocated because of the destruction of arable land. Of the 27 reported fatalities (McClelland et al. 1989 (p. 33)), most were caused by indirect effects such as accidents and illnesses. With international assistance, a well-staffed and equipped volcano observatory was established having capabilities for monitoring seismicity, ground deformation and gas emissions. After the activity subsided in 1983, much of the staff and equipment was transferred to service at other volcanoes, with the added benefit of experience gained during the emergency. The principal continuing hazard for the next few decades is likely to be rain-induced lahars, although a growing lake in the now-enclosed crater will pose a serious hazard during the next eruptive episode (Lubis et al. 1987). The 1982–83 Galunggung experience provides an example of an eruption at a volcano with known historical activity in a country where people have experienced many eruptions. In Indonesia, the ability to respond scientifically to volcanic activity is limited because of small staff and inadequate resources. However, good use is made of the resources that are available, and the interaction with local civil officials is generally very effective.

The volcanic emergency that developed rapidly at Colo volcano (Tables 14.2 & 3) illustrates how effective a response can be even with limited resources. A swarm of earthquakes began in July 1983 and increased quickly. Even though monitoring of this volcano was only rudimentary, and little was known of the geology, VSI scientists recognized the precursors, notified authorities and an evacuation was ordered of the entire island, about 7,000 people. The evacuation was completed just hours before devastating pyroclastic flows swept the island on 23 July; this outbreak, and those that followed during succeeding weeks, destroyed all the vegetation, killed the animals that remained behind, and destroyed all habitations and buildings. Locally, it was a major economic disaster but, because it occurred in a distant, remote region, and because no lives were lost, little news of this remarkably successful life-saving effort reached the outside world. Pragmatic knowledge of volcanic behavior by experienced scientists, coupled with prompt, effective response by civil officials and full co-operation by citizens, averted what might have become a major volcanic disaster.

Unfortunately, the emergency at Nevado del Ruiz (Tables 14.2 & 3) in 1985 led to a very different outcome. A complex series of interactions between the newly restless volcano, scientists and government authorities began in 1984; because detailed accounts of these interactions are given in the papers cited in Table 14.3, we give only a brief summary here. Precursory signs beginning in November 1984 signalled that the volcano might be ending its rest of more than a century. Local scientists had little experience in interpreting such signals, and foreign volcanologists were called on for help. All visiting scientists, over a span of several months in 1985, expressed a high degree of concern; they urged increased monitoring and an assessment of potential hazards, but only a few measures were actually adopted. However, a hazards map was prepared and delivered to government authorities in early October. Response by both officials and the general public was highly varied. Some officials were seriously concerned and urged prompt action, whereas others felt that the danger was overstated and merely warranted further study. With such conflicting official views, the news media and the general public were, of course, confused.

On 13 November, explosions and pyroclastic flows abruptly melted large amounts of snow and ice in the summit area, which in turn generated huge lahars that rapidly descended the rivers draining the volcano. The lahars were detected while in progress and the authorities were notified, but for reasons still not fully understood the information was never translated into effective action. No general evacuation was undertaken, and within a few hours approximately 25,000 people died, mostly in the town of Armero; this was the fourth largest death toll from a volcanic eruption in recorded history. A poignant assessment of the Ruiz tragedy has been provided by Voight (1990, p. 185):

> *The catastrophe was not produced by technological ineffectiveness or defec-*
> *tiveness, nor by an overwhelming eruption of unprecedented character, nor*
> *by an improbable run of bad luck. Armero was caused, purely and simply,*
> *by cumulative human error – by misjudgment, indecision, and bureaucratic*
> *shortsightedness. In the end, the authorities were unwilling to bear the*
> *economic or political costs of early evacuation or a false alarm, and they*
> *delayed action to the last possible minute. Catastrophe was the calculated*
> *risk, and nature cast the die. And so the lessons from Armero are not new*
> *lessons; they are old lessons forged in human behavior that once again*
> *require the force of catastrophe to drive them home. Armero could have*
> *produced no victims, and therein dwells its immense tragedy.*

A different set of events and response occurred at Oshima volcano in 1986 (Tables 14.2 & 3). Oshima has long been recognized as one of the most frequently active volcanoes in Japan, its geology and eruptive frequency are well known, and it was well monitored although plans for even further improvements were underway. The local officials and the general public were well aware of the volcanic character of their island and of the frequent eruptions, and a well-planned and practised civil defense organiza-tion was in place. Seismicity began to increase in April 1986, and had reached fairly high levels by October. In mid-November vigorous fountaining and voluminous lava flows began, which, in a few days, spread and intensified. However, of greater concern, strong earthquakes from several parts of the island far from the active vents suggested that even stronger, more extensive activity might develop. In accordance with prescribed procedures, scientists met with public officials, and on 21 November the decision was made to evacuate the island. Within 2 hours of the decision, evacuation had begun, and 11 hours later 10,000 people had been safely removed from the island. Only a small staff of scientists and government and law-enforcement officials remained on the island. Activity continued at intense levels for a few more days, but then subsided, and, fortunately, the threatened accelerated level of eruption did not materialize. After about a month of low-level activity, the decision was made to allow the people of the island to return to their homes. Oshima is an example of an emergency at a well-studied and monitored volcano, where both the government and the public are well informed on volcanic characteristics and are fully responsive to warnings.

The preceding examples are of volcanoes that involved either a single event or a series of specific events through a period of time. At some of them, activity has subsided and returned to background levels, whereas at others the threat of further activity persists but not as an immediate impending crisis (as of mid-1990). Two further examples follow, where volcanic activity has persisted for years at a low to moderate level and the population has had to adjust to continuous or repeated hazards that have caused inconvenience, damage, or destruction. These volcanoes are Kilauea and Sakurajima (Tables 14.2 & 3).

Kilauea, as one of the most frequently active volcanoes in the world, is also one of the best monitored and best studied. The Hawaiian Volcano Observatory was established in 1912, and its staff members have been pioneers in developing new and improved monitoring techniques. The frequent eruptions of Kilauea and its sister volcano, Mauna Loa, have accustomed island residents to the possibility of occasional property losses. Before 1960, the more obviously vulnerable areas, such as downslope from the rift zones, were not used for expensive or elaborate developments. However, as the population began to increase more rapidly in the 1960s, residential and commercial projects encroached into lands more susceptible to damage, and the potential for losses increased. Scientists, concerned about this trend, prepared a map and report to illustrate and explain the hazards (Mullineaux & Peterson 1974), which, in accord with then-prevailing policies, remained confidential until the time of release. When the report was released, some civil authorities and commercial executives were outraged, fearing losses in business, tourism, and investment. They claimed, "We all know about the hazards; we don't need this report! Why tell everyone else about them?" This antagonistic encounter exemplifies the consequences that arise when adequate communications are not maintained between scientists and civil officials. Had the officials been briefed in advance about the map and report, and had discussions been held about the potential problems, even though some disagreements might have resulted, the exchanges would likely have been less acrimonious.

A marked change in attitude has subsequently developed, in large measure a direct benefit of improved communications between scientists and civil authorities. The eruption that began on Kilauea's east rift zone in 1983 has continued unabated into 1992. Although the activity has been non-explosive and has caused no loss of human life, lava flows have covered about 71 km^2 of land and destroyed about 185 structures. Continuing problems caused by the spreading lava produced close co-operation between the scientists and the Hawaii County Civil Defense Agency, and cordial and productive relations have prevailed throughout the repeated crises. The character of the eruptions and the paths of the lava flows have been in accord with the hazard zones outlined in 1974, and subsequent revisions (Mullineaux et al. 1987) and a new general-interest publication on hazards (Heliker 1990) have been well accepted by the community.

The population in the vicinity of Sakurajima must cope with what is likely the most persistent inconvenience and potential hazard from volcanic emissions of any metropolitan area on Earth. Explosive emissions of ash occur frequently at variable intervals, typically ranging from only a few explosions per month to several per day; over the longer term, since 1955 the explosions have varied from fewer than 100 to more than 400 per year. Scientists have estimated that, on average, 1.4×10^9 kg of ash is ejected every year (Kamo 1988). The ejecta fall not only on the slopes of the volcano, but are also carried downwind to fall on nearby cities, villages, farms and forests. Ash must be repeatedly cleaned from residences, businesses, roads, etc., and operation of heavy equipment for ash removal is part of the way of life. Some of the strongest outbursts hurl fragments large enough to injure people and cause damage; for safety, children are required to wear protective helmets on their way to and from school. The unconsolidated volcanic debris on the steep slopes of the volcano is readily eroded, and during heavy rainfall lahars are generated. To mitigate the effects of lahars, elaborate engineering works have been erected along stream courses draining the volcano.

Sakurajima's activity is monitored by scientists at some of the best staffed and equipped volcano observatories in the world, and the geology of the volcano and surrounding region is well understood. The scientists at the observatories not only study the virtually continuous activity but also scrutinize the extensive volcano-monitoring data, seeking possible longer-term precursory patterns that might presage the buildup towards the next catastrophic eruption, such as those that devastated the region in 1471–76, 1779 and 1914. The scientists are in regular communication with the civil authorities, who have established comprehensive emergency procedures that are regularly rehearsed and involve the participation of all concerned agencies as well as the general population. Indeed, the situation at Sakurajima represents a societal adaptation to a troublesome volcano that, we believe, is as close to the ideal scenario (pp. 345–347) as has likely been achieved anywhere. The advanced state of scientific studies, the productive interactions between scientists and civil authorities, the responsive and responsible management practices of the authorities, and the level of knowledge and preparation of the entire population can serve as an excellent model towards which scientists and civil authorities can aspire.

In addition to the eruptions summarized in Table 14.2, episodes of major unrest occurred in the 1980s at three other volcanoes, which aroused great scientific and public concern. None of these volcanic crises culminated in an eruption, and at each the unrest eventually subsided. These episodes took place at three different calderas at widely separated parts of the world: Rabaul, on New Britain Island in Papua New Guinea; Campi Flegrei (the Phlegrean Fields), near Naples, Italy; and Long Valley, California, USA (see Table 14.4 for references). The unrest at all three localities consisted of dramatically increased seismicity, including damaging earthquakes, accelerated rates of ground deformation, and increased gas emissions and fumarolic activity. Statements issued by scientists at each volcano included the customary wide range of possible outcomes, ranging from a return to quiescence, through small and moderate eruptions, to major activity, but all assessments were necessarily qualified by considerable uncertainty. Public response was varied, but partial evacuations were carried out at both Rabaul and Campi Flegrei. Some news reporters sensationalized the possibilities, contributing to widespread public confusion and alarm; commercial activity declined, property values dropped and those affected began to complain. As the unrest continued at each location with no eruption, dissatisfaction with the scientists grew, both from officials and the public and, finally, when the unrest gradually subsided, antagonism became intense. At all three places scientists worked very hard to alleviate the discontent and were partly successful, but residual problems remain at all three. Episodes of unrest continue to recur at Long Valley, and scientists realize, of course, that it is just a matter of time until new, more intense unrest will resume, and they hope that official and public response will be appropriate. The references cited in Table 14.4 provide information on the crises and the scientific and public response.

Table 14.4 Selected references pertinent to the volcanic unrest and public response at the calderas at Campi Flegrei, Rabaul and Long Valley during the 1980s

Caldera and dates of unrest	Geological and historical background	Descriptions of the crises	Public response
Campi Flegrei, Italy, (1982–1984)	Lirer et al. (1987) Newhall & Dzurisin (1988, pp 95–98, 103–109) Rosi & Santacroce (1984)	Barberi et al. (1984b) Newhall & Dzurisin (1988), pp 98–100, 109–117) McClelland et al. (1989 pp 41–49)	Barberi et al. (1984a)
Rabaul, Papua New Guinea (1983–1985)	Heming (1974) Walker et al. (1981) Johnson & Threlfall (1985) Newhall & Dzurisin (1988, pp 219–222, 231–234)	McKee et al. (1984, 1985) Newhall & Dzurisin (1988, pp 222–228, 234–240) McClelland et al. (1989, pp 180–189)	McKee et al. (1985) Lowenstein (1988)
Long Valley, California, USA, (1980–1984, and intermittent to present)	Hill et al. (1985a,b) Bailey (1989) Newhall & Dzurisin (1988, pp 739–741, 747–748)	Miller et al. (1982) Hill (1984) Hill et al. (1990) Newhall & Dzurisin (1988, pp 741–743, 748–754) McClelland et al. (1989, pp 399–403)	Mader & Blair (1987) Hill et al. (1991)

Conclusions

The foregoing comparison of case histories of volcanoes in the 1980s illustrates the diversity in the nature of human response to different kinds of activity and in the caliber of interrelations between scientists, civil authorities and the populace. Several conclusions emerge from the comparison:

(a) Small, frequent eruptions induce regular contacts between scientists and authorities, leading to better acquaintance with one another's needs and to optimal interrelations between them. Examples where good relations prevail include Sakurajima, Oshima, and Kilauea. Improved relations were attained at Mount St Helens after the volcano began a pattern of small, repeated extrusive episodes, and at Nevado del Ruiz after the volcano had demonstrated – all too tragically – the validity of the scientists' warnings.

(b) Conversely, uncertainty about the outcome of volcanic unrest, especially if major violence is among the possibilities, seems to induce poor interrelations. Better relations develop when an ominous threat is immediate and definite, rather than vague. Examples of antagonistic or ineffective response to scientists' indefinite warnings include Mount St Helens prior to 18 May 1980, Ruiz prior to 13 November 1985, Long Valley, Rabaul, and Campi Flegrei. In contrast, effective responses to definite warnings occurred at Galunggung, Colo and Oshima.

(c) Volcanoes that show unrest after prolonged inactivity are particularly vexing to diagnose and have especially dangerous potential. Examples in the 1980s include Mount St Helens, El Chichón and Ruiz, all of which had been quiescent for more than a century. Still, as if to confirm the difficulty of diagnosis, the equally ominous episodes of unrest at Long Valley, Rabaul and Campi Flegrei were not followed by eruptions, and these so-called "false alarms" caused seriously strained relations between scientists and the population. Scientists must help the public learn tolerance for these "aborted eruptions" (Banks et al. 1989), because to ignore any future episode of unrest could be a grave mistake.

(d) An important goal should be to carry out adequate geological studies – *before* a volcanic crisis develops – of all potentially dangerous volcanoes in populated regions, and these, in turn, can be used as a basis to design monitoring networks and to construct hazards maps (Crandell et al. 1984). Scientists generally must work patiently with authorities to help them see how the hazards maps can be used to develop suitable land-use zones for guiding current and future development, for making plans for dealing with emergencies, and for planning the conduct of emergency drills. Such activities are most effective if carried out when no emergency is in progress, although human nature seems to dictate that such projects are invariably resisted until an actual emergency is imminent. Scientists can help their cause by preparing general-interest publications, films, videotapes and by giving programs and public lectures on the nature of volcanic hazards.

In summary, we reiterate some of the conclusions expressed by Peterson (1988, p. 4168). Scientists must draw deeply from reserves of creativity and ingenuity to find the most effective ways to convey the essential facts about volcanic hazards to civil authorities and the general public, and to convince them to take suitable action. This task must be treated with the same determination as in addressing a stubborn but intriguing and vital scientific problem. The task can be aided by utilizing the findings and techniques of sociologists (Sorensen & Mileti 1987, Lundgren 1988). Depending on the course taken by a restless volcano, the degree of understanding achieved and action taken by officials and the public can be of critical importance to society.

Acknowledgements

The ideas and the opinions stated here have been developed during work at, and visits to, several of the volcanoes discussed in this chapter. Our scientific colleagues, civil defense officials, other community leaders, members of the news media, and vast numbers of the general public have all played essential roles in forming and shaping the authors' views. The ideas have been further refined by the participation of the authors in various meetings of United Nations committees, conferences and field excursions held by commissions of the International Association of Volcanology and Chemistry of the Earth's Interior, and by various other organizations. Thoughtful reviews of earlier drafts of the paper by R. A. Bailey, T. L. Wright and D. K. Chester, and helpful comments by D. A. Swanson, have substantially improved the final product.

Note added in press.

Following completion of this manuscript, destructive eruptions occurred at two long-quiescent volcanoes: Unzen (Kyushu, Japan) and Pinatubo (Luzon, Philippines), both in June 1991. The principles reviewed in this paper were applicable at both localities.

References

Alexander, D. 1986. Comment on Smith, 1985. *Geology* **14**, 443.

Aramaki, S., K. Kamo, M. Kamada (eds) 1988. A guide book for Sakurajima Volcano. *Kagoshima International Conference on Volcanoes*. Kagoshima Prefectural Government.

Bailey, R. A. 1989. Geologic map of Long Valley caldera, Mono-Inyo craters volcanic chain and vicinity, eastern California. US Geological Survey Miscellaneous Investigations Series, Map I-1933.

Banks, N. G., R. I. Tilling, D. H. Harlow, J. W. Ewert 1989. Volcano monitoring and short-term forecasts. In *American Geophysical Union, Short Course in Geology*. Vol. 1, *Volcanic hazards*, R. I. Tilling (ed.), 51–80. Washington, DC: American Geophysical Union.

Barberi, F., G. Corrado, F. Innocenti, G. Luongo 1984a. Phlegraean Fields 1982–1984: brief chronicle of a volcano emergency in a densely populated area: *Bulletin Volcanologique* **47**, 176–185.

Barberi, F., D. Hill, F. Innocenti, G. Luongo, M. Treuil (eds.) 1984b. The 1982–1984 bradyseismic crisis at Phlegraean Fields (Italy). *Bulletin Volcanologique* **47**, 173–411.

Barberi, F., M. Martini, M. Rosi 1990. Nevado del Ruiz volcano (Colombia): pre-eruption observations and the November 13, 1985 catastrophic event. *Journal of Volcanology and Geothermal Research* **42**, 1–12.

Blong, R. J. 1984. *Volcanic hazards: a sourcebook on the effects of eruptions*. Orlando, Florida: Academic Press.

Canul, R. F. & V. S. Roch 1981. *Informe geológico de la zona geotérmica de "El Chichónal," Chiapas*. Geotherm. Dep. de la Com. Fed. de Electr., Morelia, Michoacán, Mexico, Report, 1–38.

Chester, D. K., A. M. Duncan, J. E. Guest, C. R. J. Kilburn 1985. *Mount Etna: The anatomy of a volcano*. London: Chapman & Hall.

Crandell, D. R. & D. R. Mullineaux 1978. Potential hazards from future eruptions of Mount St. Helens volcano, Washington. *US Geological Survey Bulletin* **1383-C**, 1–26.

Crandell, D. R., D. R. Mullineaux, M. Rubin 1975. Mount St. Helens volcano: recent and future behavior. *Science* **187**, 438–41.

Crandell, D. R., B. Booth, K. Kusumadinata, D. Shimozuru, G. P. L. Walker, D. Westercamp, D. 1984. *Source book for volcanic-hazard zonation*. Paris: UNESCO.

Decker, R. W. 1986. Forecasting volcanic eruptions. *Annual Reviews in Earth and Planetary Sciences* **14**, 267–91.

Decker, R. W., T. L. Wright, P. H. Stauffer (eds) 1987. *Volcanism in Hawaii*. US Geological Survey Professional Paper 1350, 1–1667.

Duffield, W. A., R. I. Tilling, R. Canul 1984. Geology of El Chichón volcano, Chiapas, Mexico. *Journal of Volcanology and Geothermal Research* **20**, 117–32.

Endo, K., T. Chiba, H. Taniguchi, M. Sumita, S. Tachikawa, T. Miyahara, R. Uno, N. Miyaji 1988. Izu-Oshima 1986–1987 eruptions and the eruptive products. *Proceedings of the Kagoshima International Conference on Volcanoes, Kagoshima, Japan*, 119–22.

Fiske, R. S. 1984. Volcanologists, journalists, and the concerned local public: a tale of two crises in the eastern Caribbean. In *Explosive volcanism: inception, evolution, and hazards*, Geophysics Study Committee & National Research Council, 170–6. Washington, DC: National Academy Press.

Fiske, R. S. 1988. Volcanoes and society – challenges of coexistence. *Proceedings of the Kagoshima International Conference on Volcanoes, Kagoshima, Japan*, 14–21.

Fournier d'Albe, E. M. 1979. Objectives of volcanic monitoring and prediction. *Geological Society of London, Journal* **136**, 321–6.

Foxworthy, B. L. & M. Hill 1982. *Volcanic eruptions of 1980 at Mount St. Helens – the first 100 days*. US Geological Survey Professional Paper 1249, 1–125.

Hall, M.L. 1990. Chronology of the principal scientific and governmental actions leading up to the November 13, 1985 eruption of Nevado del Ruiz, Colombia. *Journal of Volcanology and Geothermal Research* **42**, 101–15.

Hall, M. L. 1992. The 1985 Nevado del Ruiz eruption: scientific, social, and governmental response and interaction prior to eruption. In *Geohazards*. Association of Geoscientists for International Development Report 15, G. J. H. McCall, D. J. C. Laming, S. Scott (eds), 43–52. London: Chapman & Hall.

Havskov, J., S. De la Cruz-Reyna, S. Singh, P. Medina, C. Gutierrez 1983. Seismic activity related to the March–April 1982 eruptions of El Chichón volcano, Chiapas, Mexico. *Geophysics Research Letters* **10**, 293–6.

Hays, W. W. & C. F. Shearer 1981. *Suggestions for improving decision-making to face geologic and hydrologic hazards*. US Geological Survey Professional Paper 1240-B, B103–8.

Heliker, C. 1990. *Volcanic and seismic hazards on the Island of Hawaii*. US Geological Survey General Interest Publication, 1–48.

Heming, R. F. 1974. Geology and petrology of Rabaul caldera, Papua New Guinea. *Geological Society of America, Bulletin* **85**, 1253–64.

Herd, D. G. & Comité de Estudios Vulcanológicos 1986. The 1985 Ruiz volcano disaster. *Eos, Transactions of the American Geophysical Union* **67**, 457–60.

Hill, D. P. 1984. Monitoring unrest in a large silicic caldera, the Long Valley–Inyo craters volcanic complex in east-central California. *Bulletin Volcanologique* **47**, 371–96.

Hill, D. P., R. A. Bailey, A. S. Ryall 1985a. Active tectonic and magmatic processes beneath Long Valley caldera, eastern California: an overview. *Journal of Geophysical Research* **20**, 11,111–20.

Hill, D. P., R. E. Wallace, R. S. Cockerham 1985b. Review of evidence on the potential for major earthquakes and volcanism in the Long Valley–Mono craters–White Mountains regions of eastern California. *Earthquake Prediction Research* **3**, 571–94.

Hill, D. P., W. L. Ellsworth, M. J. S. Johnston, J. O. Langbein, D. H. Oppenheimer, A. M. Pitt, P. A. Reasenberg, M. L. Sorey, S. R. McNutt 1990. The 1989 earthquake swarm beneath Mammoth Mountain, California: an initial look at the 4 May through 30 September activity. *Seismological Society of America, Bulletin* **80**, 325–39.

Hill, D. P., M. J. S. Johnston, J. O. Langbein, S. R. McNutt, C. D. Miller, C. E. Mortensen, A. M. Pitt, S. Rojstaczer, S. 1991. *Response plans for volcanic hazards in the Long Valley caldera and Mono Craters area, California*. US Geological Survey Open-File Report 91-270, 1–64.

Hodge, D., V. Sharp, M. Marts 1979. Contemporary responses to volcanism: case studies from the Cascades and Hawaii. In *Volcanic activity and human ecology*, P. D. Sheets & D. K. Grayson (eds), 221–48. Orlando: Academic Press.

Housner, G. W. (ed.) 1987. *Confronting natural disasters: an international decade for natural hazard reduction*, 1–60. Washington, DC: US National Academy of Sciences.

Johnson, R. W. & N. A. Threlfall 1985. *Volcano town: the 1937–43 eruptions at Rabaul*. Bathurst: Robert Brown.

Kamo, K. 1988. A dialogue with Sakurajima volcano. *Proceedings of the Kagoshima International Conference on volcanoes, Kagoshima, Japan*, 3–13.

Katili, J. A. & A. Sudradjat 1984a. *Galunggung: the 1982–1983 eruption*. Bandung: Volcanological Survey of Indonesia.

Katili, J. A. & A. Sudradjat 1984b. The devastating 1983 eruption of Colo Volcano, Una-Una Island, central Sulawesi, Indonesia. *Geologisch Jahrbuch* **A75**, 27–47.

Katili, J. A., A. Sudradjat, K. Kusumadinata (eds) 1986. *Letusan Galunggung 1982–83*. Bandung: Volcanological Survey of Indonesia (some chapters in Indonesian).

Lipman, P. W. & D. R. Mullineaux (eds) 1981. *The 1980 eruptions of Mount St. Helens, Washington*. US Geological Survey Professional Paper 1250, 1–844.

Lirer, L., G. Luongo, R. Scandone 1987. On the volcanological evolution of Campi Flegrei. *Eos, American Geophysical Union, Transactions* **68**, 226–34.

Lowenstein, P. L. 1988. The Rabaul seismo-deformational crisis of 1983–85: monitoring, emergency planning and interaction with the authorities, the media and the public. *Proceedings of the Kagoshima International Conference on Volcanoes, Kagoshima, Japan*, 580–82.

Lubis, H., S. Hamidi, T. J. Casadevall 1987. Volcanic hazards at Galunggung volcano, West Java, Indonesia, since the 1982–1983 eruption. *Abstract Volume, Hawaii Symposium on How Volcanoes Work, Hilo, Hawaii, 19–25 January, 1987*, 160 (abstract).

Luhr, J. F., & J. C. Varekamp (eds) 1984. El Chichón Volcano, Chiapas, Mexico. *Journal of Volcanology and Geothermal Research* **23**, 1–191.

Lundgren, L. 1988. Comment on Smith, 1985. *Geology* **16**, 760–1.

Macdonald, G. A. 1972. *Volcanoes*. Englewood Cliffs, New Jersey: Prentice-Hall.

Macdonald, G. A. 1975. Hazards from volcanoes. In *Geological hazards*, B. A. Bolt, W. L. Horn, G. A. Macdonald, R. F. Scott (eds), 63–131. New York: Springer.

McClelland, L., T. Simkin, M. Summers, E. Nielsen, T. C. Stein 1989. *Global volcanism 1975–1985*. Englewood Cliffs, New Jersey: Prentice-Hall.

McKee, C. O., P. L. Lowenstein, P. de St. Ours, B. Talai, I. Itikarai, J. J. Mori 1984. Seismic and ground deformation crises at Rabaul caldera: prelude to an eruption? *Bulletin Volcanologique* **47**, 397–411.

McKee, C. O., R. W. Johnson, P. L. Lowenstein, S. J. Riley, R. J. Blong, P. de St. Ours, B. Talai 1985. Rabaul caldera, Papua New Guinea: volcanic hazards, surveillance, and eruption contingency planning. *Journal of Volcanology and Geothermal Research* **23**, 195–237.

Metropolitan Tokyo 1988. *November 1986 record of the volcanic eruption on Izu-Oshima*, 1–121. Tokyo: Geological Survey Center, Industrial Technology Institute.

Mader, G. G. & M. L. Blair 1987. *Living with a volcanic threat: response to volcanic hazards, Long Valley, California*. Portola Valley, California: William Spangle.

Marts, M. E. 1978. *Social implications of volcano hazard case studies in the Washington Cascades and Hawaii*. NSF-RANN Project Report ENV-76-20735, 1–364. Seattle: Department of Geography, University of Washington.

Miller, C. D., D. R. Mullineaux, D. R. Crandell 1981. *Hazards assessments at Mount St. Helens*. US Geological Survey Professional Paper 1250, 789–802.

Miller, C. D., D. R. Mullineaux, D. R. Crandell, R. A. Bailey 1982. *Potential hazards from future volcanic eruptions in the Long Valley–Mono Lake area, east-central California and southwest Nevada – a preliminary assessment*. US Geological Survey Circular 877, 1–10.

Mullineaux, D. R. & D. W. Peterson 1974. *Volcanic hazards on the island of Hawaii*. US Geological Survey Open-File Report 74-239, 1–61.

Mullineaux, D. R., D. W. Peterson, D. R. Crandell 1987. *Volcanic hazards in the Hawaiian islands*. US Geological Survey Professional Paper 1350, 599–621.

Murton, B. J. & S. Shimabukuro 1974. Human adjustment to volcanic hazards in Puna district, Hawaii. In *Natural hazards, local, national, and global*, G. F. White (ed.), 151–9. New York: Oxford University Press.

Newhall, C. G. & D. Dzurisin 1988. Historical unrest at large calderas of the world. *US Geological Survey Bulletin* 1855, 1–1108.

Parra, E. & H. Cepeda 1990. Volcanic hazard maps of the Nevado del Ruiz volcano, Colombia. *Journal of Volcanology and Geothermal Research* **42**, 117–27.

Perry, R. W. & M. R. Greene 1983. *Citizen response to volcanic eruptions*. New York: Irvington.

Peterson, D. W. 1986a. Volcanoes: tectonic setting and impact on society. In *Active Tectonics, Geophysics Study Committee, National Research Council*, 231–46. Washington, DC: National Academy Press.

Peterson, D. W. 1986b. Mount St. Helens and the science of volcanology: a five-year perspective. In *Mount St. Helens: five years later*, S. A. C. Keller (ed.), 3–19. Cheney, Wash.: Eastern Washington University Press.

Peterson, D. W. 1988. Volcanic hazards and public response. *Journal of Geophysical Research* **93**, 4161–70.

Podesta, B. & R. S. Olson 1988. Science and the state in Latin America – decision making in uncertainty. In *Managing disasters, strategies and policy perspectives*, L. K. Comfort (ed.), 296–311. Durham, N. Carolina: Duke University Press.

Punongbayan, R. S. & R. I. Tilling 1989. Some recent case histories. In *American Geophysical Union*,

Short Course in Geology. Vol. 1, *Volcanic hazards,* R. I. Tilling (ed.), 81–101. American Geophysical Union.

Rosi, M. & R. Santacroce 1984. Volcanic hazard assessment in the Phlegraean Fields: a contribution based on stratigraphic and historical data. *Bulletin Volcanologique* **47**, 359–70.

Saarinen, T. F. & J. L. Sell 1985. *Warning and response to the Mount St Helens eruption.* Albany: State University of New York Press.

Sheets, P. D. & D. K. Grayson (eds) 1979. *Volcanic activity and human ecology.* New York: Academic Press.

Smith, J. V. 1985. Protection of the human race against natural hazards (asteroids, comets, volcanoes, earthquakes). *Geology* **13**, 675–8.

Smith, J. V. 1986. Reply to comment by Alexander, 1986. *Geology* **14**, 443–4.

Smith, J. V. 1988. Reply to comment by Lundgren, 1988. *Geology* **16**, 761–2.

Sorensen, J. H. & D. S. Mileti 1987. *Public warning needs.* US Geological Survey Open-File Report 87-269, 9–75.

Sudradjat, A. & R. I. Tilling 1984. Volcanic hazards in Indonesia: the 1982–83 eruption of Galunggung. *Episodes* **7**, 13–9.

Swanson, D. A., W. A. Duffield, D. B. Jackson, D. W. Peterson 1979. *Chronological narrative of the 1969–71 Mauna Ulu eruption of Kilauea volcano, Hawaii.* US Geological Survey Professional Paper 1056, 1–55.

Swanson, D. A., T. J. Casadevall, D. Dzurisin, S. D. Malone, C. G. Newhall, C. S. Weaver 1983. Predicting eruptions at Mount St. Helens, June 1980 through December 1982. *Science* **221**, 1369–76.

Swanson, D. A., T. J. Casadevall, D. Dzurisin, R. T. Holcomb, C. G. Newhall, S. D. Malone, C. S. Weaver 1985. Forecasts and predictions of eruptive activity at Mount St. Helens, USA: 1975–1984. *Journal of Geodynamics* **3**, 397–423.

Swanson, D. A., D. Dzurisin, R. T. Holcomb, E. Y. Iwatsubo, W. W. Chadwick, T. J. Casadevall, J. W. Ewert, C. C. Heliker 1987. Growth of the lava dome at Mount St. Helens, Washington, (USA), 1981–1983. In *The emplacement of silicic domes and lava flows,* J. H. Fink (ed.). *Geological Society of America Special Paper* 212, 1–16.

Tilling, R. I. 1985. A lesson from the 1985 Ruiz volcanic disaster. *Episodes* **8**, 230 (guest editorial).

Tilling, R. I. 1989a. Volcanic hazards and their mitigation: progress and problems. *Reviews of Geophysics* **27**, 237–69.

Tilling, R. I. (ed.) 1989b. *American Geophysical Union, Short Course in Geology.* Vol. 1, *Volcanic hazards,* 1–123. American Geophysical Union.

Tilling, R. I., R. L. Christiansen, W. A. Duffield, E. T. Endo, R. T. Holcomb, R. Y. Koyanagi, D. W. Peterson, J. D. Unger 1987a. *The 1972–1974 Mauna Ulu eruption, Kilauea volcano: an example of quasi-steady-state magma transfer.* US Geological Survey Professional Paper 1350, 405–69.

Tilling, R. I., C. Heliker, T. L. Wright 1987b. *Eruptions of Hawaiian volcanoes: past, present, and future.* US Geological Survey General Interest Publication, 1–54.

Tilling, R. I., L. Topinka, D. A. Swanson 1990 *Eruptions of Mount St. Helens: past, present, and future.* US Geological Survey General Interest Publication, 1–56 (revised edn; 1st edn 1984). Washington, DC: American Geophysical Union.

UNDRO & UNESCO 1985. *Volcanic emergency management.* New York: United Nations.

Voight, B. 1988. Countdown to catastrophe. *Earth and Mineral Sciences (Pennsylvania State University)* **57**, 17–30.

Voight, B. 1990. The 1985 Nevado del Ruiz volcano catastrophe: anatomy and retrospection. *Journal of Volcanology and Geothermal Research* **42**, 151–88.

Walker, G. P. L. 1974. Volcanic hazards and the prediction of volcanic eruptions. In *Prediction of geological hazards,* B. M. Funnell (ed.). Geological Society of London, Miscellaneous Paper 3, 23–41.

Walker, G. P. L., R. F. Heming, T. J. Sprod, T. J., H. R. Walker 1981. Latest major eruptions of Rabaul volcano. In *The Cooke–Ravian volume of volcanological papers,* R. W. Johnson (ed.). Geological Survey of Papua New Guinea Memoir 10, 181–93.

Watanabe, H. 1988. The 1986–1987 eruption of Izu-Oshima volcano. *Proceedings of the Kagoshima International Conference on Volcanoes, Kagoshima, Japan*, 37–40.

Williams, H. & A. R. McBirney 1979. *Volcanology*. San Francisco: Freeman, Cooper.

Williams, S. N. (ed.) 1990a. Nevado del Ruiz Volcano, Colombia, I. *Journal of Volcanology and Geothermal Research* **41**, 1–377.

Williams, S. N. (ed.) 1990b. Nevado del Ruiz Volcano, Colombia, II. *Journal of Volcanology and Geothermal Research* **42**, 1–224.

Wolfe, E. W. (ed.) 1988. *The Puu Oo eruption of Kilauea volcano, Hawaii: episodes 1 through 20, January 3, 1983, through June 8, 1984*. US Geological Survey Professional Paper 1463, 1–251.

Wolfe, E. W., M. O. Garcia, D. B. Jackson, R. Y. Koyanagi, C. A. Neal, A. T. Okamura 1987. *The Puu Oo eruption of Kilauea volcano, episodes 1–20, January 3, 1983, to June 8, 1984*. US Geological Survey Professional Paper 1350,

Index

For Product Safety Concerns and Information please contact our
EU representative GPSR@taylorandfrancis.com Taylor & Francis
Verlag GmbH, Kaufingerstraße 24, 80331 München, Germany